PLASMA PHYSICS
IN THEORY AND APPLICATION

PLASMA PHYSICS
IN THEORY AND APPLICATION

edited by

WULF B. KUNKEL
UNIVERSITY OF CALIFORNIA, BERKELEY

McGRAW-HILL BOOK COMPANY
NEW YORK ST. LOUIS SAN FRANCISCO TORONTO LONDON SYDNEY

Plasma Physics in Theory and Application

Library of Congress Catalog Card Number 65-28728

35629

1234567890 MP 7321069876

PREFACE

In the fall semester, 1961, the University Extension of the University of California organized a statewide lecture series under the title Plasma Physics in Theory and Application. As in other such series, these lectures were offered as a course with university credit by the Engineering and Sciences Extension, University of California, Berkeley, jointly with the Engineering Extension and Physical Sciences Extension, University of California, Los Angeles. The course was intended to provide for active research workers and other scientists desiring a general knowledge of plasma physics a comprehensive introduction to this new and fascinating field, to give them an insight both into theory and into practice, and to bring them up to date with the most recent developments. The subject matter was divided into a number of separate topics, and lecturers were found (mostly from the West Coast), all of whom were known as experts in the particular specialties.

The advisory committee charged with the planning of this series consisted of M. N. Rosenbluth (San Diego); A. Baños, C. Bell, W. D. Hershberger (Los Angeles); G. S. Kino (Stanford); H. M. Heming, R. V. Pyle, A. W. Trivelpiece, C. M. Van Atta, and myself (Berkeley). Following the example of many previous such lecture series, this committee decided that it would be very desirable if the lectures were subsequently published in book form. While most of the speakers agreed to work up their notes into self-contained chapters, a few of them were unable to participate, for a variety of reasons. It fell to me as coordinator of the lectures and editor of the volume to look for substitutes wherever possible and to rearrange the material in the interest of better continuity. Unfortunately, although excellent new coauthors were found for the most essential topics, considerable delays could not be avoided.

In the meantime a fair number of texts dealing with various aspects of plasma physics have appeared in print, and it may be questioned whether or not the need for publication of this material still exists. A survey showed, however, that none of the available books was as broadly based as our lecture series, and in no case, of course, were the authors experts in the many different aspects and applications of plasma physics simultaneously. We have therefore attempted to bring our presentation up to date wherever possible and to concentrate on those features of the field which have not yet been extensively covered in the existing literature. For example, subjects such as single-particle motion and plasma kinetic theory are treated in a rather condensed manner and on a fairly advanced level, since they have already been discussed adequately in several existing texts. On the other hand, the treatment of hydromagnetic shocks or of various applications of plasma physics is kept more elementary and more detailed because these topics have not yet been covered very thoroughly in the literature. For the same reason the discussion of waves in plasmas has been limited to a brief survey of laboratory applications. It is quite apparent that the subject of waves in plasma has been given exhaustive attention elsewhere. The actual coverage offered in this volume is therefore not intended to be completely thorough, nor the level of presentation completely uniform. Nevertheless, since most omissions are adequately compensated for by extensive references to the pertinent literature, this collection of articles may be considered as a reasonably complete summary, or survey, of the entire field of plasma physics (excluding plasmas existing in solid-state materials) as it has developed so far.

We frequently complain that theory and experiments in physics research are too widely separated. The same certainly holds for plasma research, although perhaps a little less so than for some of the other branches of physics. The schism became particularly apparent when it turned out to be impossible to enforce a single set of units for this text. We therefore agreed to admit two types of units: all the fundamental chapters make use of the symmetrical CGS (Gaussian) system of units, while most of the applied topics are described in terms of the practical

MKSA (Giorgi) system.† The comparison of the two systems proves to be instructive rather than confusing, and no difficulties stemming from this short-coming are anticipated.

It also proved impossible to maintain a consistent set of symbols for all chapters. Depending on the field of specialization, the same Greek letter, for example, may be used to denote completely different physical quantities. Conversely, certain quantities are customarily identified by different symbols, depending on the field of specialization. We have not attempted an artificial unification here, and the reader is cautioned to rely on consistent notation only within each chapter. Although at first sight this feature represents an inconvenience for the reader, it is in keeping with the notion that each chapter should, where possible, serve as a proper introduction to the specialized literature in the field.

Finally, I wish to thank my able associates at the Lawrence Radiation Laboratory of the University of California at Berkeley for their assistance in reading and commenting on several of the manuscripts. Their suggestions made this volume more valuable. I also wish to thank Miss Margaret Thoman for help in the preparation of the index.

Wulf B. Kunkel

†See *Physics Today,* **15**, 6, pp. 29-30 (June, 1962).

CONTENTS

CONTRIBUTORS

George R. Brewer   Manager, Ion Propulsion Department
Hughes Research Laboratories
Malibu, California

Julian D. Cole   Professor of Aeronautics
California Institute of Technology
Pasadena, California

Malcolm R. Currie   Vice President and Director
Electron Dynamics Laboratory
Hughes Research Laboratories
Malibu, California

Burton D. Fried   Professor of Physics
University of California
Los Angeles, California

Hans R. Griem   Professor of Physics
University of Maryland
College Park, Maryland

Arthur Kantrowitz

Vice President and Director
Avco Corporation and Director,
Avco-Everett Research Laboratory
Everett, Massachusetts

Allan N. Kaufman

Associate Professor of Physics
University of California
Berkeley, California

Ronald C. Knechtli

Head, Plasma Physics Department
Hughes Research Laboratories
Malibu, California

Wulf B. Kunkel

Senior Physicist, Lawrence Radiation
Laboratory and Lecturer in Physics
University of California
Berkeley, California

Theodore G. Northrop

Theoretical Division
Goddard Space Flight Center
Greenbelt, Md.

Harry E. Petschek

Principal Research Scientist
Avco-Everett Research Laboratory
Everett, Massachusetts

Robert V. Pyle

Senior Physicist, Lawrence Radiation Labora-
tory and Lecturer in Nuclear Engineering
University of California
Berkeley, California

Richard J. Rosa

Principal Research Scientist
Avco-Everett Research Laboratory
Everett, Massachusetts

Marshall N. Rosenbluth

Professor of Physics
University of California
San Diego, California

Norman Rostoker

Manager of Fusion and Plasma Physics
General Atomic Division of
General Dynamics Corporation
John Jay Hopkins Laboratory
for Pure and Applied Science
San Diego, California

Alvin W. Trivelpiece

Associate Professor of Electrical Engineering
University of California
Berkeley, California

Charles B. Wharton

Staff Member
General Atomic Division of
General Dynamics Corporation
John Jay Hopkins Laboratory
for Pure and Applied Science
San Diego, California

1

INTRODUCTION TO PLASMA PHYSICS

KUNKEL and ROSENBLUTH

WULF B. KUNKEL, *Senior Physicist, Lawrence Radiation Laboratory, and Lecturer in Physics, University of California, Berkeley, California*

MARSHALL N. ROSENBLUTH, *Professor of Physics, University of California, San Diego, California*

1

1.1 PLASMA PHYSICS AS INTERDISCIPLINARY SCIENCE

Plasma physics had its beginning in the studies both of electrical discharges through gases and of certain problems in astrophysics. In recent years interest in the field has grown primarily because of its many applications in electrical and aeronautical engineering, in geophysics and space science, and, last but not least, in the possible development of thermonuclear power generation. To the physicist, however, the principal attraction of plasma research perhaps lies in the fact that it combines so many of the disciplines of physics into a single study. For example, if we restrict ourselves to low-density high-temperature plasmas in which quantum effects are unimportant, then we find that almost all aspects of classical physics can be brought to bear in our studies. And in fact quite different points of view have resulted when the same problem has been considered by experts in different fields. It is our hope that these will become apparent in this collection of basic topics in plasma research.

For instance, classical Hamiltonian mechanics is involved in the study of single-particle orbits. In the course of this study interesting new theorems have been developed on the adiabatic invariants of particle motion.

Many types of plasma motion resemble hydrodynamic flows. A large number of the stability studies parallel hydrodynamic problems.

2

In some cases, however, novel features emerge. Thus large-amplitude electrostatic oscillations present many of the features of turbulence in an essentially one-dimensional system.

Insofar as the plasma may be considered as a dielectric medium whose dielectric constant is a function of frequency, wave number, and direction of propagation, a new class of radiation problems emerges, as, for example, in the emission of synchrotron radiation from a plasma.

Another important range of plasma problems is related to kinetic theory and statistical mechanics. Thus several authors have recently shown how to rigorously derive the Boltzmann and Fokker-Planck equation in the case of Coulomb forces.

1.2 BASIC DEFINITION

All the principal phenomena in plasma physics can be traced to the simple fact that charged particles interact with each other by long-range forces. For instance, the electric field due to a point charge decreases only as the inverse square of the distance, whereas, at a given density, the total number of charges increases as the cube of the linear dimension. Thus it is possible for the electric (and magnetic) fields of an assembly of charged particles to add together in a coherent way. This feature gives rise to collective modes of plasma behavior which do not exist in ordinary gases, where molecules interact with short-range forces only. Plasma physics is a many-body problem.

In a basic and most general sense the word *plasma* may then be taken to refer to any collection of charged particles sufficiently dense that space-charge effects can result in strongly coherent behavior. It should be emphasized that this definition restricts us neither to fully ionized gases nor to quasineutral assemblies. There exist many examples where unneutralized electron beams exhibit typical plasmalike properties.

A quantitative criterion may be obtained by considering the following artificial situation: Two plane-parallel conducting surfaces are separated by a distance $2x$ and are held at the same potential φ_0. Let the space between the surfaces be occupied by a uniform number density n of particles carrying an identical charge q each. The potential at the midplane is then $\varphi_m = \varphi_0 + 2\pi nqx^2$ so that the energy required to transfer a particle from the surface to the midplane is $q(\varphi_m - \varphi_0) = 2\pi nq^2x^2$. Suppose now that the particles have thermal motion characterized by a temperature T. The collective (space charge) effects then dominate over the particles' random behavior if the above energy is larger than the one-dimensional mean energy of the particles' random motion, $\frac{1}{2}kT$, i.e., if

$$x > \left(\frac{kT}{4\pi q^2 n}\right)^{1/2} \equiv \lambda_D \tag{1.1}$$

The quantity λ_D is the most fundamental unit of length in plasma physics. As it appeared in the theory of electrolytes developed by P. Debye, it has been given the name *Debye length*. It is obvious, then, that an ionized gas (or any collection of charged particles) can be called a plasma only if its dimensions (L) are much larger than a Debye length,

$$L \gg \lambda_D \qquad (1.2)$$

Numerically, we find, for $q = e$, the electron charge,

$$\lambda_D = 740 \left(\frac{kT}{n}\right)^{1/2} \quad \text{cm} \qquad (1.3)$$

where kT expresses the energy of random motion in electron volts ($kT = 1$ at $T = 11,400°\text{K}$) and n is the number per cubic centimeter. From this one sees, for example, that a low-current accelerator might not exhibit plasma behavior, while in a typical thermonuclear reactor the Debye length would be only a fraction of a millimeter. Note that λ_D is independent of the particle mass. Also, no assumption has been made about the absence of neutral particles: the term plasma is independent of the degree of ionization.

1.3 FIRST CONSEQUENCES

1.3.1. Quasineutrality

Although our basic definition of plasma includes systems with substantial net charge density, these may be considered exceptional rather than typical. A true one-component plasma satisfying condition (1.2) cannot exist in a static equilibrium. Except for gravity, which we can usually neglect, there is no static force available to counteract the electric repulsion. Poorly neutralized plasmas therefore exist only in dynamic situations that represent extreme deviations from thermal equilibrium. They appear either as brief transients or in situations involving rapid streaming. All such "plasmas" are characterized by very large internal electric fields. The electric forces are always in a direction that tends to reduce the space-charge density.

When charges of both signs are present, these are strongly attracted to each other and are forced to intermix intimately. This tendency toward gross neutrality is particularly striking in all ionized gases which satisfy condition (1.2) and in which the charged species do not deviate drastically from statistical equilibrium. In such cases the difference in potential energy of particles at different points in the space-charge field can at most be of the order of the mean random energy, i.e., for electrons

$-e\Delta\varphi \lesssim kT_e$.   But from Poisson's equation we obtain

$$\Delta\varphi \approx L^2\nabla^2\varphi = -4\pi e \left(\sum_j Z_j n_j - n_e\right) L^2$$

where n_e is the average electron density and n_j denotes the density of positive ions with charge $Z_j e$ per particle.† The inequality (1.2), with λ_D referring to the electrons (as is usually the case), thus leads to the conclusion that now

$$\left| n_e - \sum_j Z_j n_j \right| \ll n_e \tag{1.4}$$

Condition (1.4) is often called *quasineutrality* and is sometimes used as a more limited definition of the term plasma. In particular it states that in macroscopic motion of plasma the electrons and ions are constrained to move without separating appreciably: they are bound to each other collectively by the space-charge forces.

In analytic work on plasmas use is made of Eq. (1.4) by setting $n_e = \sum_j Z_j n_j$ everywhere except in Poisson's equation.

1.3.2. Debye Shielding

It is implicit in Eq. (1.4) that the densities n_e and n_j must be averaged over volumes larger than $\lambda_D{}^3$. This requirement can be given a more precise meaning by considering the potential distribution around a particle of charge q, at rest, embedded in a plasma in thermal equilibrium. Here

$$\nabla^2\varphi = -4\pi \left[q\delta(\mathbf{r}) - en_e \exp\left(\frac{e\varphi}{kT}\right) + e\sum_j Z_j n_j \exp\left(-\frac{Z_j e\varphi}{kT}\right) \right] \tag{1.5}$$

The last terms in Eq. (1.5) contain the Boltzmann factors giving the probability that an electron with $q = -e$ or an ion with $q = +Z_j e$ will be found at a point with potential φ.

If Eq. (1.5) is linearized, it can easily be solved. Using Eq. (1.4) and introducing the mean ionic charge $Z = \sum_j Z_j{}^2 n_j / n_e$, one obtains

$$\varphi = \frac{q}{r} \exp\left(\frac{-r}{\lambda_D \sqrt{1+Z}}\right) \tag{1.6}$$

where λ_D refers to the plasma electrons alone. Thus a shielding cloud with radius of the order of λ_D forms around each particle in the plasma, at least statistically. The effect of this shielding cloud is to reduce the electric potential around each particle so that at distances large com-

† If negative ions are present, these must be added to the electrons, of course.

pared with λ_D its Coulomb field is no longer felt. This more detailed presentation of condition (1.4) points up the fact that, in spite of the long range of the Coulomb force, charged particles in a plasma do not interact with each other individually at distances much larger than the Debye length. Without this feature the calculation of effective "Coulomb-collision" rates of the particles in a plasma would not be possible.

1.3.3. The Plasma Parameter

The arguments of Sec. 1.3.2 are sensible of course only if there are many charged particles within a *Debye sphere*. Thus we must require that

$$N_D \equiv \frac{4\pi}{3} n\lambda_D{}^3 \gg 1 \tag{1.7}$$

if our plasma description is to be statistically meaningful. Fortunately, in highly ionized gases near equilibrium Eq. (1.7) is usually satisfied (except at very high pressures) as can be seen on comparison with Saha's equation (cf. Chap. 3 or 10). In fact we can express condition (1.7) in terms of the mean kinetic and interparticle potential energy,[†]

$$\frac{\langle \text{KE} \rangle}{\langle \text{PE} \rangle} \approx \frac{kT}{q^2 n^{1/3}} = 4\pi\lambda_D{}^2 n^{2/3} = (36\pi)^{1/3}N_D{}^{2/3} \gg 1 \tag{1.8}$$

If the ionized gas is not to recombine, at least in a classical approximation, we must of course have KE > PE.

Introducing the critical distance $r_c \equiv q^2/kT$, that is, the distance at which the potential energy equals the kinetic energy, and the simple Rutherford scattering free path (90° deflections) $\lambda_c = (4\pi r_c{}^2 n)^{-1}$, we find that the Debye length is given by

$$\lambda_D{}^2 = \lambda_c r_c \tag{1.9}$$

Upon combining (1.9) and (1.7) or (1.8) it follows that in our gaseous plasmas we usually have

$$r_c \ll n^{-1/3} \ll \lambda_D \ll \lambda_c \tag{1.10}$$

For use in further discussion it is most convenient to introduce the ratio

$$\Lambda = \frac{\lambda_D}{r_c} = \frac{\lambda_c}{\lambda_D} \tag{1.11}$$

as a new dimensionless parameter describing the statistical quality of the plasma. It is easily verified that this plasma parameter really indicates

---

[†] A more refined calculation of this ratio, averaged over the velocity distribution and taking the Debye screening into account, shows that actually $\langle \text{KE} \rangle / \langle \text{PE} \rangle \propto N_D$ instead of $N_D{}^{2/3}$.

the number of particles in a Debye sphere, since $\Lambda = 3N_D$. There is no question that Λ is the most important dimensionless number in plasma kinetic theory; it appears naturally in many derivations and expressions. The parameter Λ may be interpreted as a measure of the degree to which collective (plasma) effects dominate over individual particle phenomena.

1.3.4. The Plasma Frequency

From the fundamental length λ_D and the basic particle speed $v = (kT/m)^{1/2}$ we can construct a fundamental frequency

$$\omega_p = \frac{v}{\lambda_D} = \left(\frac{4\pi n q^2}{m}\right)^{1/2} \tag{1.12}$$

This is the so-called plasma frequency. Here the particle energy has canceled out, but the mass (inertia) enters in the denominator. The quantity ω_p is the characteristic oscillation rate for electrostatic disturbances in the plasma. In the vast majority of problems we are concerned primarily with the dynamics of the plasma electrons. Therefore we usually refer to the electron component in all the quantities appearing in Eqs. (1.7) to (1.12). The numerical value of ω_p in this case is given by

$$\omega_p = 5.6 \times 10^4 n_e^{1/2} \qquad \text{rad/sec} \tag{1.13}$$

where n_e is expressed in cm^{-3}. In laboratory plasmas this frequency usually lies in the microwave region. The plasma frequency is of such fundamental importance that the plasma density is frequently described by the quantity ω_p rather than by n itself.

It is instructive to display a variety of important plasmas on a plot of ω_p (or n_e) versus kT as shown in Fig. 1.1, in which lines of constant λ_D as well as of constant N_D are also entered. It is seen that conditions (1.2) and (1.7) are both usually satisfied very well indeed.

1.4 MICROSCOPIC DESCRIPTION

1.4.1. Orbits

As we have indicated, in an idealized plasma the collective effects dominate completely over individual particle interactions. In this spirit the many-body problem of plasma physics may be most crudely approximated by an analysis of single-particle motion in macroscopic "smooth" fields, which, in turn, are themselves determined by an averaging process over all particles. Any development of a fundamental plasma theory therefore presupposes an understanding of the behavior of charged particles in

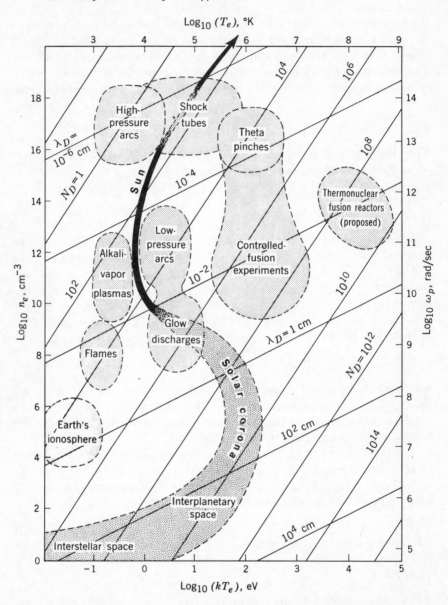

Fig. 1.1 Various approximate plasma domains on an n_e versus T_e plot. Lines of constant λ_D and of constant N_D are also shown.

prescribed electric and magnetic fields. This facet is frequently called *orbit theory*. Any resemblance to the kinetic theory of unionized gases has practically disappeared in this approximation.

The pertinent fundamental equation is simply the equation of motion of charged particles of mass m_i and charge q_i,

$$m_i \ddot{\mathbf{r}}_i = q_i \left(\mathbf{E} + \dot{\mathbf{r}}_i \times \frac{\mathbf{B}}{c} \right) \qquad\qquad (1.14)$$

In the absence of a magnetic field the resulting trajectories are relatively simple. On the other hand, if $\mathbf{B}$ is uniform and constant and if $\mathbf{E} = 0$, one sees that the particle follows a helical path along a field line, moving at constant speed and performing circles of radius $a_i = m_i v_\perp c / q_i B$ with frequency $\Omega_i = q_i B / m_i c$. This so-called gyroradius and gyrofrequency introduce another fundamental length and time of great importance in the plasma properties. It is certainly not surprising that the plasma properties depend critically on the value of a_i in relation to λ_D and λ_c (or the value of Ω_i in relation to ω_p and the collision frequency).

In the general case when $\mathbf{E}$ or $\mathbf{B}$ are not constant or not uniform, the particle orbits may of course be quite complicated, and only very few special problems of this type have been worked out rigorously. Fortunately there is a large class of situations in which the fields vary only slightly during a gyroperiod or over distances of the order of the gyroradius so that workable approximations may be introduced. In these cases the orbits are nearly periodic (adiabatic), permitting a separation of the motion into a circular gyration and a general drift of the so-called guiding center. Several approximate constants of the motion (adiabatic invariants) may then be defined which permit considerable simplification of the analysis. Thus in any slow motions of the plasma the particles remain organized in the sense that all the particles originally orbiting along a given field line remain orbiting along the displaced field lines as the motion proceeds. Thus the field acts to preserve, at least two dimensionally, the ordering of the particles. Hence it plays much the same role as collisions in ordinary hydrodynamics, which also tend to make the particles in a fluid element move together. This is the basic reason why the low-frequency hydromagnetic disturbances are so strikingly similar to hydrodynamic motions, the principal differences arising from the free flow of heat along the lines.

1.4.2. Collisions

In reality the fields in a plasma cannot be completely "smooth" in either space or time because of the discrete nature of the charges. For example, in Sec. 1.3.2 we have already shown that the Coulomb field of each particle makes itself felt over finite distances of the order of the Debye length. The effect of these "microscopic" irregularities and fluctuations on the particle orbits are commonly included under the generalized term *collisions* in analogy to ordinary gas kinetics. These are to be added to close encounters with neutral atoms or molecules if the gas is not fully ionized.

It is clear that even on a microscopic scale a charged particle in a plasma interacts (i.e., "collides") with a large number of other particles simultaneously. Nevertheless, a useful estimate of the particle scattering rate can be obtained by considering the microscopic interaction as a superposition of many two-body collisions. In other words, in a first approximation each particle is treated as experiencing multiple Coulomb scattering. However, the momentum-transfer rate calculated in this way would be infinite for an infinite plasma and in general would depend on the size of the plasma, since the differential cross section $d\sigma/d\Omega$ for Rutherford scattering diverges very rapidly for small deflections θ,

$$\frac{d\sigma}{d\Omega} = \left(\frac{q_1 q_2}{2mv^2}\right)^2 \csc^4 \frac{\theta}{2} \tag{1.15}$$

Here m is the reduced mass of the two charges q_1 and q_2 and v is their relative velocity. $d\Omega$ is the solid angle between θ and $\theta + d\theta$. This difficulty is removed by making use of a many-body aspect of the plasma, i.e., the Debye screening. For instance, the screened potential given in Eq. (1.6) could be used instead of the unmodified Coulomb potential. In that case expression (1.15) does not apply of course, and we are actually no longer dealing with simple binary interactions. More easily, the Debye length may be used as an upper limit on the impact parameter. Both procedures lead to the same finite momentum-transfer "frequency." For collisions between electrons and stationary protons, when averaged over a Maxwellian distribution with temperature T, this frequency is

$$\nu = n_e \langle \sigma v \rangle = \frac{4}{3} \left(\frac{2\pi}{m_e}\right)^{1/2} \frac{e^4 n_e}{(kT)^{3/2}} \ln \Lambda \tag{1.16}$$

where Λ is defined in Eq. (1.11). All other types of collision rates are readily expressed in terms of the ν defined by (1.16). We see that as a result of the many small-angle deflections ν exceeds the frequency of individual close-encounter collisions by a factor of order $\ln \Lambda$. For most plasmas of interest according to Fig. 1.1 one has $5 < \ln \Lambda < 20$. If n_e is again given in cm^{-3} and kT is expressed in electron volts per particle, the numerical value of ν is

$$\nu \approx 3 \times 10^{-6} n (kT)^{-3/2} \ln \Lambda \qquad \text{sec}^{-1} \tag{1.17}$$

Using the definition (1.12) for the plasma frequency, we can express Eq. (1.16) in the compact form

$$\frac{\nu}{\omega_p} = \left(\frac{2}{\pi}\right)^{1/2} \frac{\ln \Lambda}{3\Lambda} \tag{1.18}$$

Equation (1.18) illustrates clearly that if the parameter Λ is large, collective phenomena in the plasma dominate over individual particle effects.

In the treatment so far we have separated the behavior of charged particles into two distinct domains. The first has been the motion of the particles in the fluctuation-free average fields mentioned in Sec. 1.4.1. In some sense this motion also can of course be considered as a form of "collision" since each particle in this case interacts with many or all others simultaneously. By definition this interaction does not produce randomization and therefore does not result in dissipation.

The second class of interaction has been handled in exactly the opposite approximation. The particle scattering has been considered as an accumulation of quasibinary Coulomb collisions which occur only for impact parameters smaller than λ_D. Conceptually, these may be replaced by relatively infrequent equivalent large-angle deflections. In this way a correspondence to the kinetics of ideal gases may be established.

This drastic separation into two distinctly different types of interactions is of course only a crude approximation. Greater precision is obtained in more refined and more satisfying theories in which the effect of the plasma fluctuations and waves on the individual particles (and vice versa), as a form of generalized "collision" phenomenon, is being considered instead of the unrealistic binary Coulomb scattering. In this case the evaluation has to make use of the methods of statistical mechanics.

1.4.3. Plasma Kinetic Theory

We are now in a position to sketch the path by which the kinetic theory has been developed. One may without difficulty write down the complete set of equations of motion for the particles moving under the Lorentz force and the set of Maxwell's equations for the evaluation of the fields generated by the charges and currents of the particles.

Thus, in addition to Eq. (1.14) for each particle, we have

$$\nabla \cdot \mathbf{B} = 0 \qquad\qquad c\nabla \times \mathbf{B} = \frac{\partial \mathbf{E}}{\partial t} + 4\pi \sum_i q_i \dot{\mathbf{r}}_i \delta(\mathbf{r} - \mathbf{r}_i)$$

$$c\nabla \times \mathbf{E} = -\frac{\partial \mathbf{B}}{\partial t} \qquad\qquad \nabla \cdot \mathbf{E} = 4\pi \sum_i q_i \delta(\mathbf{r} - \mathbf{r}_i) \tag{1.19}$$

Needless to say, it is quite hopeless to consider solving this exact set of equations. Perhaps the most unpleasant feature of these equations is represented by the δ function, in other words, by the fact that we are dealing with a set of discrete particles. However, as we have remarked earlier, the individuality of the particles is really not too important— collisional processes are quite slow compared with collective ones. Let us consider a limiting process in which

$$q, m \to 0$$
while $v, nq, nm \to$ const

In other words we subdivide each charged particle into smaller subparticles, gradually smearing out the δ functions occurring in (1.19) so that the properties of a continuous fluid are approached. It is easy to check that in this smearing process the plasma frequency, Debye length, gyrofrequency, and gyroradius remain fixed, while the collision mean-free path becomes infinite. Formally this corresponds to an expansion in Λ^{-1} of the exact Liouville equation for the plasma.

Each species may now be conveniently described by the *collisionless Boltzmann equation* (often called the *Vlasov equation*), which is the continuity equation in phase space for the single-particle distribution function $f(\mathbf{r},\mathbf{v},t)$

$$\frac{\partial f}{\partial t} + \mathbf{v} \cdot \nabla f + \frac{q}{m} \left(\mathbf{E} + \frac{\mathbf{v}}{c} \times \mathbf{B} \right) \cdot \nabla_v f = 0 \tag{1.20}$$

Equation (1.20) merely expresses Liouville's theorem on the incompressibility of phase space. In Maxwell's equation the sum over δ functions is now replaced by an integral over velocity space so that the current density, for instance, is given by

$$\mathbf{j} = \sum_i \frac{q_i}{c} \int f_i \mathbf{v} \, d^3\mathbf{v} \tag{1.21}$$

If we were to carry out the expansion to the next order in Λ^{-1}, collisional-type terms would appear and we would obtain the Fokker-Planck equation, which has been developed in connection with the analysis of Brownian motion. These are the fundamental equations of plasma physics.

It can be seen that (1.20) describes the constancy of the distribution function as we move along a particle trajectory. The degree of complexity is determined primarily by the complexity of the orbits. Thus progress to date has been largely made by studying motions which move the plasma only slightly away from situations in which simple particle orbits describe the motion, as already mentioned in Sec. 1.4.1.

The dispersion relations for waves in such plasma fluids have been studied in great detail. As a result the properties of plasmas near equilibrium, and in certain situations even far from equilibrium, are now well known, and the response to small perturbations can be calculated. For instance, the formation of the Debye clouds described in Sec. 1.3.2 is readily derived by this procedure.

Most important, however, it can be shown that the fluid has peculiar dielectric properties which permit the propagation of longitudinal (compressional) electrostatic waves, i.e., where $\nabla \times \mathbf{E} = 0$. These waves or propagating plasma oscillations have a certain similarity to acoustic waves in ordinary gases, except that the pressure is transmitted via the collective space-charge field instead of via individual particle collisions. A point charge moving through the fluid represents an electric perturbation and therefore continually excites some of these oscillations; i.e., the particle emits waves and loses energy. Therefore, at a finite temperature, plasma oscillations are constantly generated by the individual particles. Conversely, the waves are also continuously washed out (damped) by particles that in their random motion gain energy from the wave fields, i.e., the particles absorb waves.

We thus end up with a model that describes the plasma as a dual medium, consisting of particles with a distribution of random velocities on the one hand and a continuous fluid on the other, supporting a spectrum of oscillations with which the particles interact and whose amplitudes are determined by the relative rates of wave emission and absorption. The net effect of the wave-particle interaction is a diffusion of $f(\mathbf{r},\mathbf{v},t)$ in velocity space in a manner resembling collisions between particles. It should be noted that this interaction does not disappear completely when Λ^{-1} becomes negligible. It apparently contains a contribution which is additional to the ordinary so-called Coulomb interaction, i.e., to the Fokker-Planck terms describing the scattering produced by field perturbations associated with the individual charges. In its most satisfying form the model even obviates the need to define a demarcation between individual particle fields and high-frequency plasma waves. Near thermal equilibrium the effect of the waves is minor. However, in certain nonequilibrium situations the wave amplitudes can build up by orders of magnitude, and the resulting effect on the particles can represent a considerably enhanced dissipation rate which then completely overshadows the ordinary collisions. This property of plasmas, which has no counterpart in the kinetics of unionized gases, has not yet been fully explored.

The most important problems in the microscopic theory of plasmas that still require major efforts thus are connected with nonlinear phenomena resulting from large deviations from thermal equilibrium, e.g. large-amplitude waves and fluctuations.

1.5 MACROSCOPIC PROPERTIES

1.5.1. The Single-fluid Approximation

There are a large number of problems in plasma physics for which a microscopic treatment is not necessary. In such cases it is usually advantageous to consider each of the constituent species as a continuous fluid with only macroscopic properties derived from appropriate averages over the particles. Furthermore, as we have seen in Sec. 1.3.1, in the usual quasineutral plasmas the electrons and ions are intimately coupled together. It is therefore reasonable to attempt a macroscopic description of a plasma as a single medium with well-defined characteristics. This is certainly justified for fully ionized two-component plasmas. If the gas is partially ionized, the neutral component can be considered as a separate fluid coupled to the plasma proper by friction and by collisional-energy transfer. Sometimes the friction is so effective that the relative motion between ions and neutrals is negligible; the neutral component can then be considered as a part of the plasma.

In any event, our medium has all the properties that characterize ordinary compressible fluids except that the viscosity is not isotropic if a magnetic field is present. Likewise, the energy-transport coefficients must be replaced by tensors. In addition, of course, the fluid is electrically conducting so that a new body force $\mathbf{j} \times \mathbf{B}$ appears in the equation of motion,

$$\rho \frac{d\mathbf{u}}{dt} + \nabla \cdot \mathbf{P} = \mathbf{F} + \mathbf{j} \times \mathbf{B} \tag{1.22}$$

Here ρ is the density, $d\mathbf{u}/dt$ is the comoving derivative of the velocity $\mathbf{u}$ of the fluid, $\mathbf{P}$ is the material stress tensor, and $\mathbf{F}$ represents the usual body forces such as gravity or an electrostatic force on any net space-charge density. The latter are usually negligible. At the same time a new energy source $\mathbf{j} \cdot \mathbf{E}$ enters in the energy equation. The terms containing $\mathbf{j}$ represent the coupling between the plasma and the electromagnetic field.

In highly ionized gases the large discrepancy in mass between the electrons and ions has interesting consequences. The viscosity is always dominated by the ions, whereas the thermal-transport coefficients, except for transport across strong magnetic fields, are controlled by the electrons. Quite in general the complications arising from the two-fluid aspect of the macroscopic plasma model appear in the transport coefficients. For instance, in many respects the most important and at the same time the most complex transport equation is the generalized Ohm's law connecting the current (charge transport) with the gradients of the electric and the fluid stresses.   After a Fourier transformation in time this relation can be written

$$\mathbf{j} - \alpha \cdot \nabla T = \sigma(\omega) \cdot \left(\mathbf{E}' + \frac{\nabla p_e}{n_e e} \right) \tag{1.23}$$

where α is the thermoelectric tensor.   In the frame of the fluid the electric current is carried predominantly by the electrons.   The conductivity σ is therefore a tensor relating the current density $\mathbf{j}$ to the electric field in the fluid frame $\mathbf{E}' = \mathbf{E} + \mathbf{u} \times \mathbf{B}/c$, reduced by the electron pressure gradient per electron.   The components of this tensor must describe not only the dissipation (electron-ion collisions) and the Hall effect (Lorentz force on the electrons) but also the high-frequency response (electron inertia).   Thus Ohm's law expresses the properties of the plasma both as a conductor at low frequencies ($\omega < \nu$) and as a dielectric for very rapid oscillations of the electric field ($\omega > \omega_p$).   In the intermediate domain, and particularly in the presence of a magnetic field, the electric properties of the plasma are more complex, and no such simple terms are applicable.

1.5.2. The Role of the Macroscopic Electric Field

As pointed out in Sec. 1.3.1 the quasineutrality of a plasma does not permit the a priori assumption that $\nabla \cdot \mathbf{E} = 0$.   It follows that the value of $\mathbf{E}$ cannot be derived directly from boundary conditions.   In a complete description of the dynamics of the various charge species in the plasma the electric field is of course an essential variable and must be determined in a self-consistent manner.   In particular it should be noted that the positive- and negative-charge species of a plasma interact with each other primarily by means of space-charge fields and only secondarily through individual collisions.   In the grossly neutral single-fluid approximation it is therefore more logical at low frequencies to consider the dynamics of the entire fluid as primary.   This means that the macroscopic behavior of the plasma is determined by Maxwell's equations and by the equation of motion (1.22), in which $\mathbf{E}$ is not an essential variable.   Ohm's law (1.23) is then used either to eliminate the electric field or to determine it, if it is needed for any reason (e.g., to fit boundary conditions of electric potentials).

As an example, let us consider a sphere of hot plasma freely expanding into vacuum, and let $\mathbf{B} = 0$ everywhere.   The total (ion plus electron) pressure in Eq. (1.22) forces the mass to expand at a rate somewhat higher than the ion thermal speed.   The physical reason is that the electrons with their large thermal speeds try to expand rapidly, leaving the slow ions behind.   The resulting charge separation sets up an electric field which holds the electrons back and accelerates the ions outward. The net expansion, which carries no net current, is sometimes called *ambipolar*.   In the interior of the plasma this "ambipolar" electric field is precisely predicted by Eq. (1.23); i.e., it opposes the electron pressure gradient.

It is interesting to consider also what happens to this expansion if a quenching surface is interposed, such as the walls of a plasma container, onto which the ionized gas transmits pressure without reflection of the charged particles.   It is still true that no net current can be collected by this surface.   Thus a negative charge layer is set up on the boundary by which most of the electrons are turned back into the plasma and ions are accelerated toward the surface.   The thickness of the resulting space-charge *sheath* is of the order of the Debye length, and the potential across it corresponds to a few times the mean electron energy.   The structure of the sheath cannot of course be analyzed on the basis of a one-fluid model.

Another instructive and very important situation involving a macroscopic electric field exists whenever the Lorentz term $\mathbf{j} \times \mathbf{B}$ in Eq. (1.22) is not zero.   If the plasma is treated as a single fluid, the tensor conductivity σ in Eq. (1.23) requires that in the frame of this fluid an electric-field component appears which is given by $\mathbf{E}_H = \mathbf{j} \times \mathbf{B}/en_e$.   This "Hall field" simply reflects the fact that in the frame of the fluid the Lorentz force is exerted primarily on the moving electrons, and these can transmit the momentum to the ions only by means of a macroscopic electric field.   Above all it should be noted that this electric field is present even when the resistivity of the fluid is zero.†   It is therefore essential that the Hall effect be taken into consideration, or its neglect be carefully justified, whenever the plasma is to be considered as a unified electrically conducting medium.

Quite in general we note that the most significant function of the self-consistent electric field in a plasma is the macroscopic coupling of the electron and the ion fluids so that the quasineutrality of Sec. 1.3.1 is preserved at all times.

1.5.3. Magnetohydrodynamics

The current density $\mathbf{j}$ in Eq. (1.22) can be expressed in terms of the magnetic field by means of Ampère's law.   If the electric field and other forces can be neglected in the equation of motion, as is usually justified, we can

---

† Some analyses of macroscopic plasma behavior in the past have been in error because the neglected Hall effect rendered the solutions inconsistent with, for instance, the existing electrical boundary conditions.

then write Eq. (1.22) in the form

$$\rho \frac{d\mathbf{u}}{dt} + \nabla \cdot \left(\mathbf{P} + \frac{B^2}{8\pi} \mathbf{I} - \frac{\mathbf{BB}}{4\pi} \right) = 0 \tag{1.24}$$

where $\mathbf{I}$ denotes the identity matrix and $\mathbf{BB}$ is a diadic. This relation merely expresses the fact that the magnetic components of the Maxwell stress tensor affect the dynamics of a conducting fluid in exactly the same way as the material stresses. This is the basis for the merger between fluid dynamics and electrodynamics that has become known under the name *magnetohydrodynamics*, or, better perhaps, *magnetofluiddynamics*. In particular, when the conductivity is so high that the magnetic field can be considered as practically "frozen" into the fluid, we note that the well-known elastic properties of the magnetic field are very effectively communicated to the fluid. The scalar $B^2/8\pi$ is at once recognized as an equivalent pressure, whereas $\mathbf{BB}/4\pi$ must be interpreted as a tension along the field lines. Thus Eq. (1.24) leads to several very important conclusions.

The elasticity mentioned above is responsible for the propagation of compressional waves across the magnetic field. These are very similar to acoustic waves, but they require neither a finite particle pressure nor interparticle collisions for the transport of momentum. In strict analogy to the speed of sound the propagation speed of these disturbances is given by $V^2 = (\gamma p + B^2/4\pi)/\rho.$† More interesting even is the existence of transverse (shear) waves that can travel along the magnetic field direction because of the finite tensile strength of the magnetized medium. These so-called Alfvén waves travel with the velocity $V = B/\sqrt{4\pi\rho}$. The latter can also be interpreted as genuine electromagnetic waves propagating at a speed $V = c/\sqrt{K}$, where $K = 1 + 4\pi\rho c^2/B^2$ is the low-frequency transverse dielectric constant of the plasma. Neglecting unity as compared with K is identical with neglecting the displacement current in going from Eq. (1.22) to Eq. (1.24). The description is restricted to frequencies well below the ion gyrofrequency, because these simple results are obtained only if the inertial effects contained in $\sigma(\omega)$ of Eq. (1.23) are negligible. As indicated in Sec. 1.4.1, these phenomena can also be perfectly well understood on an individual-particle orbit model.

Equation (1.24) can of course also be used to describe the acceleration of plasma by means of magnetic forces such as are used in plasma-propulsion devices or the inverse process, i.e., the production of electromagnetic energy in magnetohydrodynamic generators.

Finally, when $\mathbf{u} = 0$, Eq. (1.24) expresses the condition for magnetohydrostatic equilibrium. In the presence of finite gradients this

† Note that the magnetic field acts like a compressible medium with 2 degrees of freedom, so that $\gamma_M = 2$.

solution is of course consistent with Ohm's law (1.23) only if σ does not contain any dissipation, i.e., if the resistivity is zero.   The search for such equilibria in finite geometries that are stable against small perturbations constitutes the first problem in the control of thermonuclear-energy production: the magnetic containment of high-temperature plasmas. The stability requirement is a serious restriction on the acceptable confinement systems: magnetohydrostatic equilibria are stable only if the field-energy density, at least when integrated along the field lines, increases in all directions with distance from the plasma.   This condition is a direct consequence of the fact that plasma is a diamagnetic medium; i.e., it experiences a net force in the direction of decreasing magnetic field.

The diamagnetic behavior is already apparent in Eq. (1.24).   Consider, for instance, the equilibrium $d\mathbf{u}/dt = 0$ if the magnetic field lines are straight and parallel to each other.   In this special case Eq. (1.24) can be integrated directly, yielding the simple result

$$P_\perp + \frac{B^2}{8\pi} = \text{const} \tag{1.25}$$

where $P_\perp$ denotes the pressure exerted in the direction transverse to the field lines.   This says that the field must be weaker inside the plasma, where $P_\perp > 0$.   The origin of this plasma diamagnetism can be readily understood with the help of a particle-orbit picture (see Fig. 1.2).   The

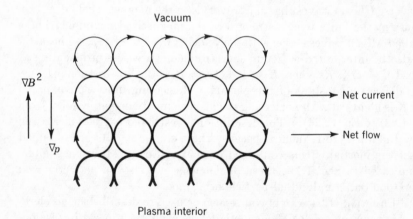

Fig. 1.2   Schematic sketch of gyration orbits, showing origin of plasma diamagnetism.   Heavy lines indicate regions of large microscopic current circulation.

gyrating motion of each charge represents a microscopic current which is always in such a direction that it reduces the magnetic field.   In the interior of a plasma these currents overlap, however, and cancel each other except in regions of gradients where the resultant unbalance exactly reproduces condition (1.25). The current density j in this case is thus caused not by guiding-center drifts but by a phenomenon that might be termed *net circulation*, given by the curl of the magnetic moment per unit volume.

It should be noted that this result is not inconsistent with the statements of Sec. 1.5.2 concerning the Hall effect, although in Fig. 1.2 we have implied that $E \times B = 0$. Actually, it is easily verified that the same unbalance in orbit density that is responsible for the current also represents a mass flow $u \times B \neq 0$. In a frame of reference in which $u = 0$ the orbits are distorted because there $E \times B \neq 0$, and the guiding-center motion exactly cancels the mass flow shown in Fig. 1.2. Quite in general, when the approximations made are equivalent, the analysis of orbits in self-consistent average fields gives of course the same result as the macroscopic description. The latter is usually much simpler, but the former contains more detailed information.

In this chapter we have attempted to present a condensed survey of the field of plasma physics by briefly discussing several of the most outstanding features of this subject. It is our hope that the remaining chapters in this volume will serve to amplify and deepen these matters, particularly those areas which have not yet been adequately treated in existing texts.

2

ADIABATIC CHARGED PARTICLE MOTION

NORTHROP

THEODORE G. NORTHROP, *Theoretical Division, Goddard Space Flight Center, Greenbelt, Maryland*

2

2.1 INTRODUCTION

The motion of each particle in a collection of charged particles is determined by the fields of the other particles, plus the fields due to external sources. To understand the behavior of the collection, it is necessary to understand first the behavior of a single particle in prescribed fields. If self-consistency conditions are applied to the fields instead of regarding them as prescribed, equations for the collective behavior of the plasmas result. Solution of these equations will give the fields, which can then be regarded as prescribed in interpreting the role of the particle motions in producing the collective behavior which has been calculated.

If the prescribed electric and magnetic fields are zero or uniform in space and constant in time, the particle motion is elementary. If the fields are not so simple, an exact integration of the equations of motion generally is impossible. Nevertheless, some approximate description of the motion is needed to deal with plasmas analytically. The adiabatic approximation is such an approximation and is the subject of this chapter.

This work was performed under the auspices of the U.S. Atomic Energy Commission.

2.2 THE GUIDING-CENTER MOTION OF NONRELATIVISTIC PARTICLES

In a uniform magnetic field that is constant in time a charged particle moves in a helical path.   The motion may be described exactly as motion about a circle whose center is moving along a line of force.   If the field is not quite uniform and not quite time-independent, one expects that the motion will not be quite helical; one also expects that something approximating helical motion will still be discernible and therefore that a good approximation will contain gyration about a center that now may move at right angles to the line of force as well as along it.   This expectation is indeed correct, and the equations governing this *guiding-center motion* can be derived by following one's physical intuition.   To do this, let $\mathbf{r} = \mathbf{R} + \boldsymbol{\varrho}$, where the vectors are defined in Fig. 2.1.   To correspond to the picture of

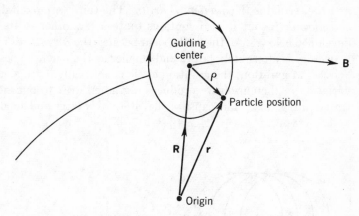

Guiding center

ρ

B

Particle position

R   r

Origin

Fig. 2.1   The charged particle gyrates about its guiding center.

rapid gyration about the guiding center, let $\boldsymbol{\varrho} = \rho(\hat{\mathbf{e}}_2 \sin \omega t + \hat{\mathbf{e}}_3 \cos \omega t)'$ where ω is the angular frequency of gyration $eB(\mathbf{R})/mc$, $B(\mathbf{R})$ is the magnetic field at $\mathbf{R}$, and $\hat{\mathbf{e}}_2(\mathbf{R})$ and $\hat{\mathbf{e}}_3(\mathbf{R})$ are unit vectors perpendicular to $\mathbf{B}(\mathbf{R})$ and to each other.   If $\mathbf{R} + \boldsymbol{\varrho}$ is now substituted into the equation of motion for the particle

$$m\ddot{\mathbf{r}} = \frac{e\dot{\mathbf{r}}}{c} \times \mathbf{B}(\mathbf{r},t) + e\mathbf{E}(\mathbf{r},t) \tag{2.1}$$

and an average is taken over a period of the gyration, the result after a little algebra with the unit vectors is [1,2]

$$\ddot{\mathbf{R}} = \frac{e}{m}\left[\mathbf{E}(\mathbf{R}) + \frac{\dot{\mathbf{R}}}{c} \times \mathbf{B}(\mathbf{R}) \right] - \frac{M}{m}\nabla B(\mathbf{R})$$

$$+ \text{ terms proportional to } \frac{m}{e} \tag{2.2}$$

Here M is the well-known magnetic moment $e\rho^2\omega/2c = mv_\perp{}^2/2B$, where $v_\perp$ is the particle velocity perpendicular to $\mathbf{B}(\mathbf{R})$. In (2.2) only terms through zero order in m/e have been kept; m/e can be used as the expansion parameter because, if (2.1) is written in suitable dimensionless form, the dimensionless parameter that appears is the gyration radius divided by the dimensions of the system, and m/e is proportional to this ratio.

The component of $\dot{\mathbf{R}}$ perpendicular to $\mathbf{B}(\mathbf{R})$ in (2.2) is the guiding-center velocity perpendicular to $\mathbf{B}(\mathbf{R})$. It is the so-called drift velocity and is obtained by taking the vector product of (2.2) with $\mathbf{B}$. We have

$$\dot{\mathbf{R}}_\perp = \frac{c\mathbf{E} \times \hat{\mathbf{e}}_1}{B} + \frac{Mc}{e}\frac{\hat{\mathbf{e}}_1 \times \nabla B}{B} + \frac{mc}{e}\frac{\hat{\mathbf{e}}_1 \times \ddot{\mathbf{R}}}{B} + 0(\varepsilon^2) \tag{2.3}$$

where ε is m/e, $\hat{\mathbf{e}}_1$ is $\mathbf{B}/B$, and all field quantities are evaluated at $\mathbf{R}$. There are three drift terms here. The first is the well-known $\mathbf{E} \times \mathbf{B}$ *drift*, and the second is the *gradient B drift*. The third term contains the *line-curvature drift*, but it also contains quite a few other drifts, as will be developed below. All the drifts occur because the curvature of the particle trajectory is alternately larger and smaller as the particle goes around its "circle" of gyration; the gyration circle is not really quite a circle. This variation in the curvature produces a gradual drift to one side, as illustrated in Fig. 2.2. The cause of the alternately large and small curvature

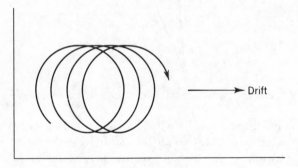

Fig. 2.2   A drift.

is different for each of the drifts. The $\mathbf{E} \times \mathbf{B}$ and ∇B drifts have been frequently described before.[3,4] The six drifts that are contained in the last term of (2.3) also can be given geometric interpretations. That term may be expanded by writing $\ddot{\mathbf{R}}$ as

$$\frac{d}{dt}(\dot{\mathbf{R}}_\perp + \hat{\mathbf{e}}_1\dot{\mathbf{R}} \cdot \hat{\mathbf{e}}_1) = \hat{\mathbf{e}}_1\frac{dv_\parallel}{dt} + v_\parallel\frac{d\hat{\mathbf{e}}_1}{dt} + \frac{d\dot{\mathbf{R}}_\perp}{dt}$$

where $v_\parallel$ is $\dot{\mathbf{R}} \cdot \hat{\mathbf{e}}_1(\mathbf{R})$, the component of guiding-center velocity parallel to the line of force at $\mathbf{R}$. We need $d\dot{\mathbf{R}}_\perp/dt$ only to zero order in ε, since the entire term is multiplied by ε in (2.3). By iteration of (2.3) we obtain $d\dot{\mathbf{R}}_\perp/dt = d\mathbf{u}_E(\mathbf{R})/dt + 0(\varepsilon)$, where $\mathbf{u}_E$ is $c\mathbf{E} \times \hat{\mathbf{e}}_1/B$. Also, $d\hat{\mathbf{e}}_1(\mathbf{R})/dt$ is needed. It is the rate of change of the unit vector as one follows the guiding center. This unit vector changes direction in a time-dependent magnetic field even in the absence of guiding-center motion. In addition the guiding center sees a change in $\hat{\mathbf{e}}_1$ as it moves in a field whose direction in space is not constant. Consequently, the total derivative

$$\frac{d\hat{\mathbf{e}}_1}{dt} = \frac{\partial \hat{\mathbf{e}}_1}{\partial t} + v_\parallel \frac{\partial \hat{\mathbf{e}}_1}{\partial s} + \mathbf{u}_E \cdot \nabla \hat{\mathbf{e}}_1 + 0(\varepsilon)$$

where s is distance along the line of force. Similarly

$$\frac{d\mathbf{u}_E}{dt} = \frac{\partial \mathbf{u}_E}{\partial t} + v_\parallel \frac{\partial \mathbf{u}_E}{\partial s} + \mathbf{u}_E \cdot \nabla \mathbf{u}_E$$

With these substitutions, the total drift velocity becomes

$$\dot{\mathbf{R}}_\perp = \frac{\hat{\mathbf{e}}_1}{B} \times \left(-c\mathbf{E} + \frac{Mc}{e} \nabla B + \frac{mc}{e} v_\parallel \frac{d\hat{\mathbf{e}}_1}{dt} + \frac{mc}{e} \frac{d\mathbf{u}_E}{dt} \right)$$

$$= \frac{\hat{\mathbf{e}}_1}{B} \times \left[-c\mathbf{E} + \frac{Mc}{e} \nabla B + \frac{mc}{e} \left(v_\parallel \frac{\partial \hat{\mathbf{e}}_1}{\partial t} + v_\parallel^2 \frac{\partial \hat{\mathbf{e}}_1}{\partial s} \right. \right.$$

$$\left. \left. + v_\parallel \mathbf{u}_E \cdot \nabla \hat{\mathbf{e}}_1 + \frac{\partial \mathbf{u}_E}{\partial t} + v_\parallel \frac{\partial \mathbf{u}_E}{\partial s} + \mathbf{u}_E \cdot \nabla \mathbf{u}_E \right) \right] + 0(\varepsilon^2) \quad (2.4)$$

where all quantities are evaluated at $\mathbf{R}$. The term proportional to $\partial \hat{\mathbf{e}}_1/\partial s$ is the well-known *line-curvature drift*. However, the other five terms in the second parentheses, although possibly less familiar, should not be overlooked. In practical cases the electric fields are often so small that the four terms containing $\mathbf{u}_E$ are negligible, and the field lines may change direction so slowly that the $\partial \hat{\mathbf{e}}_1/\partial t$ drift is small. But these five terms in the second parentheses are not necessarily small, and situations where each is of primary importance are known in plasma physics. For example, the term proportional to $\mathbf{u}_E \cdot \nabla \mathbf{u}_E$ is responsible for the shear, or Helmholtz, instability of a plasma.[2,5] Shears occur at the solar wind–geomagnetic field interface, where the solar plasma slides over the geomagnetic field.

The $\partial \hat{\mathbf{e}}_1/\partial t$ drift is an easy one to understand geometrically. If the direction of the magnetic field changes without a change in the particle velocity, then some of what was "parallel" velocity will become "perpendicular," and vice versa. In other words, if there is a change in the reference direction, with respect to which one defines parallel and perpendicular, then the respective components of velocity will change. It is easy to work out the details and see that there is a periodic variation (at the gyration frequency) in the curvature of the particle trajectory while the line of force changes direction.   This leads to a drift, just as in the more familiar case of the $\mathbf{E} \times \mathbf{B}$ and ∇B drifts.

The component of (2.2) parallel to the magnetic field gives the parallel acceleration of the guiding center.   The scalar product of (2.2) with $\hat{\mathbf{e}}_1(\mathbf{R})$ is

$$\ddot{\mathbf{R}} \cdot \hat{\mathbf{e}}_1 = \frac{e}{m} E_{\parallel} - \frac{M}{m} \hat{\mathbf{e}}_1 \cdot \nabla B + 0(\varepsilon) \tag{2.5}$$

where $E_{\parallel}$ is $\mathbf{E}(\mathbf{R}) \cdot \hat{\mathbf{e}}_1(\mathbf{R})$.   The parallel acceleration $dv_{\parallel}/dt$ is $(d/dt)(\dot{\mathbf{R}} \cdot \hat{\mathbf{e}}_1)$, which differs from $\ddot{\mathbf{R}} \cdot \hat{\mathbf{e}}_1$ by $\dot{\mathbf{R}} \cdot d\hat{\mathbf{e}}_1/dt$; and since the latter equals $(\hat{\mathbf{e}}_1 v_{\parallel} + \mathbf{u}_E) \cdot (d\hat{\mathbf{e}}_1/dt) + 0(\varepsilon)$, then

$$\frac{dv_{\parallel}}{dt} = \frac{e}{m} E_{\parallel} - \frac{M}{m} \frac{\partial B}{\partial s} + \mathbf{u}_E \cdot \frac{d\hat{\mathbf{e}}_1}{dt} + 0(\varepsilon) \tag{2.6}$$

The term $v_{\parallel}\hat{\mathbf{e}}_1 \cdot d\hat{\mathbf{e}}_1/dt$ vanished because $\hat{\mathbf{e}}_1$ is a unit vector.   The term $-(M/m)(\partial B/\partial s)$ is the usual mirror effect that produces reflection of particles and makes them oscillate north and south in the geomagnetic field, thus trapping them.   The total time derivative $d\hat{\mathbf{e}}_1/dt$ may be expanded to $(\partial \hat{\mathbf{e}}_1/\partial t) + (v_{\parallel} \partial \hat{\mathbf{e}}_1/\partial s) + \mathbf{u}_E \cdot \nabla \hat{\mathbf{e}}_1$, just as in the drift equation.   This $\mathbf{u}_E \cdot d\hat{\mathbf{e}}_1/dt$ term is another example of an effect caused by a change in the reference direction.   If the electric field is small, the term may be negligible.

To conclude this section, expressions for the drift velocity and parallel acceleration [Eqs. (2.4) and (2.6)] are presented in a form useful for computational purposes. Z and A have their usual meanings of atomic number and mass in atomic mass units.   For negatively charged particles Z is to be taken as negative. $\mathbf{E}$ is in volts per centimeter, $\mathbf{B}$ is in gauss, velocities are in centimeters per second, time is in seconds, and gradients are in cm^{-1}.

$$\dot{\mathbf{R}}_{\perp} = \frac{\hat{\mathbf{e}}_1}{B} \times \left\{ -10^6\mathbf{E} + 5.19 \times 10^{-5} \frac{A}{Z} \frac{v_{\perp}^2}{B} \nabla B \right.$$
$$+ 1.039 \times 10^{-4} \frac{A}{Z} \left[v_{\parallel} \frac{\partial \hat{\mathbf{e}}_1}{\partial t} + v_{\parallel}^2 \frac{\partial \hat{\mathbf{e}}_1}{\partial s} + v_{\parallel}\mathbf{u}_E \cdot \nabla \hat{\mathbf{e}}_1 \right.$$
$$\left. \left. + \frac{\partial \mathbf{u}_E}{\partial t} + v_{\parallel} \frac{\partial \mathbf{u}_E}{\partial s} + \mathbf{u}_E \cdot \nabla \mathbf{u}_E \right] \right\}$$

where $\mathbf{u}_E = 10^6 \dfrac{\mathbf{E} \times \hat{\mathbf{e}}_1}{B}$

$$\frac{dv_{\parallel}}{dt} = 0.963 \times 10^{12} \frac{Z}{A} E_{\parallel} - \frac{v_{\perp}^2}{2B} \frac{\partial B}{\partial s}$$
$$+ \mathbf{u}_E \cdot \left(\frac{\partial \hat{\mathbf{e}}_1}{\partial t} + v_{\parallel} \frac{\partial \hat{\mathbf{e}}_1}{\partial s} + \mathbf{u}_E \cdot \nabla \hat{\mathbf{e}}_1 \right)$$

For a 100-Mev proton at one-half an earth radius from the equator in the inner Van Allen belt, these formulas lead to a precessional time about the earth of approximately 1 min.

2.3 ENERGY CHANGES

The kinetic energy W of a particle, averaged over a gyration, is $(mv_\parallel^2/2) + (mu_E^2/2) + MB$. This can be demonstrated, but it is really obvious: the first two terms are the energy of the guiding-center motion, and MB is the energy of rotation about the guiding center.   The parallel energy $W_\parallel$ is $mv_\parallel^2/2$, and the average perpendicular energy $W_\perp$ is $(mu_E^2/2) + MB$. The rate of change dW/dt of total kinetic energy, averaged over a gyration, can be deduced in a formal fashion, but the result is so intuitively correct that the procedure will be omitted here.   The result is

$$\frac{1}{e}\frac{dW}{dt} = \dot{\mathbf{R}} \cdot \mathbf{E}(\mathbf{R},t) + \frac{M}{e}\frac{\partial B}{\partial t}(\mathbf{R},t) + 0(\varepsilon^2) \tag{2.7}$$

where $\dot{\mathbf{R}}$ is $\hat{\mathbf{e}}_1 v_\parallel + \dot{\mathbf{R}}_\perp$.   The first term on the right side is energy increase resulting from the average particle motion in the electric field, while the second term is the induction effect, or *betatron acceleration*, caused by the curl of $\mathbf{E}$ acting about the circle of gyration.   Part of the energy increase given by (2.7) is fed into the parallel energy and the rest into perpendicular energy.   Simultaneously, energy is exchanged between parallel and perpendicular components by the mirror effect, the exchange occurring without a change in total kinetic energy.   The process may be visualized as in Fig. 2.3, where the partition of dW/dt between $dW_\perp/dt$ and $dW_\parallel/dt$ comes from the formal analysis.   Note that $M\,\partial B/\partial t$ is only part of the

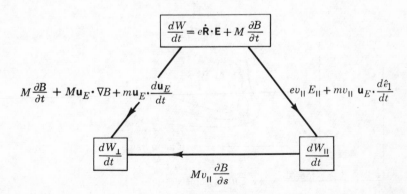

Fig. 2.3   Energy changes in a time-dependent field.

perpendicular-energy increase; $e\dot{\mathbf{R}} \cdot \mathbf{E}$ contains the rest of the perpendicular-energy increase plus the entire rate of increase of parallel energy.

2.4 FERMI ACCELERATION

Fermi acceleration[6–10] is a special case of the adiabatic-energy change of the preceding section. Fermi suggested that repeated collisions between a charged particle and moving clumps of magnetized plasma in space would accelerate a few particles to extreme energies. In effect, the clumps act as massive particles with which the high-energy particles attempt to establish kinetic equilibrium. The many particles in a clump, although of low energy, give it a very large mass. Thus, at thermal equilibrium, the high-energy particles will have very high energies indeed. The statistics of these collisions will not be discussed here (see Teller[8]); instead, details of a single Fermi-type collision will be interpreted in the light of the preceding section.

Equation (2.7) applies to any adiabatic situation, but Fermi had in mind special cases, namely, those where there is a frame of reference (that of the clump) in which the magnetic field is static and there is no electric field. In the frame of the clump there is therefore no energy gain or loss by the particle. The collision is elastic, and its net effect is to alter the velocity of the guiding center. In the earth's frame, with respect to which the clump is in motion, there may be an energy change, somewhat in analogy to a ball struck by a baseball bat. A particle will lose energy if the clump is overtaken by the particle, and it will gain if the clump overtakes the particle.

Suppose that the earth is fixed at 0 in Fig. 2.4 and that the clump is fixed in a frame 0* moving at velocity $\mathbf{u}$ with respect to the earth.

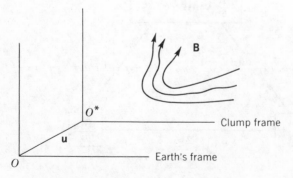

Fig. 2.4   Fermi acceleration.

The rate of energy gain, from (2.7) and (2.4), is

$$\frac{dW}{dt} = ev_\parallel E_\parallel + e\dot{\mathbf{R}}_\perp \cdot \mathbf{E} + M\frac{\partial B}{\partial t}$$

$$= ev_\parallel E_\parallel + M\mathbf{u}_E \cdot \nabla B + mv_\parallel \mathbf{u}_E \cdot \frac{d\hat{\mathbf{e}}_1}{dt}$$

$$+ m\mathbf{u}_E \cdot \frac{d\mathbf{u}_E}{dt} + M\frac{\partial B}{\partial t} \qquad (2.8)$$

Quantities in (2.8) must now be expressed in terms of $\mathbf{u}$. For example, the electric field seen in 0 is

$$-\frac{\mathbf{u}}{c} \times \mathbf{B}^* \cong -\frac{\mathbf{u}}{c} \times \mathbf{B}$$

The magnetic fields $\mathbf{B}$ and $\mathbf{B}^*$ are equal through order $\mathbf{u}/c$, that is, non-relativistically. The actual cosmic problem may have relativistic clump velocities, and relativistic energy for the colliding particle. Relativistic adiabatic motion will be reviewed in the next section, but the non-relativistic case is adequate here for illustrative purposes.

The following relations also hold, as seen from the earth's frame of reference:

$$\frac{\partial B}{\partial t} = -\mathbf{u} \cdot \nabla B$$

$$\frac{d\hat{\mathbf{e}}_1}{dt} = (v_\parallel - u_\parallel)\frac{\partial \hat{\mathbf{e}}_1}{\partial s}$$

and $$\mathbf{u}_E \cdot \frac{d\mathbf{u}_E}{dt} = -(v_\parallel - u_\parallel)u_\parallel \mathbf{u}_\perp \cdot \frac{\partial \hat{\mathbf{e}}_1}{\partial s}$$

Substitution into (2.8) gives

$$\frac{1}{e}\frac{dW}{dt} = -\frac{M}{e}u_\parallel \frac{\partial B}{\partial s} + \frac{m}{e}(v_\parallel - u_\parallel)^2 \mathbf{u}_\perp \cdot \frac{\partial \hat{\mathbf{e}}_1}{\partial s} + 0(\varepsilon^2) \qquad (2.9)$$

If the magnetic field in the clump is such that the guiding center moves along a straight line of force, the last term in (2.9) is zero and one then has what Fermi named *type a acceleration*. As seen from the clump frame, the particle moves into an increasing magnetic field (magnetic mirror) along a straight line of force and reflects with no energy change. As viewed from the earth's frame, there will be an energy change.

On the other hand, if the field line along which the guiding center moves is curved, and if the magnitude of the field is constant along the line, the first term on the right side of (2.9) vanishes. The last term is then Fermi's *type b acceleration*. In either case, (2.9) may be integrated with respect to time to give the total energy change produced by the particle's collision with the clump. Types a and b really differ only in the mechanism whereby the guiding-center velocity is reversed in the clump frame. In either case the energy change seen by the observer on the earth is $2mu_\parallel(v_\parallel - u_\parallel)$, where $v_\parallel$ is the component of guiding-center velocity parallel to the magnetic field after the collision (i.e., far from the clump) and $u_\parallel$ is the component of $\mathbf{u}$ parallel to that field. This energy change is naturally more easily obtained from the fact that the velocity in the static frame is merely reversed by the collision. But our purpose here has been to apply (2.7) in the frame of reference in which there is an energy change. Equation (2.9) can also be integrated over a collision without breaking it up into the special cases a and b.

Fermi acceleration and betatron acceleration are sometimes invoked as distinct processes whereby a particle gains energy. However, they are not distinct. If one follows the fate of the $(M/e)\,\partial B/\partial t$ term in the transition from (2.8) to (2.9), he finds that the term contributes to $-(M/e)u_\parallel(\partial B/\partial s)$, which is the type a acceleration. Consequently, betatron acceleration should not be viewed as a process distinct from Fermi acceleration, since it is part of type a. It *is* correct to distinguish between betatron acceleration and acceleration resulting from guiding-center motion in the electric field, since these appear as distinct terms in (2.8).

Pure betatron acceleration in space is improbable, since if there is a $\partial B/\partial t$, there will usually be an electric field at the guiding center, and the $\dot{\mathbf{R}} \cdot \mathbf{E}$ term in (2.7) will be nonvanishing.

2.5 RELATIVISTIC ADIABATIC MOTION

If the particle has relativistic energy, (2.1) is replaced by

$$\frac{d\mathbf{p}}{dt} = \frac{d}{dt}\frac{m_0\dot{\mathbf{r}}}{(1 - \beta^2)^{1/2}} = \frac{e}{c}\dot{\mathbf{r}} \times \mathbf{B}(\mathbf{r}) + e\mathbf{E}(\mathbf{r}) \qquad (2.10)$$

where $\mathbf{p}$ is the momentum, $\beta = v/c$, and m_0 is the rest mass. Three cases can be distinguished: when the electric field is zero, when its component $\mathbf{E}_\perp$ perpendicular to $\mathbf{B}$ is small, and when $\mathbf{E}_\perp$ is large.

If there is no electric field, the force on the particle is always at right angles to the velocity, with the result that the energy is constant. Then $m_0/(1 - \beta^2)^{1/2}$ can be removed from under the d/dt in (2.10), and the equation is identical with the nonrelativistic one for a particle of mass $m_0/(1 - \beta^2)^{1/2}$. All the preceding nonrelativistic theory, with $\mathbf{E}$ set equal to zero, now applies. In the following two equations the nonrelativistic guiding-center equations are rewritten with $m_0/(1 - \beta^2)^{1/2}$ replacing m. The drift velocity is

$$\dot{\mathbf{R}}_\perp = \frac{1}{(1 - \beta^2)^{1/2}} \frac{m_0 v_\perp{}^2}{2B} \frac{c}{e} \frac{\hat{\mathbf{e}}_1 \times \nabla B}{B}$$

$$+ \frac{1}{(1 - \beta^2)^{1/2}} \frac{m_0 c}{e} v_\parallel^2 \frac{\hat{\mathbf{e}}_1}{B} \times \frac{\partial \hat{\mathbf{e}}_1}{\partial s} \quad (2.11)$$

and the parallel force is

$$\frac{m_0}{(1 - \beta^2)^{1/2}} \frac{dv_\parallel}{dt} = - \frac{1}{(1 - \beta^2)^{1/2}} \frac{m_0 v_\perp{}^2}{2B} \frac{\partial B}{\partial s} \quad (2.12)$$

Nonrelativistically, the magnetic moment is $M = m v_\perp{}^2/2B$. Relativistically, the corresponding invariant is $M_r = m_0 v_\perp{}^2/2B(1 - \beta^2) = p_\perp{}^2/2m_0 B$. It is not obvious that this is the correct generalization of M for relativistic energy. It is easy enough to verify for the simple case of a particle in a uniform azimuthally symmetric field that changes with time. The general case is not so easy to prove. The adiabatic invariants will be further studied in the next section.

The parallel force in (2.12) is now larger by $(1 - \beta^2)^{-1/2}$ than would be predicted by the nonrelativistic equation for the same rest mass. Similarly, the drifts in (2.11) are faster by the same factor. These effects are caused by the increased gyration radius resulting from the relativistic mass increase. For example, the increased gyration radius increases the amount of field inhomogeneity sampled by the particle, hence increases the ∇B drift. Similarly, the parallel force increases because the larger gyration radius subjects the particle to a greater convergence of the field lines, and it is this convergence which produces the mirror effect. As illustrated in Fig. 2.5 it is the product of $v_\perp$ and the radial component of $\mathbf{B}$ that results in a parallel force.

If the electric field is sufficiently small (formally, of order ε), the four terms containing $\mathbf{u}_E$ in (2.4) become of order ε^2 and may be dropped. The drift proportional to $\partial \hat{\mathbf{e}}_1/\partial t$ will also probably be negligible, since $\nabla \times \mathbf{E}$ and $\partial \mathbf{B}/\partial t$ are related by the Maxwell equation. Then only the three familiar drifts remain. One may surmise that the correct relativistic modification is obtained by adding $c\mathbf{E} \times \mathbf{B}/B^2$ to (2.11) and $eE_\parallel$ to the parallel force in (2.12). This does in fact turn out to be the correct procedure, but it is not a deductive one, since (2.11) and (2.12) were derived by assuming no electric field. The relativistic case has been studied by Hellwig[1] and by Vandervoort[11] for $\mathbf{E}_\perp$ large (i.e., of order 1), and the small $\mathbf{E}_\perp$ results are a special case.

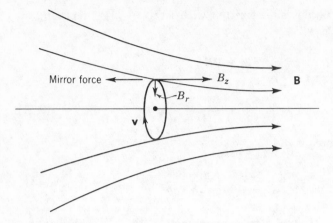

Fig. 2.5 The mirror effect.

The relativistic rate of energy change for $\mathbf{E}_\perp$ small is

$$\frac{dW}{dt} = e\dot{R} \cdot \mathbf{E} + M_r(1 - \beta^2)^{1/2}\frac{\partial B}{\partial t} \tag{2.13}$$

Only the betatron term has been altered, a comparison with (2.7) shows.

The complete guiding-center equations for large $\mathbf{E}_\perp$ are rather long and will not be repeated here.[11,12] Their principal features are corrections to existing terms of the small-$\mathbf{E}_\perp$ relativistic expressions above. Additionally, two new drift terms that are in the *direction* of $\mathbf{E}_\perp$ appear. They are pure relativistic effects that have no analog in the small-$\mathbf{E}_\perp$ relativistic case. One of these two drifts can be explained by the change in direction of $\mathbf{B}$ when a Lorentz transformation is made in the presence of an electric field. Basically, the drift is a result of the change in the reference direction with respect to which parallel and perpendicular are defined. Some of what was parallel velocity is converted to perpendicular velocity.

2.6 THE ADIABATIC INVARIANTS

2.6.1. The Magnetic Moment

The emphasis so far has been on the guiding-center motion and on energy changes. Not only are the guiding-center equations useful, but also valuable are quantities that are constant over long periods of guiding-center motion—i.e., any invariants of the adiabatic motion, or *adiabatic invariants*. They are not exact invariants of the particle motion, any more than the guiding-center equations are exact equations for the particle motion. Formal analysis[13,14] shows that there are at most three adiabatic invariants for the charged particle.   Each one is really an asymptotic series in a smallness parameter ε—a series of the form

$$\text{const} = a_0 + \varepsilon a_1 + \varepsilon^2 a_2 + \cdots$$

Systematic analysis[13,15] is essential for obtaining higher-order terms in the series.   Historically, however, the forms of the lowest-order invariants (i.e., the a_0's) were deduced by physical insight and by consideration of special cases.[3,14,16]

The connection with more formal theory was made later.   Such an evolutionary history is common in physical science.   In this chapter only invariance to lowest order (the a_0's) will be proved.

The formal theories also show that the adiabatic-invariant series are not the action integrals of the form $\oint p \, dq$, where p and q are canonical variables, but are instead Poincaré integral invariants of the form $\sum_i \oint p_i \, dq_i$, where the number of terms in the sum is the number of degrees of freedom of the canonical system.   However, the number of adiabatic invariants may vary from one to three, depending on the field geometry, as will become apparent shortly.   In general, the number of invariants is less than or equal to the number of degrees of freedom of the system.[13]

The first invariant is the magnetic moment, defined previously as $mv_\perp^2/2B$ for the nonrelativistic case; $mv_\perp^2/2B$ is really M_0 of the magnetic-moment series, $\text{const} = M_0 + \varepsilon M_1 + \varepsilon^2 M_2 + \cdots$.   The definition of $v_\perp$ was glossed over slightly at the beginning of this discussion.   If the component of $\mathbf{E}$ perpendicular to $\mathbf{B}$ is small, the $\mathbf{E} \times \mathbf{B}$ drift is much less than the particle velocity and the particle trajectory will be as in Fig. 2.2.   The motion is almost circular, and the $v_\perp$ to be used in the magnetic moment is the velocity about the circle.   When $\mathbf{E}_\perp$ is this small, the last four drifts in (2.4) will probably be negligible.   Suppose that $\mathbf{E}_\perp$ is now increased.   Eventually the trajectory will resemble a prolate cycloid as in Fig. 2.6.   There is no resemblance to circular motion in the laboratory

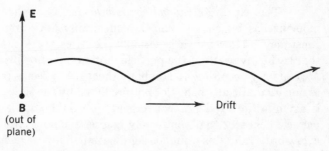

Fig. 2.6   Particle trajectory when $\mathbf{E}_\perp$ is large.

frame, but in the frame moving at $\mathbf{E} \times \mathbf{B}$ the motion is approximately circular again, as in Fig. 2.2.   It is the $v_\perp$ in this drifting frame that should be used in $mv_\perp{}^2/2B$.   Adiabatic theory therefore can still apply even when the perpendicular electric field is so large that the particle trajectory in the observer's frame shows no looping or resemblance to circular motion. One must only be careful to use the complete expressions in (2.4) and (2.6) and to define $v_\perp$ properly.

The invariance of M is easy to demonstrate for simple cases, like a time-dependent magnetic field with azimuthal symmetry and straight lines of force.   A proof for the most general situation (general time-dependent magnetic field and large electric field) seems to be rather long.[12,15,17]   The most general case for which a simple proof seems to exist is the static one, where the energy is constant; a large curl-free electric field may be present.   By conservation of energy,

$$\frac{\overset{.}{d}}{dt}\left(\frac{mv_\parallel^2}{2} + \frac{mu_E{}^2}{2} + MB + e\phi\right) = 0 \tag{2.14}$$

where ϕ is the electrostatic potential.   Recall that the invariance of M was not invoked in deriving the guiding-center equations.   Thus the value of $dv_\parallel/dt$ from (2.6) can be used to convert (2.14) to

$$\frac{d(MB)}{dt} = -\frac{e\,d\phi}{dt} - m\mathbf{u}_E \cdot \frac{d\mathbf{u}_E}{dt} - mv_\parallel\left(\frac{e}{m}E_\parallel - \frac{M}{m}\frac{\partial B}{\partial s} + \mathbf{u}_E \cdot \frac{d\hat{\mathbf{e}}_1}{dt}\right) \tag{2.15}$$

The total derivative $d\phi/dt$ equals $v_\parallel(\partial\phi/\partial s) + \dot{\mathbf{R}}_\perp \cdot \nabla\phi$, where $\dot{\mathbf{R}}_\perp$ is given by (2.4).   Putting it all together and doing a little vector algebra gives

$$\frac{d(MB)}{dt} = M\mathbf{u}_E \cdot \nabla B + Mv_\parallel\frac{\partial B}{\partial s} \equiv M\frac{dB}{dt} \tag{2.16}$$

or $\dfrac{dM}{dt} = 0$

The next two higher terms in the magnetic-moment series have also been derived.[17,18]   They are rather complicated.

The expression *nonadiabatic behavior* as applied to the magnetic moment has by custom come to mean any deviation of $mv_\perp{}^2/2B$ from constancy.   However, it is actually the series $M_0 + \varepsilon M_1 + \varepsilon^2 M_2 + \cdots$ that is the invariant of the particle motion, and not just M_0.   Therefore, M_0 can vary according to adiabatic theory.   It seems preferable to define as nonadiabatic any behavior not predicted by the series.   Since the series is asymptotic[19] and not convergent, it would not be surprising to see particle behavior that completely ignores the adiabatic predictions, even in low order, and this would be genuine nonadiabatic behavior.   Examples of such motion are known[12,20] for the magnetic moment.

2.6.2. The Second, or Longitudinal, Invariant

Another invariant of the particle motion, or really of the guiding-center motion, is

$$J = \oint p_\parallel \, ds \qquad (2.17)$$

where $p_\parallel$ is $mv_\parallel$, the guiding-center momentum parallel to the line of force. The invariant J exists if there is a mirror-type geometry, such that the guiding center oscillates back and forth along the lines of force while drifting slowly at right angles to them, as illustrated in Fig. 2.7.   For J

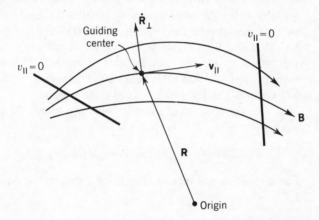

Fig. 2.7   Mirror geometry needed for existence of second adiabatic invariant.

to be constant, it is necessary that the drift be slow compared with $v_\parallel$, that is, that $\mathbf{E}_\perp$ be of order ε.   The integral is taken over a complete oscillation, the deviation of the guiding center from a line due to the drift during one oscillation being negligible if $\mathbf{E}_\perp$ is small.

The earliest suggestion that J is an invariant appears to have come from Rosenbluth.[16]   A proof of the invariance of J and some applications to laboratory magnetic field configurations was given by Kadomtsev[21] for a nonrelativistic particle in a static magnetic field.   A proof that remains valid at relativistic energies and that includes time-dependent fields has been given by Northrop and Teller[14] along with applications to the Van Allen radiation.   The proof of the invariance of J given below is for a nonrelativistic particle in a static field with no electric field; inclusion of nonstatic fields greatly increases the length of the proof.   Therefore, only the results will be given for the time-dependent case.   The time-dependent results will be needed to discuss the third invariant. Relativistic modifications do not seem materially to complicate the proofs.

To begin the proof of J, a curvilinear coordinate system will now be introduced. The three coordinates will be denoted by α, β, and s, where α and β are two parameters specifying the line of force and s denotes position along the line. (Distinguish this β from v/c in a previous section.) A system of nonintersecting lines can be generated as the intersections of two families of surfaces $\alpha(\mathbf{r}) =$ const and $\beta(\mathbf{r}) =$ const, where $\alpha(\mathbf{r})$ and $\beta(\mathbf{r})$ are two different functions of position. It is apparent that for a given system of lines the functions $\alpha(\mathbf{r})$ and $\beta(\mathbf{r})$ are not unique. Consider the simple example of straight lines of force. They can be generated by the intersections of two families of planes, by a family of planes and one of cylinders, etc. Among the many possible pairs of functions $\alpha(\mathbf{r})$ and $\beta(\mathbf{r})$ for a given magnetic field, there is a subclass for which the vector potential $\mathbf{A}$ is $\alpha\nabla B$ and $\mathbf{B}$ therefore is $\nabla\alpha \times \nabla\beta$. That such a subclass exists is not quite obvious, but it is not difficult to prove. The utility of the subclass is that for it $|\nabla\alpha \times \nabla\beta|/B$ is constant everywhere, being unity, and this fact reduces the algebra involved in the proof.

In the absence of electric fields the energy W equals $mv_\parallel^2/2 + MB$, so that

$$J(\alpha,\beta,M,W) = \oint \{2m[W - MB(\alpha,\beta,s)]^{1/2}\, ds\} \tag{2.18}$$

The instantaneous rate of change of J due to the particle drift $\dot{\mathbf{R}}_\perp$ in Fig. 2.7 is

$$\frac{dJ}{dt} = \frac{\partial J}{\partial \alpha}\frac{d\alpha}{dt} + \frac{\partial J}{\partial \beta}\frac{d\beta}{dt} \tag{2.19}$$

Differentiation of the integral in (2.18) gives

$$\frac{\partial J}{\partial \alpha} = -mM \oint \frac{ds}{[2m(W - MB)]^{1/2}} \frac{\partial B(\alpha,\beta,s)}{\partial \alpha}$$
$$\text{and} \quad \frac{\partial J}{\partial \beta} = -mM \oint \frac{ds}{[2m(W - MB)]^{1/2}} \frac{\partial B(\alpha,\beta,s)}{\partial \beta} \tag{2.20}$$

Because α and β are constant on a line of force, they are changed only by the drift velocity, and not by the parallel velocity. Therefore, $d\alpha/dt = \dot{\mathbf{R}}_\perp \cdot \nabla\alpha(\mathbf{R})$ and $d\beta/dt = \dot{\mathbf{R}}_\perp \cdot \nabla\beta(\mathbf{R})$. Substituting $\dot{\mathbf{R}}_\perp$ from (2.4), with the electric field zero, gives

$$\frac{d\alpha}{dt} = \frac{\hat{\mathbf{e}}_1}{B} \times \left(\frac{Mc}{e}\nabla B + \frac{mc}{e}v_\parallel^2\frac{\partial\hat{\mathbf{e}}_1}{\partial s}\right) \cdot \nabla\alpha \tag{2.21}$$

Consider now the quantity $(\partial\mathbf{R}/\partial\beta) \times \mathbf{B}$, where the guiding-center position

$\mathbf{R}$ is a function of (α,β,s),

$$\frac{\partial \mathbf{R}}{\partial \beta} \times \mathbf{B} = \frac{\partial \mathbf{R}}{\partial \beta} \times (\nabla \alpha \times \nabla \beta) = -\left(\nabla \alpha \cdot \frac{\partial \mathbf{R}}{\partial \beta}\right)\nabla \beta + \left(\nabla \beta \cdot \frac{\partial \mathbf{R}}{\partial \beta}\right)\nabla \alpha$$
(2.22)

By implicit differentiation of $\alpha = \alpha[\mathbf{R}(\alpha,\beta,s)]$ one finds that

$$\nabla \alpha \cdot \frac{\partial \mathbf{R}}{\partial \beta} = 0 \qquad \text{and} \qquad \nabla \beta \cdot \frac{\partial \mathbf{R}}{\partial \beta} = 1$$

Thus, $(\partial \mathbf{R}/\partial \beta) \times \mathbf{B} = \nabla \alpha$, and (2.21) becomes

$$\frac{d\alpha}{dt} = \frac{\hat{\mathbf{e}}_1}{B} \times \left(\frac{Mc}{e}\nabla B + \frac{mc}{e}v_\parallel^2 \frac{\partial \hat{\mathbf{e}}_1}{\partial s}\right) \cdot \left(\frac{\partial \mathbf{R}}{\partial \beta} \times \mathbf{B}\right)$$
(2.23)

Interchanging the dot and cross, and expanding the triple-vector product $\hat{\mathbf{e}}_1 \times [(\partial \mathbf{R}/\partial \beta) \times \mathbf{B}]$ gives

$$\frac{d\alpha}{dt} = -\left(\frac{Mc}{e}\nabla B + \frac{mc}{e}v_\parallel^2 \frac{\partial \hat{\mathbf{e}}_1}{\partial s}\right) \cdot \left(\frac{\partial \mathbf{R}}{\partial \beta} - \hat{\mathbf{e}}_1 \hat{\mathbf{e}}_1 \cdot \frac{\partial \mathbf{R}}{\partial \beta}\right)$$

$$= -\frac{Mc}{e}\frac{\partial \mathbf{R}}{\partial \beta} \cdot \nabla B - \frac{mc}{e}v_\parallel^2 \frac{\partial \mathbf{R}}{\partial \beta} \cdot \frac{\partial \hat{\mathbf{e}}_1}{\partial s} + \frac{Mc}{e}\hat{\mathbf{e}}_1 \cdot \frac{\partial \mathbf{R}}{\partial \beta}\hat{\mathbf{e}}_1 \cdot \nabla B$$

$$= -\frac{Mc}{e}\frac{\partial B(\alpha,\beta,s)}{\partial \beta} - \frac{mc}{e}v_\parallel^2 \frac{\partial \mathbf{R}(\alpha,\beta,s)}{\partial \beta} \cdot \frac{\partial \hat{\mathbf{e}}_1(\alpha,\beta,s)}{\partial s}$$

$$\qquad\qquad\qquad + \frac{Mc}{e}\hat{\mathbf{e}}_1 \cdot \frac{\partial \mathbf{R}}{\partial \beta}\frac{\partial B(\alpha,\beta,s)}{\partial s} \quad (2.24)$$

In the second term on the right-hand side of (2.24) we have

$$\frac{\partial \mathbf{R}}{\partial \beta} \cdot \frac{\partial \hat{\mathbf{e}}_1}{\partial s} = \frac{\partial}{\partial s}\left(\hat{\mathbf{e}}_1 \cdot \frac{\partial \mathbf{R}}{\partial \beta}\right)$$

since $\quad \hat{\mathbf{e}}_1 \cdot \dfrac{\partial}{\partial s}\dfrac{\partial \mathbf{R}(\alpha,\beta,s)}{\partial \beta} = \hat{\mathbf{e}}_1 \cdot \dfrac{\partial}{\partial \beta}\dfrac{\partial \mathbf{R}}{\partial s}$

and $\quad \hat{\mathbf{e}}_1 \cdot \dfrac{\partial}{\partial \beta}\dfrac{\partial \mathbf{R}}{\partial s} = \hat{\mathbf{e}}_1 \cdot \dfrac{\partial \hat{\mathbf{e}}_1}{\partial \beta}$

which is zero.   Therefore, the second term becomes

$$-\frac{mc}{e}v_\parallel^2 \frac{\partial}{\partial s}\left(\hat{\mathbf{e}}_1 \cdot \frac{\partial \mathbf{R}}{\partial \beta}\right) = -\frac{mc}{e}v_\parallel \frac{d}{dt}\left(\hat{\mathbf{e}}_1 \cdot \frac{\partial \mathbf{R}}{\partial \beta}\right)$$

From (2.6) $\quad \dfrac{\partial B}{\partial s} = -\dfrac{m}{M}\dfrac{dv_\parallel}{dt}$

so that the last two terms in (2.24) combine to

$$\frac{mc}{e}\frac{d}{dt}\left(v_\parallel \hat{\mathbf{e}}_1 \cdot \frac{\partial \mathbf{R}}{\partial \beta}\right)$$

The instantaneous rate of change of α finally is

$$\frac{d\alpha}{dt} = -\frac{Mc}{e}\frac{\partial B(\alpha,\beta,s)}{\partial \beta} - \frac{mc}{e}\frac{d}{dt}\left(v_{\parallel}\hat{\mathbf{e}}_1 \cdot \frac{\partial \mathbf{R}}{\partial \beta}\right) \tag{2.25}$$

By a similar analysis, we have

$$\frac{d\beta}{dt} = \frac{Mc}{e}\frac{\partial B}{\partial \alpha} + \frac{mc}{e}\frac{d}{dt}\left(v_{\parallel}\hat{\mathbf{e}}_1 \cdot \frac{\partial \mathbf{R}}{\partial \alpha}\right) \tag{2.26}$$

If $\partial B/\partial \beta$ from (2.25) is substituted into $\partial J/\partial \beta$ from (2.20), the result is

$$\begin{aligned}
\frac{\partial J}{\partial \beta} &= \oint \frac{m\,ds}{[2m(W - MB)]^{1/2}}\frac{e}{c}\left[\frac{d\alpha}{dt} + \frac{mc}{e}\frac{d}{dt}\left(v_{\parallel}\hat{\mathbf{e}}_1 \cdot \frac{\partial \mathbf{R}}{\partial \beta}\right)\right] \\
&= \frac{e}{c}\oint \frac{ds}{v_{\parallel}}\frac{d\alpha}{dt}
\end{aligned} \tag{2.27}$$

The integral of $m(d/dt)[v_{\parallel}\hat{\mathbf{e}}_1 \cdot (\partial \mathbf{R}/\partial \beta)]$ has vanished because $ds/v_{\parallel}$ is dt, and $v_{\parallel}$ is zero at the reflection points. Equation (2.27) can be written as

$$\frac{\partial J}{\partial \beta} = \frac{e}{c}\,T\langle\dot{\alpha}\rangle \tag{2.28}$$

where T is the time for a longitudinal oscillation and the broken brackets denote the time average over an oscillation. Similarly,

$$\frac{\partial J}{\partial \alpha} = -\frac{e}{c}\,T\langle\dot{\beta}\rangle \tag{2.29}$$

Equation (2.19) can then be written as

$$\frac{dJ}{dt} = \frac{e}{c}\,T(\langle\dot{\alpha}\rangle\dot{\beta} - \langle\dot{\beta}\rangle\dot{\alpha}) \tag{2.30}$$

Now this quantity is not zero except under very special circumstances, so that J is not instantaneously being conserved by the guiding-center motion. However, the rate of change of J averaged over a longitudinal oscillation is

$$\begin{aligned}
\oint \frac{ds}{v_{\parallel}}\frac{dJ}{dt} &= \frac{eT}{c}\left(\langle\dot{\alpha}\rangle\oint \frac{ds}{v_{\parallel}}\dot{\beta} - \langle\dot{\beta}\rangle\oint \frac{ds}{v_{\parallel}}\dot{\alpha}\right) \\
&= \frac{eT}{c}\left(\langle\dot{\alpha}\rangle\langle\dot{\beta}\rangle - \langle\dot{\beta}\rangle\langle\dot{\alpha}\rangle\right)
\end{aligned} \tag{2.31}$$

which is identically zero, and this is the important fact for the long-term motion.

Equations (2.28) and (2.29) are new equations of motion, with the guiding-center oscillation averaged out; they are the analog of the guiding-center equations of motion, which are the particle equations of motion with the particle gyration averaged out.

When Eqs. (2.28) and (2.29) are solved for $\langle \dot{\alpha} \rangle$ and $\langle \dot{\beta} \rangle$, they are at first sight suggestively canonical in form, with $J(\alpha,\beta,M,W)$ playing the role of Hamiltonian. But they are not quite canonical. In the first place the time of oscillation T is also a function of (α,β,M,W). Furthermore there are the time averages of $\dot{\alpha}$ and $\dot{\beta}$, rather than the instantaneous values. The first difficulty can be overcome by differentiating $J = J(\alpha,\beta,M,W)$ implicitly with respect to α and β to yield

$$\partial J(\alpha,\beta,M,W)/\partial \beta = -(\partial J/\partial W)\, \partial W(\alpha,\beta,M,J)/\partial \beta$$

etc., for $\partial J/\partial \alpha$. The factor $\partial J/\partial W$ is simply T, as can be verified from (2.18). Then

$$\langle \dot{\alpha} \rangle = -\frac{c}{e} \frac{\partial W(\alpha,\beta,M,J)}{\partial \beta}$$

and $\quad \langle \dot{\beta} \rangle = \frac{c}{e} \frac{\partial W}{\partial \alpha}$ \hfill (2.32)

Except for the time averages, these are now canonical. It would seem that the matter of the time averages could be overlooked if one is interested only in the average guiding-center position and therefore that the equations of motion can be regarded as canonical. If this is the case, any theorems in classical mechanics that come from the canonical equations should have an analog in the (α,β) space. Liouville's theorem comes to mind immediately, and it is possible to derive it[14] for the density in (α,β) space by disregarding the time averages. To dispel any lingering doubts about the time averages, a more direct derivation can be made using the expressions for the instantaneous values of $\dot{\alpha}$ and $\dot{\beta}$. The Liouville theorem can be used to find equilibria in so-called minimum-B geometries.[22-24]

2.6.3. The Third Adiabatic Invariant

As a guiding center oscillates between mirror points, it gradually changes lines of force. During its motion along a line, it instantaneously drifts toward a variety of adjacent lines, but there is one line toward which it moves on the average, and this line is specified by (2.32). Thus a surface composed of lines on which J is constant is gradually traversed by the guiding center. Now it may happen that this surface is closed, so that the particle eventually returns to a line it traversed earlier. If so, there is a third periodicity, and a third adiabatic invariant is to be expected. The surfaces seem to be closed for particles in the inner Van Allen belt. Such a surface (idealized) is sketched in Fig. 2.8.

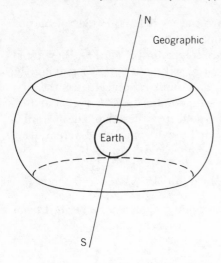

Fig. 2.8 An invariant surface for a particle trapped in the earth's field.

If the motion along the field is not periodic (i.e., oscillatory or unidirectional on a closed field line), there is not even a second adiabatic invariant, nor is there a third. Only the magnetic moment exists. This illustrates the fact that the number of adiabatic invariants depends on the geometry and is less than or equal to the number of degrees of freedom.

The third adiabatic invariant is the flux Φ of **B** enclosed by the surface of Fig. 2.8. That this flux should be constant in a static situation is a trivial statement, much as the invariance of the magnetic moment in a uniform field is trivially true. But the flux is also invariant if the field is time-dependent, and this is the significant fact. The surface about which the particle precesses is not even well defined unless the particle traverses it in a time small compared with the time scale for fields to change. It is not surprising therefore that this rapid precession assumption is necessary to prove the invariance of Φ. From a practical standpoint, the time scale of field fluctuations must be slowest to conserve Φ; they can be faster and still conserve J, and fastest of all without disturbing M, since the time scale then need be long only compared with the gyration period.

Proof of the invariance of Φ is reminiscent of the proof for J. It is necessary to extend Eqs. (2.32) to include time-dependent fields. When the fields are time-dependent, it is appropriate to generalize the quantity W used previously to a quantity K, defined by

$$K = \frac{mv_\parallel^2}{2} + MB + e\left(\phi + \frac{\alpha}{c}\frac{\partial\beta}{\partial t}\right) \tag{2.33}$$

where ϕ is the scalar potential for the electric field, so that **E** is equal to

$-\nabla\phi - [(1/c)(\partial(\alpha\nabla\beta)/\partial t)]$. In a time-dependent field α and β are functions of both time and position. The second invariant is now defined by

$$J(\alpha,\beta,M,K,t) = \oint \left\{ 2m \left[K - e \left(\frac{\alpha}{c} \frac{\partial\beta}{\partial t} + \phi \right) - MB \right] \right\}^{1/2} ds$$

(2.34)

where $\partial\beta/\partial t$ is to be expressed as a function of (α,β,s,t). The generalization of Eqs. (2.32) turn out to be

$$\langle\dot\alpha\rangle = - \frac{c}{e} \frac{\partial K}{\partial\beta} (\alpha,\beta,J,M,t)$$

$$\langle\dot\beta\rangle = \frac{c}{e} \frac{\partial K}{\partial\alpha}$$

$$\langle\dot K\rangle = \frac{\partial K}{\partial t}$$

(2.35)

and $1 = T \dfrac{\partial K}{\partial J}$

The quantity $\langle\dot K\rangle$ is related to the gain in energy averaged over a longitudinal oscillation.

The details of the proof that Φ is invariant will not be given here.[14] One finds that $d\Phi/dt$ is not zero as the particle drifts around the surface by the invariance of J (that is, as it precesses around the earth); the average motion from line to line as given by Eqs. (2.35) does not conserve Φ. But if $d\Phi/dt$ is averaged over a complete precession, the time average is zero. This is analogous to the situation with dJ/dt. The instantaneous rate of change of Φ is

$$\frac{d\Phi}{dt} = \frac{cT_p}{e} [\langle\dot K\rangle - \langle\langle\dot K\rangle\rangle] \neq 0$$

(2.36)

where $\langle\langle\dot K\rangle\rangle$ means $\langle\dot K\rangle$ averaged over a precession and T_p is the time for the particle to precess once around the surface. The right side of (2.36) obviously vanishes when averaged over the period T_p.

Before leaving the subject of the third invariant, several points should be discussed concerning motion of lines of force and the average (over a longitudinal oscillation) guiding-center drift. The "velocity" of a line of force in a time-dependent field is not physically observable. We cannot see lines of force. One is therefore free to define line velocity, and it should be defined so as to enhance our visualization of how the magnetic field pattern changes with time. One ordinarily uses the picture in which a magnetic field has an intensity proportional to the line density one draws. As the field changes with time, this picture remains valid if the lines are moved around at a *flux-preserving velocity*. To define this velocity, suppose that an arbitrary closed curve is drawn in space; now let each element of the curve move at a velocity $\mathbf{U}(\mathbf{r},t)$. If the flux through the curve remains constant as the curve distorts, $\mathbf{U}$ is said to be *flux-preserving*. As shown,[25] $\mathbf{U}$ must satisfy $\nabla \times (\mathbf{E} + \mathbf{U} \times \mathbf{B}/c) = 0$. This limits $\mathbf{U}$ but does not determine it uniquely. One often chooses $\mathbf{U}$ as $c\mathbf{E} \times \mathbf{B}/B^2$, which is acceptable if $\nabla \times \mathbf{E}_\parallel$ is zero.

A more general definition of line velocity that is always acceptable (but not unique) is

$$\mathbf{U}(\mathbf{r},t) = \left(\frac{\partial \beta}{\partial t}\nabla \alpha - \frac{\partial \alpha}{\partial t}\nabla \beta\right) \times \frac{\hat{\mathbf{e}}_1}{B} \qquad (2.37)$$

It is not difficult to show that $\nabla \times (\mathbf{E} + \mathbf{U} \times \mathbf{B}/c)$ is zero for this choice of $\mathbf{U}$. Moreover, this choice has the advantage that $\partial \alpha/\partial t + \mathbf{U} \cdot \nabla \alpha$ is zero, and likewise for β. The significance of this is that, as an observer moves at the line velocity, the (α,β) label on the line he is following remains unchanging with time.

A convenient space in which to visualize the invariant surfaces is a cartesian (α,β,s) space, in which the field lines are straight and parallel to the s axis, as in Fig. 2.9. The choice of $\mathbf{U}$ in (2.37) makes the lines of force

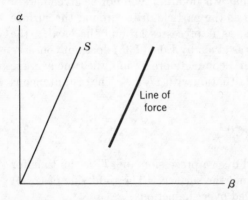

Fig. 2.9 A line of force in (α,β,s) space.

fixed in this space; by contrast, a particle for which J (but not necessarily Φ) is invariant moves in (α,β) space in accordance with Eqs. (2.35) and consequently does not remain attached to a line of force.

The picture developed so far of line motion is very appealing but is not unique. To illustrate, suppose that $\mathbf{U}$ is defined by

$$\mathbf{U}(\mathbf{r},t) = \frac{c}{eB}\hat{\mathbf{e}}_1 \times \nabla K(\mathbf{r},M,J,t) + \left(\frac{\partial \beta}{\partial t}\nabla \alpha - \frac{\partial \alpha}{\partial t}\nabla \beta\right) \times \frac{\hat{\mathbf{e}}_1}{B} \qquad (2.38)$$

where K is to be regarded as a function of the specified variables via (2.34).

This velocity can also be proved flux-preserving. However, for it, we have

$$\frac{\partial \alpha}{\partial t} + \mathbf{U} \cdot \nabla \alpha = \frac{c}{eB} \left(\hat{\mathbf{e}}_1 \times \nabla K \right) \cdot \nabla \alpha = - \frac{c}{e} \nabla K \cdot \frac{\partial \mathbf{R}}{\partial \beta}$$

$$= - \frac{c}{e} \frac{\partial K}{\partial \beta} \left(\alpha, \beta, M, J, t \right) = \langle \dot{\alpha} \rangle \tag{2.39}$$

and similarly for $\langle \dot{\beta} \rangle$. With this definition of line velocity, the line of force consequently moves at exactly the average particle drift velocity, and the particle remains attached to the line.

Either of the two pictures is acceptable, though definition (2.37) seems preferable since it does not depend on any particle parameters, while definition (2.38) depends on J and M. It is a little unappealing to use a definition of line velocity that depends on the particle under observation. One prefers to visualize the motion of field lines as being intrinsic to the field and not dependent on particles. Furthermore, if two particles with different J and M are on the same line of force, there will be an ambivalence in the line velocity. Finally, if the electric fields are so large that all the drift terms in (2.4) must be retained, J is not conserved. Definition (2.38) is still flux-preserving, but J is now a time-dependent parameter. And because the guiding center no longer shows a slow average drift, governed by Eqs. (2.35), it is not possible to say that the particle follows the line of force on the average. The guiding center follows a trajectory in Fig. 2.9 determined by (2.4) and (2.6). Under these circumstances, definition (2.37) for the line velocity certainly is superior to (2.38).

2.7 APPLICATION OF ADIABATIC THEORY TO PLASMAS

In the previous sections the motion of a single particle in a prescribed field has been studied. The adiabatic model may also apply to a plasma, where the density of positively and negatively charged particles is so large that their interactions are important in determining their motions. The field each particle moves in is the sum of (1) any "external" field and (2) those fields due to the motions and positions of all other particles. For the particle motion to be adiabatic, close collisions between charged particles must be infrequent (high plasma temperature and low density) so that a particle at no time feels a sudden force. Such self-consistent calculations are necessary to analyze the stability of plasma confinement in a given field configuration.

Newcomb[26] has developed a method for using the first two adiabatic invariants in studying plasma stability. The change in energy of an equilibrium plasma under a prescribed displacement $\xi(\mathbf{r})$ of the element of plasma at $\mathbf{r}$ can be obtained from invariance of the magnetic moment and longitudinal invariant. If this energy change is positive for all possible $\xi(\mathbf{r})$, the plasma is stable. If the change is negative for any $\xi(\mathbf{r})$, it is unstable. It is plausible that the change in particle energies should be derivable from the first two invariants. The magnetic moment is associated with perpendicular energy, while the longitudinal invariant is associated with parallel velocity and energy. Changes in field energy under the perturbation must also be accounted for in obtaining the total change in energy.

The mechanism of these instabilities can be explained in terms of the adiabatic particle drifts. In the presence of the perturbation the drifts lead to charge accumulations whose electric fields drive the perturbation farther in a typically regenerative fashion (see Rosenbluth and Longmire,[27] Lehnert,[28] and Northrop[2] for examples and the discussion in Chap. 5).

We can also apply adiabatic motion to the current density in a collisionless plasma. Each component (i.e., ions or electrons) of the plasma obeys the macroscopic momentum-conservation equation

$$nm \frac{d\mathbf{V}}{dt} = -\nabla \cdot \mathbf{P} + ne \frac{\mathbf{V}}{c} \times \mathbf{B} + ne\mathbf{E} \tag{2.40}$$

where $\mathbf{V}$ is the average (over the velocity distribution) of the particle velocity $\mathbf{v}$ and $\mathbf{P}$ is the pressure tensor defined as $\langle nm(\mathbf{v} - \mathbf{V})(\mathbf{v} - \mathbf{V}) \rangle$, where the broken brackets mean an average over the particle velocity distribution. The current density $\mathbf{J}$ of that component is $ne\mathbf{V}$, where n is the particle density. Solving (2.40) for $\mathbf{V}$, we obtain

$$\mathbf{V} = \mathbf{V}_{\parallel} + \frac{c\hat{\mathbf{e}}_1 \times \nabla \cdot \mathbf{P}}{neB} + \frac{c\mathbf{E} \times \hat{\mathbf{e}}_1}{B} + \frac{mc}{eB^2} \hat{\mathbf{e}}_1 \times \frac{d\mathbf{V}}{dt} \tag{2.41}$$

Consider now a steady situation, where there is no electric field. Then $\mathbf{V}_{\perp}$ is $c\hat{\mathbf{e}}_1 \times \nabla \cdot \mathbf{P}/neB$. This is just the east-west asymmetry effect of mirroring protons observed by Heckman and Nakano.[29] They observed that, at the inner edge of the inner Van Allen belt, more high-energy protons are moving east than west; there is an average proton velocity $\mathbf{V}$ toward the east. The pressure gradient is caused by the atmospheric density gradient, there being fewer particles at lower altitudes owing to the greater loss to the atmosphere. At the outer edge of a radiation belt, where the density decreases with increasing radius (for whatever reason), the reverse asymmetry should appear, with more particles moving west than east.

The divergence of the pressure tensor can be expanded in the adiabatic case as[30]

$$\nabla \cdot \mathbf{P} = \hat{\mathbf{e}}_1 \left(\frac{\partial P_\parallel}{\partial s} - \frac{P_\parallel - P_\perp}{B} \frac{\partial B}{\partial s} \right) + \left[(P_\parallel - P_\perp) \frac{\partial \hat{\mathbf{e}}_1}{\partial s} + \nabla P_\perp \right] \tag{2.42}$$

where $P_\parallel$ is $nm\langle(v_\parallel - V_\parallel)^2\rangle$ and $P_\perp$ is $\frac{1}{2}nm\langle v_\perp^2\rangle$. In the east-west asymmetry experiment there would be a small contribution from the line curvature $\partial\hat{\mathbf{e}}_1/\partial s$ in addition to the one from the pressure gradient $\nabla P_\perp$.

It is possible to prove from the Vlasov (collisionless Boltzmann) equation that

$$\mathbf{J}(\mathbf{r},t) \equiv ne\mathbf{V} = Ne\langle \dot{\mathbf{R}}_\perp + \hat{\mathbf{e}}_1 v_\parallel \rangle + c\nabla \times \mathbf{M} \tag{2.43}$$

where N is the number of guiding centers per unit volume at $(\mathbf{r},t)$ and $\mathbf{M}$ is the total magnetic moment per unit volume of particles with guiding centers at $\mathbf{r}$ and is antiparallel to $\mathbf{B}$. The brackets mean the average over particles with guiding centers at $\mathbf{r}$. The perpendicular component of (2.43) is easily derived from (2.41) and the guiding-center equations. However, the parallel component is rather difficult to prove formally,[12] even though the entire expression (2.43) is intuitively correct. It says that the total current density in a plasma is the sum of the guiding-center current and the current that results from the curl of the magnetic moment per unit volume. This current is diamagnetic in the sense that it partially shields the interior of a plasma from an externally produced magnetic field.

2.8 NONADIABATIC EFFECTS

The application of adiabatic theory and the lowest-order invariants to the Van Allen radiation has been outlined in previous sections. According to the theory, in the absence of collisions, particles would remain indefinitely in the geomagnetic field and repeatedly precess about their invariant surfaces. In practice all three invariants may not hold sufficiently well for this permanent trapping to occur. There is low-temperature plasma permeating the magnetosphere about the earth, and the solar wind may produce disturbances that are propagated through this plasma. These disturbances in turn may be fast enough to affect one or more of the lowest-order invariants. Even if the deviation of one of them from a constant is very small, this very small effect can operate over very long times in the geophysical case. The question becomes whether these effects are cumulative or whether they are oscillatory and self-canceling over a long period. If the motion is truly nonadiabatic, in the sense defined in Sec. 2.6, the effects may be cumulative and the particle may become lost from the geomagnetic field.   For example, if the magnetic moment decreases continuously, the particle will eventually become lost in the atmosphere.   However, if the motion is adiabatic, in the sense of being predicted by the first few terms of the invariant series, then the particle may still be permanently trapped, with the guiding center following a slightly different path from that predicted by the lowest-order invariant.   The distinction between these two possibilities, cumulative and oscillatory, may not always be sharp, though in one geometry it seemed to be quite sharp for the magnetic moment.[20]

The consequences of the earth's rotation, coupled with the azimuthal asymmetry of its field, do not seem to be known except in the limit when Φ is invariant.   In this limit a particle precesses rapidly about its invariant surface, and the surface rotates slowly and rigidly with a 24-hr period.

The extreme limit of nonadiabatic motion must be treated by different means.   Only certain special cases, for instance as discussed by Longmire,[31] can be handled analytically.   To conclude, it does not seem possible at present to make any general statements about nonadiabatic effects, other than that numerical computation is probably needed to study them.   However, these effects may be important in the dynamics of the trapped radiation and therefore merit attention.

REFERENCES

1. Hellwig, G.: Z. Naturforsch., **10A,** 508 (1955).
2. Northrop, T. G.: Ann. Phys., **15,** 79 (1961).
3. Alfvén, H.: "Cosmical Electrodynamics," Oxford University Press, Fair Lawn, N.J., 1950.
4. Spitzer, L., Jr.: Astrophys. J., **116,** 299 (1952).
5. Northrop, T. G.: Phys. Rev., **103,** 1150 (1956).
6. Fermi, E.: Phys. Rev., **75,** 1169 (1949).
7. Fermi, E.: Astrophys. J., **119,** 1 (1954).
8. Teller, E.: Rept. Progr. Phys., **17,** 154 (1954).
9. Davis, L., Jr.: Phys. Rev., **101,** 351 (1956).
10. Parker, E. N.: Phys. Rev., **109,** 1328, (1958).
11. Vandervoort, P. O.: Ann. Phys., **10,** 401, (1960).
12. Northrop, T. G.: "The Adiabatic Motion of Charged Particles," Wiley, New York, 1963.
13. Kruskal, M. D.: J. Math. Phys., **3,** 806 (1962).
14. Northrop, T. G., and E. Teller: Phys. Rev., **117,** 215 (1960).
15. Gardner, C. S.: Phys. Rev., **115,** 791 (1959).
16. Chew, G. F., M. L. Goldberger, and F. E. Low: The Individual Particle Equations of Motion in the Adiabatic Approximation, Los Alamos Sci. Lab. Rept. LA-2055, chap. T-759, p. 9, 1955.

17. Kruskal, M. D.: The Gyration of a Charged Particle, *Princeton Univ. Project Matterhorn Rept.* PM-S-33 (NYO-7903), March, 1958.

18. Gardner, C. S.: Private communication, RCA Laboratories, Princeton, N.J., 1962. The expression for M_2 in Ref. 12, eq. (5.5), should have $2v_\theta^2 v_\parallel^2$ added in the square bracket of the last term.

19. Berkowitz, J., and C. S. Gardner: *Communs. Pure Appl. Math.*, **12**, 501 (1959).

20. Garren, A., et al.: Individual Particle Motion and the Effect of Scattering in an Axially Symmetric Magnetic Field, *Proc. Second UN Intern. Conf. on Peaceful Uses of Atomic Energy*, Geneva, 1958, vol. 31, p. 65.

21. Kadomtsev, B. B.: In M. A. Leontovich (ed.), "Plasma Physics and the Problem of Controlled Thermonuclear Reactions," vol. 3, p. 340, Pergamon, 1959.

22. Northrop, T. G., and K. J. Whiteman: *Phys. Rev. Letters*, **12**, 639 (1964).

23. Furth, H. P.: *Phys. Rev. Letters*, **11**, 308 (1963).

24. Taylor, J. B.: *Phys. Fluids*, **6**, 1529 (1963).

25. Newcomb, W. A.: *Ann. Phys.*, **3**, 347 (1958).

26. Newcomb, W. A.: Stability of a Collisionless Plasma, I and II, *Ann. Phys.* To be published.

27. Rosenbluth, M. N., and C. L. Longmire: *Ann. Phys.*, **1**, 120 (1957).

28. Lehnert, B.: "Dynamics of Charged Particles," North Holland, Amsterdam, 1964.

29. Heckman, H. H. and G. H. Nakano: *J. Geophys. Research*, **68**, 2117 (1963).

30. Chew, G. F., M. L. Goldberger, and F. E. Low: *Proc. Roy. Soc. (London)*, **A236**, 112 (1956).

31. Longmire, C. L.: "Elementary Plasma Physics," chap. 5, Wiley, New York, 1963.

3

STATISTICAL MECHANICAL FOUNDATIONS

FRIED

BURTON D. FRIED, *Professor of Physics, University of California, Los Angeles, California*

3

3.1 INTRODUCTION

So long as the density n of a plasma is sufficiently small, a single-particle description of the dynamics, such as that discussed in the previous chapter, will be a good approximation. In studying denser plasmas, however, it becomes necessary to consider the total force on a particle, which also includes, besides external fields, the influence of all the other particles. To do this, one must (to some approximation) simultaneously solve the equations of motion for all the particles in the plasma. In this chapter we shall apply to the plasma the methods of statistical mechanics, which were developed to deal with just such many-body problems.[1]

To begin with, we may ask under what circumstances it becomes necessary to consider these cooperative phenomena. One measure is provided by the criterion, discussed in Chap. 1, that there be many particles within a sphere whose radius equals the Debye length, or, equivalently, that the mean kinetic energy per particle be much larger than the mean potential energy. In the case of a plasma confined by a magnetic field, we should clearly abandon the single-particle picture when the magnetic field B_i due to current flow within the plasma becomes comparable with the external field B_0. If v is a typical particle velocity and L a characteristic distance for changes in B, then

$$B_i \sim \frac{4\pi n e v L}{c} \gtrsim 4\pi \, \frac{nev}{c} \, \frac{cvm}{eB_0}$$

if we assume that L cannot be much smaller than the cyclotron radius, mvc/eB. (Here e = electron charge, m = electron mass, c = light velocity, n = electron density.) Thus, replacing $mv^2/2$ by kT, we have

$$\frac{B_i}{B_0} \gtrsim \frac{nmv^2/2}{B_0^2/8\pi} = \frac{nkT}{B_0^2/8\pi} \equiv \beta$$

The ratio β of thermal-energy density to magnetic-energy density (or of kinetic pressure to magnetic pressure) is among the most important of the several dimensionless parameters used to characterize a plasma. As we see here, the single-particle description may be appropriate for the *low-β* plasmas found in such laboratory devices as the *mirror machine* or *stellarator* but is not adequate to deal with the *high-β* plasmas observed in a *pinch* or in an electromagnetic shock tube.

In general, we may expect the single-particle picture to fail whenever the internal fields, magnetic or electric, become comparable with those imposed externally. Of course, the dynamics of the many-body problem can always, in a sense, be reduced to that of a (representative) single particle moving under the influence of fields (which must eventually be determined in a self-consistent fashion) so that single-particle concepts such as guiding center, $\mathbf{E} \times \mathbf{B}$ drift, mirror action, and the like, continue to have heuristic value even for high-β plasmas.

The simplest, most thoroughly studied, and best-understood aspects of the plasma are, naturally, those associated with the equilibrium state, which we shall consider first. We shall then study small deviations from equilibrium, i.e., the kinetic theory of the plasma. Finally, we shall discuss some features of the lowest approximation to the kinetic equations for a plasma, that which neglects correlations between particles. Our aim throughout will be to develop the basic equations, from which the various properties of a plasma needed for applications can (and in subsequent chapters will) be derived.

3.2 THE PLASMA IN EQUILIBRIUM

3.2.1. Debye Shielding and the Equation of State

The equilibrium statistical mechanics of a classical perfect gas (i.e., one composed of noninteracting particles with an average separation much greater than the de Broglie wavelength) leads to the familiar expressions

$$p = nkT$$

for the equation of state and

$$U = \tfrac{3}{2}nkT$$

for the energy per unit volume.   Interactions between the particles cause deviations from these simple relationships, which are conventionally expressed in terms of an expansion in powers of the density.   The coefficients of the terms of this series are called *virial coefficients*.

For the long-range Coulomb interaction characteristic of a plasma, this procedure encounters formal difficulties in the form of divergent integrals.   In various forms this difficulty will appear repeatedly, since it is a consequence of the peculiar character of an inverse-square force law.   This character is manifest, for instance, in the fact that the momentum transfer to a particle from interactions with other particles a distance R away increases with R, rather than falling rapidly to zero as in the case of short-range force laws.   Although we shall see how to circumvent the formal divergence difficulties [cf. the remark following Eq. (3.11) below], we should emphasize first the underlying physics, namely, the phenomenon of *Debye shielding*, for this, in a sense, provides the proper "cutoff" for the divergent integrals, both here and in other parts of the theory.

The picture (originally introduced by Debye and Hückel in the theory of electrolytes) is a very simple one: each particle of the plasma tends to attract to it particles of opposite charge and to repel particles of like charge, thereby creating a kind of "cloud" of opposite charge.   This process continues out to a radius λ_D, at which the total charge in the cloud is equal and opposite to that on the original particle.   At distances beyond λ_D, the potential due to the particle plus its cloud vanishes, and it is clear that integrals originally divergent at large r will be effectively cut off at $r = \lambda_D$.

The size and charge distribution of the cloud are determined by two competing effects: thermal agitation, which tends to produce a uniform density, and the Coulomb forces, which give rise to the cloud in the first place.   The quantitative expression of this competition is provided by the Maxwell-Boltzmann distribution law plus Poisson's equation.   If the potential at distance r from a particle of species i is $\phi(\mathbf{r})$, then the density of particles of species j (with charge $Z_j e$ and average density $\bar{n}_j$) is

$$n_j(\mathbf{r}) = \bar{n}_j \exp\left[-\beta Z_j e \phi(\mathbf{r})\right] \qquad \beta = \frac{1}{kT} \tag{3.1}$$

One the other hand, ϕ is determined by Poisson's equation,

$$\nabla^2 \phi = -4\pi e \sum_j Z_j n_j = -4\pi e \sum_j Z_j \bar{n}_j \exp\left(-\beta e Z_j \phi\right) \tag{3.2}$$

and the boundary conditions

$$\phi(\mathbf{r}) \to \frac{Z_i e}{r} \qquad \text{as } r \to 0 \tag{3.3}$$

$$\phi(\mathbf{r}) \to 0 \qquad \text{as } r \to \infty$$

The simultaneous solution of these equations determines $\phi(\mathbf{r})$, and hence $n_j(\mathbf{r})$. However, for a majority of plasmas of interest we can linearize Eq. (3.2) for ϕ, since the average potential energy of a particle is small compared with its average kinetic energy, i.e.,

$$\frac{e^2 n^{1/3}}{kT} \ll 1 \tag{3.4}$$

This represents a kind of perturbation approach, in that we consider small corrections to the perfect-gas behavior, where the potential energy is neglected and only the kinetic-energy part of the Hamiltonian is considered. We may then expand the exponential in (3.2). The condition of charge neutrality for the plasma

$$\Sigma Z_j n_j = 0 \tag{3.5}$$

eliminates the first term, and we are left, in lowest order, with the equation

$$\nabla^2 \phi - \frac{\phi}{\lambda_D{}^2} = 0 \tag{3.6}$$

with $\quad \lambda_D{}^{-2} \equiv \dfrac{4\pi e^2 (\Sigma Z_j{}^2 n_j)}{kT} \tag{3.7}$

The spherically symmetric solution of (3.6) which satisfies the boundary condition (3.3) is just

$$\phi(\mathbf{r}) = \frac{Z_i e}{r} \exp\left(-\frac{r}{\lambda_D}\right) \tag{3.8}$$

We see that the form factor of the charge cloud is just a Yukawa potential and that its size is given by λ_D. Moreover, it follows from (3.4) that there will be many particles in a sphere of radius λ_D, a condition which is clearly necessary if the underlying picture of a screening charge cloud is to make sense.

By using the simple result (3.8) it is easy to find expressions for the thermodynamic properties of the plasma. The average potential energy for the plasma is

$$U_1 = \tfrac{1}{2}\int d^3\mathbf{r}\, \rho\phi = \tfrac{1}{2}\Sigma (n_i V)(eZ_i)\phi_i \tag{3.9}$$

where ρ is the average charge density, $N_i = n_i V$ is the total number of particles of species i, and ϕ_i is the average potential experienced by an ion of species i due to all *other* particles of the system. Since (3.4) allows us to expand the exponential in (3.8), we have

$$\phi_i = \lim_{r \to 0}\left[\phi(\mathbf{r}) - \frac{Z_i e}{r} \right] = -Z_i e / \lambda_D \tag{3.10}$$

which, substituted in (3.9), gives

$$U_1 = -e^3 \left(\frac{\pi\beta}{V}\right)^{1/2} \left(\sum N_i Z_i{}^2\right)^{3/2} \tag{3.11}$$

Notice that U_1 is proportional to the three-halves power of e^2, which is the *coupling constant*, i.e., the parameter which measures the strength of the potential part of the Hamiltonian,

$$H = \sum \frac{P_a{}^2}{2m_a} + \frac{e^2}{2} \sum \frac{Z_a Z_b}{r_{ab}} \tag{3.12}$$

and also proportional to the square root of the density. From this it is clear that a simple perturbation expansion in powers of e^2 or density will lead to difficulty.

For computing thermodynamic properties the free energy

$$F = U - TS \tag{3.13}$$

is more useful than the internal energy U. Since

$$U = -T^2 \left[\frac{\partial(F/T)}{\partial T}\right]_V = \left[\frac{\partial(\beta F)}{\partial \beta}\right]_V \tag{3.14}$$

we obtain the interaction part of F by integrating (3.11) on β,

$$F_1 = \beta^{-1}\!\int d\beta \; U_1 = \tfrac{2}{3}U_1 \tag{3.15}$$

To this we add the free energy of a gas of noninteracting particles to get the total free energy, correct to order e^3,

$$F = F_0 + F_1 = \Sigma_i N_i kT[\log(\lambda_i{}^3 n_i) - 1] + \tfrac{2}{3}U_1 \tag{3.16}$$

where $\lambda_i = \left(\dfrac{2\pi\hbar^2\beta}{m_i}\right)^{1/2}$

is the thermal de Broglie wavelength of species i.

From this point, standard thermodynamic relations give the other quantities of interest, e.g., the entropy

$$S = -\left(\frac{\partial F}{\partial T}\right)_V$$

or the pressure

$$p = -\left(\frac{\partial F}{\partial V}\right)_T = \sum n_i kT - \frac{1}{3}(\pi\beta)^{1/2}e^3 \left(\sum n_i Z_i{}^2\right)^{3/2} \tag{3.17}$$

The partition function is just

$$Z = e^{-\beta F}$$

To lowest order

$$Z = \exp{(-\beta F_0)} \left(1 - \frac{2}{3} \beta U_1\right) \tag{3.18}$$

so that it, like p, U, and F, has a branch cut at the origin of the e^2 or n planes.

For the particular case of a fully ionized hydrogen plasma, with equal ion and electron densities, $n_i = n_e = n$, (3.17), for instance, becomes

$$p = 2nkT \left(1 - \frac{\sqrt{2}}{24\pi} \frac{k_D{}^3}{n}\right) \tag{3.19}$$

where $\quad k_D = \sqrt{\dfrac{4\pi n e^2}{kT}} \tag{3.20}$

is the Debye wave number for either species. It is clear that the correction term is small whenever the condition (3.4), which justifies the present derivation, is satisfied.

3.2.2. Ionization; the Saha Equation

Although we have so far considered only plasmas which are fully ionized, partial ionization occurs in almost all experiments, except for some with hydrogen and helium. The assumption of equilibrium is not always well justified, and one must frequently consider the specific rates of ionization, recombination, and ion loss to obtain an accurate calculation of the degree of ionization. Nevertheless, the predictions of the equilibrium theory provide at the very least a useful guide in many cases.

To avoid notational complications, we shall consider the case of hydrogen, the extension to more complicated plasmas being fairly straightforward. There are then three species: electrons (e), protons (i), and hydrogen atoms (o). The number of particles of each species is not specified, but we assume a neutral system in which ionization and recombination occur freely, so that the concentrations must satisfy

$$N^e = N^i \quad \text{and} \quad N^i + N^0 = N = \text{const} \tag{3.21}$$

Such a multiphase system is best treated with the grand partition function[2] defined by

$$Z_G = \sum_m \exp \{-\beta E(m) + \beta \mu^e [N^e(m) + N^0(m)]$$
$$+ \beta \mu^i [N^i(m) + N^0(m)]\}$$

where we sum over all quantum states m of the total system. Alternatively, we can employ the more symmetrical form

$$Z_G = \sum \exp\left[-\beta E(m) + \beta \sum_a \mu^a N^a(m)\right] \tag{3.22}$$

with $\mu^0 \equiv \mu^e + \mu^i$ $\tag{3.23}$

Here $E(m)$ denotes the energy of quantum state m and the chemical potentials μ^a are Lagrange multipliers which allow us to take into account the constraints (3.21).   The connection with thermodynamics is provided by the equation

$$pV = kT \log Z_G$$

and by the expression for the probability of finding the system in state m,

$$w_m = \exp\left[-\beta E(m) + \beta \sum_a \mu^a N^a(m)\right] Z_G^{-1} \tag{3.24}$$

The μ^a are fixed by requiring

$$\overline{N^e + N^0} = \sum_m [N^e(m) + N^0(m)]w(m) = N \tag{3.25}$$

and $$\overline{N^i + N^0} = \sum_m [N^i(m) + N^0(m)]w(m) = N \tag{3.26}$$

The inclusion of neutral atoms as one species already takes some account of the interaction between electrons and protons.   However, we shall ignore the remainder of the interactions, so that a state m of the system is specified by giving the numbers N_j^a of particles of species a which are in the (single-particle) state j, the corresponding single-particle energies being ε_j^a.   Then Z_G is a *sum* over all N_j^a of

$$\exp\left\{\beta\left[\sum_{j,a}(\mu^a - \varepsilon_j^a)N_j^a\right]\right\} \tag{3.27}$$

which can be written as the *product* over all a and j of

$$\sum_{N_j^a} \exp[\beta(\mu^a - \varepsilon_j^a)N_j^a] \tag{3.28}$$

For particles obeying Bose-Einstein statistics, like hydrogen, the sum in (3.28) extends from 0 to ∞ and (being just a geometric series) gives

$$\{1 - \exp[\beta(\mu^a - \varepsilon_j^a)]\}^{-1}$$

For the Fermi-Dirac particles (electrons and protons) the N_j^a range only over the values 0 and 1 (on the assumption that the label j specifies spin and other internal degrees of freedom as well as the center-of-mass momentum), and we get for (3.28)

$$1 + \exp[\beta(\mu^a - \varepsilon_j^a)]$$

Since Z_G is a product of terms like (3.28), $\log Z_G$ will be a sum,

$$\log Z_G = \sum_j \log \frac{\{1 + \exp\left[\beta(\mu^e - \varepsilon_j{}^e)\right]\}\{1 + \exp\left[\beta(\mu^i - \varepsilon_j{}^i)\right]\}}{1 - \exp\left[\beta(\mu^0 - \varepsilon_j{}^0)\right]}$$

(3.29)

The average value of any occupation number $N_j{}^a$ is

$$\overline{N_j{}^a} = \sum_m w_m N_j{}^a(m) = \frac{\partial(\log Z_G)}{\beta \partial \varepsilon_j{}^a} = \frac{\exp\left[\beta(\mu^a - \varepsilon_j{}^a)\right]}{1 \pm \exp\left[\beta(\mu^a - \varepsilon_j{}^a)\right]} \qquad (3.30)$$

where the upper sign applies to ions and electrons, the lower sign to the atoms.   Defining

$$z^a = \exp(\beta\mu^a) \tag{3.31}$$

we have

$$\overline{N_j{}^a} = [z_\alpha{}^{-1} \exp(\beta\varepsilon_j{}^a) \pm 1]^{-1} \tag{3.32}$$

Summing these equations on j, to get the $\overline{N^a}$, and substituting in (3.25) and (3.26), fixes the z^a.   At any densities for which the plasma is still a gas, z^e is essentially the number of electrons in a cube whose edge is the de Broglie wavelength,

$$\lambda_e = \left(\frac{2\pi\hbar^2\beta}{m^e}\right)^{1/2} \tag{3.33}$$

Then $z^e \ll 1$, as are $z^i \approx (m^e/m^i)^{3/2} z^e$ and $z^0 = z^e z^i$.   We may therefore drop the 1 in (3.32), thus obtaining classical Maxwell-Boltzmann distributions,

$$\overline{N_j{}^a} = z^a \exp(-\beta\varepsilon_j{}^a) \tag{3.34}$$

for each species.

The labels j include the center-of-mass momentum, the spin, and, in the case of the atom (and also the ion, for nonhydrogenic plasmas), quantum numbers specifying the internal states.   It is generally convenient to separate these, i.e., to write

$$\varepsilon_j{}^a = \frac{(p_j)^2}{2m^a} + W_j{}^a \tag{3.35}$$

and to work with occupation numbers which refer to the internal state, independent of the momentum.   Using densities rather than total occupation numbers, we have thus

$$n_j{}^a = \frac{\displaystyle\sum_{p_i} N_j{}^a}{V} = \frac{g_j{}^a \exp(-\beta W_j{}^a) z^a}{(\lambda^a)^3} \tag{3.36}$$

where $g_j{}^a$ is the degeneracy of the internal state,

$$\lambda^a = \left(\frac{2\pi\hbar^2\beta}{m^a}\right)^{1/2}$$

and the factor $(\lambda^a)^{-3}$ comes out simply from doing the integration of $\exp(-p_j{}^2/2m^a)$ over p_j. The conditions (3.23) and (3.26) which determine z^e, z^i, and $z^0 = z^e z^i$ can then be written

$$n = z^i[z^e Z^0(\lambda^0)^{-3} + Z^i(\lambda^i)^{-3}] = z^e[z^i Z^0(\lambda^0)^{-3} + Z^e(\lambda^e)^{-3}] \quad (3.37)$$

where $Z^a = \Sigma g_j{}^a \exp(-\beta W_j{}^a)$

is the partition function associated with the internal states and

$$n = \sum_j (n_j{}^0 + n_j{}^e) = \sum_j (n_j{}^0 + n_j{}^i)$$

is the total density of electrons (or ions) free or bound.

Instead of solving (3.37) for z^e and z^i and substituting these into (3.36), it is sometimes convenient to combine Eqs. (3.36) to eliminate the z^a. Thus, we have

$$n^e = \frac{2z^e}{(\lambda^e)^3} \tag{3.38a}$$

$$n_j{}^i = \frac{g_j{}^i \exp(-\beta W_j{}^i)z^i}{(\lambda^i)^3} \tag{3.38b}$$

$$n_k{}^0 = \frac{g_k{}^0 \exp(-\beta W_k{}^0)z^i z^e}{(\lambda^0)^3} \tag{3.38c}$$

Then $\dfrac{n_j{}^i n^e}{n_k{}^0} = 2\left(\dfrac{g_j{}^i}{g_k{}^0}\right)\exp[-\beta(W_j{}^i - W_k{}^0)]\left(\dfrac{\lambda^0}{\lambda^e\lambda^i}\right)^3$ \hfill (3.39)

This equation, or one of its various equivalent forms, is called the *Saha equation*. If we sum (3.38b) over j and (3.38c) over k and combine them, we have (neglecting m^e/m^i compared with 1)

$$\frac{n^e n^i}{n^0} = 2Z^i/Z^0(\lambda^e)^3 \tag{3.40}$$

In particular, for hydrogen, where $Z^i = Z^e = 2$,

$$\frac{n^e n^i}{n^0} = 4\exp(-\beta W)/\tilde{Z}^0(\lambda^e)^3 \tag{3.41}$$

where W is the ionization energy and we adopt a convention of measuring energies from the ground state in computing the partition function, i.e.,

$$\tilde{Z}^0 \equiv \exp(-\beta W)Z^0 = \sum_j g_j{}^0 \exp[-\beta(W + W_j{}^0)]$$

If we try to calculate with this formula, we encounter a divergence, since the jth level of hydrogen has degeneracy $g_j{}^0 = 4j^2$, whereas $W_j{}^0 +$

$W \to +13.6$ ev as $j \to \infty$. The origin of this difficulty lies in our neglect of interactions between the particles, and the divergence can, in principle, be removed by taking this into account. Collisions with electrons, ions, or other atoms will provide a cutoff in the sum over j, for, in a rough way, we can say that it does not make sense to speak of an atomic level whose quantum number is higher than that for which the associated orbit radius

$$r_j = \frac{\hbar^2}{me^2} j^2 \tag{3.42}$$

exceeds either the interparticle spacing $n^{-1/3}$ or the Debye length, whichever is the smaller. We should therefore terminate the sum in Z^0 at a level j^* defined by

$$\frac{\hbar^2}{me^2} (j^*)^2 = \min (n^{-1/3}, k_D{}^{-1}) \tag{3.43}$$

In practice, however, this usually causes no real difficulty, for if $kT \gtrsim W$, then the ionization is extremely close to 100 per cent and the Saha equation is not really necessary, whereas for $kT \ll W$ the first term in the sum for Z_0 dominates all the rest (at least out to j^*, which is sufficient). For many purposes, therefore, we can approximate the Saha equation as simply

$$\frac{n^i n^e}{n^0} = \exp (-\beta W)(\lambda^e)^{-3} \tag{3.44}$$

The basic result (3.44) can be cast in many other forms. A common one introduces the electron pressure $p^e = n^e kT$,

$$\frac{n^i p^e}{n^0} = \frac{kT}{(\lambda^e)^3} \exp \left(-\frac{W}{kT} \right) \tag{3.45}$$

Another introduces the degree of ionization,

$$\alpha = \frac{n^e}{n^0 + n^i} = \frac{n^i}{n^0 + n^i} \tag{3.46}$$

so that $\quad \dfrac{\alpha^2}{1 - \alpha} n = \dfrac{\exp (-W/kT)}{(\lambda^e)^3 \tilde{Z}^0/4} = \eta n \qquad \eta \equiv \dfrac{\exp (-W/kT)}{n(\lambda^e)^3 \tilde{Z}^0/4}$ (3.47)

or $\quad \alpha = \dfrac{\sqrt{\eta^2 + 4\eta} - \eta}{2}$ (3.48)

For small ionization (low T or high n) we have $\alpha \approx \sqrt{\eta}$, whereas in the opposite limit

$$\alpha \approx 1 - \eta^{-1}$$

3.3 KINETIC THEORY

3.3.1. Some Possible Descriptions of a Plasma

We have seen that the plasma in equilibrium is characterized by the usual thermodynamic variables—pressure, density, temperature, degree of ionization, etc.   However, any measurements of a plasma involve some disturbance of equilibrium, usually by electromagnetic fields, so that we must consider the dynamical behavior of a plasma which is near, but not at, equilibrium.   To be sure, many experiments (or natural phenomena) involve conditions far from equilibrium, but, just as in the simpler case of neutral gases, no general or comprehensive analysis of such situations is yet possible.

The simplest characterization of the nonequilibrium plasma uses a one-fluid hydrodynamic [more generally magnetohydrodynamic (MHD), since there will usually be a magnetic field] picture in which pressure, density, mean-flow velocity, and temperature are considered as functions of space and time.   One can write down phenomenological equations expressing, in terms of these variables, conservation of mass, momentum, and energy.   These equations are essentially those of hydrodynamics, supplemented by terms appropriate to a conducting fluid, such as a $\mathbf{j} \times \mathbf{B}/c$ term in the momentum balance, and by Maxwell's equations for the electromagnetic fields.   In addition to the hydrodynamic-transport coefficients—viscosity and thermal conductivity—these equations will involve electromagnetic constitutive parameters such as electrical conductivity.

The next level of sophistication is a multifluid MHD picture, in which similar equations are written down for each species (electrons, ions, and neutral particles).   This requires additional phenomenological parameters describing the energy and momentum transfer between different species but, in turn, yields many new physical phenomena.   (For example, the only linearized waves contained in the one-fluid picture are sound waves, Alfvén waves, and electromagnetic waves, whereas the two-fluid model also describes many other modes, including plasma oscillations and whistler propagation.)

We shall omit further discussion of such macroscopic theories, since they will be dealt with in later chapters, and shall take up the next most complicated model, that provided by kinetic theory.   Here, in place of the macroscopic variables p, ρ, T, $\mathbf{v}$, etc., we consider distribution functions $f(\mathbf{x},\mathbf{v},t)$ (one for each species) which specify the number of particles at $\mathbf{x}$ at time t which have velocity $\mathbf{v}$.   The equation for f can be written down phenomenologically or derived from first principles, and we shall discuss both points of view.   Moreover, the MHD equations discussed above can be derived from this kinetic-theory formulation. In fact, there exists a kind of hierarchy of approximations, starting with the exact dynamics (the Hamiltonian or Schrödinger equations for the entire system) and extending through kinetic theory, down to one-fluid MHD, with each formulation derivable, with suitable approximations, from its predecessor. In particular, the deduction of the MHD equations from kinetic theory provides not only a knowledge of the limits of validity of the former but also a means of calculating from first principles the values of transport coefficients.   This will be examined in detail in a subsequent chapter.

In a gas of noninteracting particles, moving in an external force field $\mathbf{F}$, the distribution function satisfies a continuity equation in six-dimensional $\mathbf{x}$-$\mathbf{p}$ space,

$$\frac{\partial f}{\partial t} + \frac{\partial}{\partial \mathbf{x}} \cdot (\dot{\mathbf{x}}f) + \frac{\partial}{\partial \mathbf{p}} \cdot (\dot{\mathbf{p}}f) = 0 \tag{3.49}$$

expressing the conservation of phase points.   Because

$$\mathbf{v} = \dot{\mathbf{x}} = \frac{\partial H}{\partial \mathbf{p}} \quad \text{and} \quad \dot{\mathbf{p}} = -\frac{\partial H}{\partial \mathbf{x}} = \mathbf{F}$$

this becomes $\quad \dfrac{\partial f}{\partial t} + \mathbf{v} \cdot \dfrac{\partial f}{\partial \mathbf{x}} + \mathbf{F} \cdot \dfrac{\partial f}{\partial \mathbf{p}} = 0 \tag{3.50}$

When the particles interact, we have

$$\frac{\partial f}{\partial t} + \mathbf{v} \cdot \frac{\partial f}{\partial \mathbf{x}} + \mathbf{F} \cdot \frac{\partial f}{\partial \mathbf{p}} = \frac{\delta f}{\delta t} \tag{3.51}$$

where $\delta f/\delta t$ represents contributions to $\partial f/\partial t$ (due to the interparticle forces) besides the streaming terms on the left-hand side.

We shall now discuss three means of evaluating $\delta f/\delta t$.

3.3.2. The Landau-Boltzmann Binary-collision Picture

On phenomenological grounds, Boltzmann gave an expression for $\delta f/\delta t$ which is appropriate for a low-density gas with short-range interactions. Since each particle spends most of its time in free straight-line motion, with occasional collisions whose duration is short compared with the mean free time between collisions, we evaluate $\delta f/\delta t$ as the rate at which collisions (idealized to be instantaneous) cause particles to appear at $(\mathbf{x},\mathbf{p})$ minus the rate at which particles are lost from $(\mathbf{x},\mathbf{p})$ owing to collisions. Then

$$\left(\frac{\delta f}{\delta t}\right)_B = \int d\mathbf{p}' \, d\mathbf{\Delta} \, w(\mathbf{p},\mathbf{p}',\mathbf{\Delta})[f(\mathbf{p} + \mathbf{\Delta})f'(\mathbf{p}' - \mathbf{\Delta}) - f(\mathbf{p})f'(\mathbf{p}')] \tag{3.52}$$

where $w(\mathbf{p},\mathbf{p}',\mathbf{\Delta})$, the transition-probability rate for a collision with initial momenta $\mathbf{p}$, $\mathbf{p}'$ and momentum transfer $\boldsymbol{\Delta}$, that is,

$$\mathbf{p}, \mathbf{p}' \rightarrow \mathbf{p} + \boldsymbol{\Delta}, \mathbf{p}' - \boldsymbol{\Delta}$$

is proportional to the cross section $\sigma(u,\theta,\phi)$ for a collision in which the relative velocity

$$\mathbf{u} = \frac{\mathbf{p}}{m} - \frac{\mathbf{p}'}{m'}$$

is rotated by polar angles (θ,ϕ) to the new value†

$$\bar{\mathbf{u}} = \mathbf{u} + \frac{\boldsymbol{\Delta}}{\mu} \qquad \mu = \frac{mm'}{m + m'}$$

Specifically, $\quad w(\mathbf{p},\mathbf{p}',\boldsymbol{\Delta}) = \dfrac{\sigma(u,\theta,\phi)\delta(E_i - E_f)}{\mu^2}$ \hfill (3.53)

E_i and E_f being the initial and final energies and μ the reduced mass of the colliding particles (which we here allow to be different), since (3.53) ensures that

$$\int w\, d\boldsymbol{\Delta} = \int d\Omega\, \sigma u$$

as required.   For future reference, we note here that invariance under space inversions and time reversal implies

$$w(\mathbf{p},\mathbf{p}',\boldsymbol{\Delta}) = w(-\mathbf{p},-\mathbf{p}',-\boldsymbol{\Delta}) = w(\mathbf{p} + \boldsymbol{\Delta}, \mathbf{p}' - \boldsymbol{\Delta}, -\boldsymbol{\Delta}) \qquad (3.54)$$

With short-range interactions, a particle is generally free, occasionally interacts with a second particle, and only very rarely feels simultaneously the influence of two or more others.   In a plasma with its long-range Coulomb forces, however, a particle is simultaneously exposed to the influence of a large number of others (at least all those within a sphere of radius λ_D), and we could scarcely expect the Boltzmann picture to be applicable.   If we use it nonetheless, we encounter the divergences characteristic of Coulomb interactions.   Each of the terms in (3.52) is quadratically divergent at small Δ [since $\sigma \propto \sin^{-4}(\theta/2)$] and their difference is logarithmically divergent.   Small Δ (or small θ) corresponds to large-impact parameters, so this divergence is clearly related to the one encountered in the equilibrium theory.

On a pragmatic basis we can deal with this by cutting off the integrations at a value of Δ corresponding to an impact parameter equal to λ_D.   In addition, as pointed out by Landau,[3] we can simplify (3.52) by exploiting the dominance of the small Δ contributions.   Expanding $f(\mathbf{p} + \boldsymbol{\Delta})$ and $f(\mathbf{p}' - \boldsymbol{\Delta})$ in powers of $\boldsymbol{\Delta}$ and using identities[4] consequent on (3.54), we find that, correct to second order in $\boldsymbol{\Delta}$, (3.52) may be written

† Energy conservation $E_f - E_i = \mathbf{u} \cdot \boldsymbol{\Delta} + \Delta^2/2\mu = 0$ implies $u = \bar{u}$.

as

$$\left(\frac{\delta f}{\delta t}\right)_{LB} = -\nabla_p \cdot \mathbf{J}(\mathbf{p}) \tag{3.55}$$

with $\mathbf{J}(\mathbf{p}) = -\frac{1}{2}\nabla_p \cdot [f(\mathbf{p})\int d\mathbf{p}'\, f(\mathbf{p}')\int d\boldsymbol{\Delta}\; w\boldsymbol{\Delta}\;\boldsymbol{\Delta}]$

$$+ f(\mathbf{p})\int d\mathbf{p}'\, f(\mathbf{p}')\int d\boldsymbol{\Delta}\; w(\mathbf{p},\mathbf{p}',\boldsymbol{\Delta})\boldsymbol{\Delta}$$

$$= \int d\mathbf{p}'\, [f(\mathbf{p})\nabla_{p'}f - f(\mathbf{p}')\nabla_p f]\mathbf{K}(\mathbf{v},\mathbf{v}')$$

$$\mathbf{K}(\mathbf{v},\mathbf{v}') = \tfrac{1}{2}\int d\boldsymbol{\Delta}\; \boldsymbol{\Delta}\boldsymbol{\Delta}w(\mathbf{p},\mathbf{p}',\boldsymbol{\Delta}) \tag{3.56}$$

If, then, we (1) use the Boltzmann form, (3.52), for $\delta f/\delta t$, ignoring the inapplicability of the two-particle collision picture to a plasma, and (2) make a power-series expansion (to second order) in the momentum transfer, we arrive at the results (3.55) and (3.56).

3.3.3. The Fokker-Planck Coefficients

The Fokker-Planck equation arose in the theory of Brownian motion, where it was used to deal with the stochastic character of the many small impulses which act on a suspended particle.   It was applied to the case of a Coulomb gas by Chandrasekhar,[5] who argued that a given particle (a so-called test particle) will undergo many small-angle (large-impact-parameter) collisions in a time small compared with that in which its position or velocity changes appreciably.   Since the Coulomb potential makes the cumulative effect of these small momentum transfers dominant, the stochastic methods of the Fokker-Planck equation should be more appropriate for the plasma than the Boltzmann approach outlined above.

Let $W(\mathbf{x},\mathbf{v},t)$ be the probability that a selected (test) particle of the plasma be at $(\mathbf{x},\mathbf{v})$ at time t if it was at $\mathbf{x}_0$, $\mathbf{v}_0$ at $t = 0$.   Assume there exists an interval δt short compared with the time for W to change $(\delta t\; \partial W/\partial t \ll W)$ but long enough so that a stochastic approach is justified. If $\psi(\mathbf{x},\mathbf{v},\delta\mathbf{v})$ is the probability that $\mathbf{x} \to \mathbf{x} + \mathbf{v}\,\delta t$, $\mathbf{v} \to \mathbf{v} + \delta\mathbf{v}$ in time δt, then

$$W(\mathbf{x},\mathbf{v},t)\;\delta t = \int d(\delta\mathbf{v})\; W(\mathbf{x} - \mathbf{v}\,\delta t, \mathbf{v} - \delta\mathbf{v}, t)$$

$$\psi(\mathbf{x} - \mathbf{v}\,\delta t, \mathbf{v} - \delta\mathbf{v}, \delta\mathbf{v}) \tag{3.57}$$

Expanding (3.57) to first order in δt and second order in $\delta\mathbf{v}$, we get the Fokker-Planck equation,

$$\frac{\partial W}{\partial t} + \mathbf{v}\cdot\nabla W + \mathbf{F}\cdot\nabla_p W = \nabla_v(\tfrac{1}{2}\nabla_v\cdot W\langle\delta\mathbf{v}\,\delta\mathbf{v}\rangle - W\langle\delta\mathbf{v}\rangle) \tag{3.58}$$

where $\langle\delta\mathbf{v}\rangle \equiv \overline{\overline{\delta\mathbf{v}}}/\delta t - \mathbf{F}/m$

and $\langle\delta\mathbf{v}\,\delta\mathbf{v}\rangle \equiv \overline{\overline{\delta\mathbf{v}\,\delta\mathbf{v}}}/\delta t \tag{3.59}$

are called the *Fokker-Planck coefficients*.   The double bar denotes the average change during the time interval δt, that is,

$$\overline{\overline{\delta \mathbf{v}}} = \int d(\delta \mathbf{v}) \, \psi(\mathbf{x}, \mathbf{v}, \delta \mathbf{v}) \, \delta \mathbf{v}, \text{ etc.}$$

Since all particles in the plasma are equivalent, $W(\mathbf{x}, \mathbf{v}, t)$ is just proportional to $f(\mathbf{x}, m\mathbf{v}, t)$ and we see that (3.58) is just (3.51) with the particular choice

$$\left(\frac{\delta f}{\delta t}\right)_{FP} = \nabla_v \cdot (\tfrac{1}{2} \nabla_v \cdot f \langle \delta \mathbf{v} \, \delta \mathbf{v} \rangle - f \langle \delta \mathbf{v} \rangle) \tag{3.60}$$

Although this procedure seems better adapted to a plasma than does the Landau-Boltzmann picture of binary collisions, the computation of ψ or of the Fokker-Planck coefficients still requires solution of a difficult many-body problem, if we really want to take account of the fact that the test particle feels the simultaneous action of many particles of the plasma (at least all those within a Debye sphere). In fact, the problem is so difficult that one generally falls back on a binary-collision picture and computes the Fokker-Planck coefficients as though the test particle experienced a number of individual scatterings with other plasma particles during the time δt. Thus, if α is any component of $\delta \mathbf{v}$ or $\delta \mathbf{v} \, \delta \mathbf{v}$ one takes

$$\langle \alpha \rangle = \int d\mathbf{p}' \, f(\mathbf{p}') \int d\mathbf{\Delta} \, w(\mathbf{p}, \mathbf{p}', \mathbf{\Delta}) \alpha \tag{3.61}$$

As is clear from a comparison of the equations, $(\delta f/\delta t)_{FP}$ given by (3.60) and (3.61) is identical with the $(\delta f/\delta t)_{LB}$ given by (3.55) and (3.56) (except for the trivial circumstance that one is expressed in terms of velocity, the other in terms of momentum). This is scarcely surprising, since both calculations make the same assumptions—binary encounters and an expansion in powers of the momentum transfer—differing only in the order in which these are introduced.

Before concluding this discussion, we record the explicit form of (3.56) which results if we substitute in (3.53) the Coulomb cross section,

$$\sigma(u, \theta, \phi) = \frac{4\mu^2 q^2 q'^2}{\Delta^4}$$

We may simplify the resulting w by assuming Δ to be small, since this has already been done in deriving (3.56). Thus, in

$$E_f - E_i = \mathbf{u} \cdot \mathbf{\Delta} + \frac{\Delta^2}{2\mu}$$

we neglect the quadratic term and find

$$\mathbf{K}(\mathbf{p}, \mathbf{p}') = 2q^2 q'^2 \int d\mathbf{\Delta} \, \frac{\mathbf{\Delta}\mathbf{\Delta}}{\Delta^4} \, \delta(\mathbf{\Delta} \cdot \mathbf{u}) \tag{3.62}$$

We note that the integral diverges, logarithmically, at both large and small Δ. The divergence at large Δ, corresponding to small-impact parameters, is not surprising since our derivation has assumed Δ small. In fact, however, the scattering kinematics gives $\Delta = 2\mu u \sin (\theta/2)$; so there is a natural upper limit on Δ, $\Delta \leq 2\mu u$, equality corresponding to the minimum-impact parameter, $b_{\min} = e^2/\mu u^2$, and to a deflection of 180°. The divergence at small Δ, corresponding to large-impact parameters, shows clearly that the naïve binary-collision picture is, as expected, not consistent with the long-range Coulomb interaction. Intuition suggests that a cutoff corresponding to an impact parameter less than or equal to the Debye-shielding distance is required, but whether it should come at the interparticle spacing $n^{-1/3}$, at λ_D, or somewhere in between is a question best resolved by the considerations of the following section.

3.4 MANY-BODY THEORY; DEDUCTION OF A KINETIC EQUATION FROM FIRST PRINCIPLES

3.4.1. General Formulation

In view of the somewhat heuristic character of the above two derivations, it is of interest to examine a procedure which starts from the exact dynamics of the complete system and to see just what approximations are involved in obtaining a kinetic equation like (3.51) with $\delta f/\delta t$ determined by f. At the present time, there exist a number of different formal methods of deriving a kinetic equation from the exact many-body description, all leading to the same results. The earliest, due to Rosenbluth and Rostoker,[6] starts with the Liouville distribution function

$$\mathfrak{D}(\mathbf{x}_1, \ldots ,\mathbf{x}_N,\mathbf{p}_1, \ldots ,\mathbf{p}_N,t)$$

which specifies the distribution of systems over a statistical ensemble. $\mathfrak{D}$ satisfies the Liouville equation,

$$\frac{\partial \mathfrak{D}}{\partial t} + \sum_{1}^{N} \left[\frac{\partial}{\partial \mathbf{x}_i} \cdot (\mathbf{v}_i\mathfrak{D}) + \frac{\partial}{\partial \mathbf{p}_i} \cdot (\mathbf{F}_i\mathfrak{D}) \right] = 0 \tag{3.63}$$

where $\mathbf{F}_i$ is the total force on particle i due to external *and* interparticle forces. Although (3.63) completely determines (in a statistical sense) the dynamics of the system, we can neither solve it (remember that $\log N$ is of the order of 10 to 20!) nor get any useful information directly from a solution if we had it. Of much greater interest than $\mathfrak{D}$, therefore, are its moments, i.e., the one-body, two-body, etc., distribution functions obtained by integrating $\mathfrak{D}$, which is a symmetric function of its arguments, over the coordinates and velocities of all but one particle, all but two, etc. Using ζ_i to denote $(\mathbf{x}_i,\mathbf{v}_i)$, we have

$$f^{(s)}(\zeta_1,\zeta_2, \ldots ,\zeta_s) \equiv V^s \int d\zeta_{s+1} \cdots d\zeta_N \mathfrak{D} \tag{3.64}$$

By integrating (3.63) over all particles but one, all but two, etc., we obtain a set of coupled equations for the $f^{(s)}$, which has been derived by so many people that it is called the *Born-Green-Kirkwood-Bogoliubov-Yvon (BGKBY) hierarchy.* The hierarchy is fully equivalent to (3.63) and hence just as difficult to solve exactly. The contribution of Rosenbluth and Rostoker was to devise an approximate technique for solving the hierarchy which is valid in those cases where we expect a plasma description to make sense, i.e., when $n\lambda_D^3 \gg 1$. Further development of their technique by Lenard and by Guernsey led to a good understanding of the rigorous basis for plasma kinetic theory, although the formalism involved was by no means simple.

A different, somewhat complementary, procedure was introduced by Klimontovich and Dupree.[7] Instead of starting with an ensemble, thus putting in the statistics first, they begin with the equations of motion for a single system. Moreover, instead of starting with a distribution function for N particles and integrating out the variables associated with N-1, N-2, etc., of the particles, they begin with a distribution function for just one particle and introduce two-particle and higher-order functions as needed. Although equivalent to the BGKBY hierarchy, the Klimontovich-Dupree formalism is somewhat simpler and often makes calculations easier and more understandable, so we shall follow it here. (For still a third, and quite different, approach see Balescu.[1])

We start with the "microscopic" distribution function for a single system,

$$\mathfrak{F}(\mathbf{x},\mathbf{v},t) = n^{-1} \sum_1^N \delta[\mathbf{x} - \mathbf{x}_i(t)]\delta[\mathbf{v} - \mathbf{v}_i(t)] = n^{-1} \sum_1^N \delta[\zeta - \zeta_i(t)] \quad (3.65)$$

where $\mathbf{x}_i(t)$, $\mathbf{v}_i(t)$ describe the exact trajectory of the ith particle as determined by the external *and* interparticle forces and $n = N/V$ is the average density. (Actually, we need such a function for each species present, but we shall suppress species labels whenever possible in the interests of notational simplicity.) Although singular, $\mathfrak{F}$ is a distribution function in the sense that its integral over any region of the $\mathbf{x}$, $\mathbf{v}$ phase space gives the number of particles present in that region for the particular system described by $\mathfrak{F}$.

From the individual-particle equations of motion,

$$\dot{\mathbf{x}}_i = \mathbf{v}_i \qquad \dot{\mathbf{v}}_i = \frac{\mathbf{F}_i}{m_i}$$

it follows that

$$\frac{\partial \mathfrak{F}}{\partial t} + \mathbf{v} \cdot \nabla \mathfrak{F} + \frac{\mathbf{F}}{m} \cdot \nabla_v \mathfrak{F} = 0 \quad (3.66)$$

where $\mathbf{F}(\zeta) = \mathbf{F}(\mathbf{x},\mathbf{v})$ is the net force a particle at $\mathbf{x}$, $\mathbf{v}$ would experience.

For a system with two-body forces, $\hat{\mathbf{F}}(\zeta,\zeta')$ being the force on a particle at ζ due to one at ζ', we have

$$\mathbf{F}(\zeta) = \sum_j \hat{\mathbf{F}}(\zeta,\zeta_j) = n\!\!\int d\zeta' \, \hat{\mathbf{F}}(\zeta,\zeta')\mathfrak{F}(\zeta') \tag{3.67}$$

This equation is in a form valid for a multispecies plasma, provided that we understand $\int dv$ or $\int d\zeta$ to denote integration over the indicated phase-space variable and also summation over an associated (but not explicitly indicated) species label. Of course, if an external force field $\mathbf{F}_{ex}$ is present, it would be added to the right-hand side of (3.67). Note that (3.66) is a deceptively simply equation, for $\mathfrak{F}$ and $\mathbf{F}$ have a stochastic character, being wildly varying, singular functions of ζ and t. As always, therefore, we are obliged to consider a statistical ensemble of systems, characterized, e.g., by a Liouville function $\mathfrak{D}$, each system of the ensemble evolving according to (3.66). The ensemble averaged quantities, such as

$$\langle \mathfrak{F} \rangle = \int d\zeta_1 \cdots d\zeta_N \mathfrak{D}(\zeta_1 \cdots \zeta_N)\mathfrak{F} \tag{3.68}$$

are assumed to correspond to macroscopic observations of a single system. The symmetry of $\mathfrak{D}$ makes (3.68) a sum of N equal terms, and, comparing (3.68) with (3.64), we see that

$$\langle \mathfrak{F}(\zeta,t) \rangle = f^{(1)}(\zeta,t) \tag{3.69}$$

That is, $\langle \mathfrak{F} \rangle$ is just the conventional one-body function $f^{(1)}$.

We obtain an equation for $f^{(1)}$ by taking the expectation value of (3.66),

$$\frac{\partial f^{(1)}}{\partial t} + \mathbf{v} \cdot \nabla f^{(1)} + \langle \mathbf{F} \cdot \nabla_p \mathfrak{F} \rangle = 0 \tag{3.70}$$

or, in view of (3.67),

$$\frac{\partial f^{(1)}}{\partial t} + \mathbf{v} \cdot \nabla f^{(1)} + n \int d\zeta' \, \hat{\mathbf{F}}(\zeta,\zeta') \cdot \nabla_p \langle \mathfrak{F}(\zeta)\mathfrak{F}(\zeta') \rangle = 0 \tag{3.71}$$

(Here, as is sometimes convenient, we have replaced the derivative with respect to $\mathbf{v}$ by m times the derivative with respect to $\mathbf{p} = m\mathbf{v}$.) Thus, we are inevitably led to consider expectation values of products of the $\mathfrak{F}$; to determine $\langle \mathfrak{F}\mathfrak{F} \rangle$, we multiply (3.66) by $\mathfrak{F}$ and then take the ensemble average. The resulting equation involves $\langle \mathfrak{F}\mathfrak{F}\mathfrak{F} \rangle$, and so we are led to a hierarchy similar to the BGKBY one. In fact, there is a simple relation between $\langle \mathfrak{F}(1) \cdots \mathfrak{F}(s) \rangle$ and $f^{(s)}$. For instance,

$$\langle \mathfrak{F}(\zeta)\mathfrak{F}(\zeta') \rangle = n^{-2} \int d\zeta_1 \cdots d\zeta_n \, \mathfrak{D}(\zeta_1 \cdots \zeta_n)$$
$$\sum_{i,j} \delta(\zeta - \zeta_i)\delta(\zeta' - \zeta_j)$$
$$= n^{-1} \delta(\zeta,\zeta') \, f^{(1)}(\zeta) + f^{(2)}(\zeta,\zeta') \tag{3.72}$$

where $\delta(\zeta,\zeta')$ denotes a Dirac-delta function $\delta(\zeta - \zeta')$ times a Kronecker delta in the corresponding species labels.   (This singular part, representing self-correlations of particles, arises from terms in the sum with $i = j$, whereas $f^{(2)}$ comes from terms with $i \neq j$.)

It is convenient to define the "fluctuations"

$$\delta\mathfrak{F} = \mathfrak{F} - \langle\mathfrak{F}\rangle \qquad \delta\mathbf{F} = \mathbf{F} - \langle\mathbf{F}\rangle \qquad \text{etc.}$$

i.e., the deviation between $\mathfrak{F}$ or $\mathbf{F}$ for a single system and the ensemble average.   Then (3.70) can be written

$$\left[\frac{\partial}{\partial t} + \mathbf{v} \cdot \nabla + \langle F\rangle \cdot \nabla_p\right] f^{(1)} = \frac{\delta f^{(1)}}{\delta t} \tag{3.73}$$

where $\dfrac{\delta f^{(1)}}{\delta t} = -\langle\delta\mathbf{F} \cdot \nabla_p\,\delta\mathfrak{F}\rangle$

$$= n\!\!\int d\zeta\,\,\hat{\mathbf{F}}(\zeta,\zeta') \cdot \nabla_p\langle\delta\mathfrak{F}(\zeta)\,\,\delta\mathfrak{F}(\zeta')\rangle \tag{3.74}$$

We see that the "collisional" term $\delta f^{(1)}/\delta t$ in the kinetic equation results directly from a correlation of the fluctuations $\langle\delta\mathfrak{F}\,\delta\mathfrak{F}\rangle$, or, equivalently, from the two-body function $f^{(2)}$.   Only to the extent that this can be (approximately) expressed in terms of $f^{(1)}$ will we be able to deduce a true "kinetic" equation involving only $f^{(1)}$.

3.4.2. Coulomb Interactions

We specialize now to the Coulomb gas, with

$$\hat{\mathbf{F}}(\zeta,\zeta') = -qq'\nabla|\mathbf{r} - \mathbf{r}'|^{-1} \tag{3.75}$$

since we obtain, thereby, the principal characteristics peculiar to a plasma. Inclusion of the nonretarded Lorentz force is straightforward, and extension to include radiation is possible[8] but is beyond the scope of this chapter.   The method used to render the equations soluble is Dupree's expansion in fluctuations: we assume that $\langle\delta\mathfrak{F}\,\delta\mathfrak{F}\rangle$ is of order ε compared with $\langle\mathfrak{F}\rangle$; that $\langle\delta\mathfrak{F}(1)\,\delta\mathfrak{F}(2)\,\delta\mathfrak{F}(3)\rangle$ is of order ε^2; etc., where ε is a small parameter. By examining the hierarchy of equations for[8] $\langle\delta\mathfrak{F}(1) \cdots \delta\mathfrak{F}(s)\rangle$, with $s = 2, 3, 4, \ldots$, or, equivalently, for[6] the functions $f^{(s)}$, one can show that this expansion procedure is indeed valid for a stable plasma, at least in the domain of the independent variables for which x is of order λ_D and t of order ω_p^{-1}.   The small parameter ε turns out to be the reciprocal of the number of particles in a Debye cube,

$$\varepsilon = (n\lambda_D{}^3)^{-1}$$

(Alternatively, we can consider a gedanken procedure[6] of subdividing the particles of the plasma, so that q, m, and $1/n$ are all proportional to a parameter ε, while q/m, nq, and nm remain fixed; the expansion in fluctuations is then justified in the limit $\varepsilon \to 0$.)

To lowest order, then, we have, from (3.73),

$$\frac{\partial f^{(1)}}{\partial t} + \mathbf{v} \cdot \nabla f^{(1)} + \langle \mathbf{F} \rangle \cdot \nabla_p f^{(1)} = 0 \tag{3.76}$$

The force $\langle \mathbf{F} \rangle$ represents a self-consistent electric field, $\langle \mathbf{F} \rangle = q\mathbf{E}$ (plus external force fields, if any), whose sources lie in the plasma, as described by $f^{(1)}$,

$$\nabla \cdot \mathbf{E} = 4\pi \!\!\!\!\!\! \oint dv \, nq f^{(1)}(\mathbf{v}) \qquad \nabla \times \mathbf{E} = 0 \tag{3.77}$$

Inclusion of the Lorentz force in $\mathbf{F}(\zeta, \zeta')$ would have given

$$\langle \mathbf{F} \rangle = q(\mathbf{E} + \mathbf{v} \times \mathbf{B}/c)$$

with

$$\nabla \cdot \mathbf{B} = 0 \qquad c\nabla \times \mathbf{B} = 4\pi \!\!\!\!\!\! \oint dv \, nq\mathbf{v} f^{(1)}(\mathbf{v}) + \dot{\mathbf{E}} \tag{3.78}$$

Finally, including also the retardation, e.g., by introducing transverse field oscillators, gives both Faraday's law ($\nabla \times \dot{\mathbf{E}} = -\dot{\mathbf{B}}/c$) and equations like (3.76) for the distribution function of each oscillator. Of course, any external fields present would be added to those given by (3.77) and (3.78).

The terms *Vlasov equation* and *collisionless Boltzmann equation* are frequently used for (3.76), (3.77), and (3.78), although *correlationless kinetic equation* (CKE) would be a better description, since collisionless implies the absence of interaction between particles. Neglect of $\langle \mathfrak{F}\mathfrak{F} \rangle$ in (3.71) would indeed correspond to a lack of interaction, but with neglect of the correlations, $\langle \delta\mathfrak{F} \, \delta\mathfrak{F} \rangle$, an interaction via the self-consistent field $\langle \mathbf{F} \rangle$ remains. In any case, these equations, which are a good approximation for $n\lambda_D{}^3 \gg 1$, i.e., for a sufficiently hot and not too dense plasma, contain a good deal of plasma physics—the effects of Debye shielding, plasma oscillations, a variety of linearized waves, instabilities, and the generalized (correlationless) dielectric constant (i.e., the linear response of the plasma to external electromagnetic fields). While we shall defer our discussion of these properties to a later section to avoid an interruption of the present development, we shall make a slight digression here to derive the solution of the linearized Vlasov equation, which will be needed in the discussion of the next order in the expansion in fluctuations.

3.4.3. Solution of the Linearized Vlasov Equation

Any time-independent homogeneous distribution function $f(\mathbf{v})$, with $\oint dv \, nqf = \oint dv \, nq\mathbf{v}f = 0$, trivially satisfies the Vlasov equation. Restricting ourselves to the case of Coulomb interactions, we consider the linearization of (3.76) about such a solution; i.e., we set

$$f^{(1)} = f(v) + \delta f(\mathbf{x},\mathbf{v},t) \tag{3.79}$$

and neglect terms quadratic in δf. (*Note:* The linear term δf in the one-body function should not be confused with the fluctuation $\delta \mathcal{F}$ in the microscopic distribution function. As used here, the δf is, like $f^{(1)}$, an ensemble averaged quantity, whereas $\delta \mathcal{F}$ is a singular, stochastic function.) It is convenient to take the Fourier-Laplace transform of δf, or, in more physical terms, to study the general initial-value problem for plane waves of wave number $\mathbf{k}$. Substituting (3.79) into (3.76) and linearizing in δf, we see that

$$\delta f(\mathbf{k},\mathbf{v},\omega) \equiv \int d\mathbf{x} \exp \left(-i\mathbf{k} \cdot \mathbf{x} \right) \int_0^\infty dt \exp \left(i\omega t \right) \delta f(\mathbf{x},\mathbf{v},t) \tag{3.80}$$

with Im $\omega > 0$, satisfies

$$i(\mathbf{k} \cdot \mathbf{v} - \omega) \, \delta f = - \frac{q}{m} \left(\delta \mathbf{E} + \delta \mathbf{E}_e \right) \cdot \nabla_v f + \delta f_0 \tag{3.81}$$

where δf_0 denotes the initial value of δf,

$$\delta f_0(\mathbf{k},\mathbf{v}) = \delta f(\mathbf{k},\mathbf{v},t)_{t=0}$$

and we have separated the self-consistent electric field

$$\delta \mathbf{E}(\mathbf{k},\omega) = - \frac{4\pi i}{k^2} \mathbf{k} \oint d\mathbf{v} \, qn \, \delta f(\mathbf{k},\mathbf{v} \, \omega) \tag{3.82}$$

from a possible external field $\delta \mathbf{E}_e$. From (3.81) and (3.82) it follows at once that

$$\delta f(\mathbf{k},\mathbf{v},\omega) = \frac{\delta f_0(\mathbf{k},\mathbf{v})}{i(\mathbf{k} \cdot \mathbf{v} - \omega)} - \frac{q \, \delta E \, \mathbf{k} \cdot \nabla_v f}{mik(\mathbf{k} \cdot \mathbf{v} - \omega)} \tag{3.83}$$

and

$$\delta E(\mathbf{k},\omega) = - \frac{4\pi}{k\epsilon(\mathbf{k},\omega)} \oint \frac{d\mathbf{v} \, nq}{\mathbf{k} \cdot \mathbf{v} - \omega} \left(\delta f_0 - \frac{q}{m} \delta \mathbf{E}_e \cdot \nabla_v f \right) \tag{3.84}$$

where

$$\epsilon(\mathbf{k},\omega) = 1 - \oint d\mathbf{v} \, \frac{\omega_p^2}{k^2} \frac{\mathbf{k} \cdot \nabla_v f}{\mathbf{k} \cdot \mathbf{v} - \omega} \tag{3.85}$$

The significance of ϵ is seen most clearly if we examine the case $\delta f_0 = 0$. Then (3.84) gives

$$\delta E = \frac{1 - \epsilon}{\epsilon} \, \delta E_e$$

and the total field is

$$\delta E + \delta E_e = \frac{1}{\epsilon} \, \delta E_e \tag{3.86}$$

Since the ratio of external field (the **D** *vector* of macroscopic electrodynamics) to total field (the $\mathbf{E}$ *vector*) is defined as the dielectric "constant," the ϵ defined by (3.85) is conventionally called the *longitudinal dielectric constant* of the plasma. [The designation longitudinal reflects our neglect of magnetic effects and the consequent longitudinal character of $\delta\mathbf{E}$. Of course, ϵ is not a constant, being a function of $\mathbf{k}$ and ω. Moreover, the electric-induction and electric-field vectors have the simple relation $\mathbf{D} = \epsilon\mathbf{E}$ *only* so long as we deal with the Fourier-Laplace transforms. In the space-time domain, they are related by a convolution whose kernel is the inverse transform of $\epsilon(\mathbf{k},\omega)$,

$$\mathbf{D}(\mathbf{x},t) = \int d\mathbf{x}' \, dt' \, \epsilon(\mathbf{x} - \mathbf{x}', t - t')\mathbf{E}(\mathbf{x}',t')$$

The nonlocal character in t is called *dispersion;* that in $\mathbf{x}$ is sometimes (especially in the Soviet literature) termed *spatial dispersion.*]

For the initial-value problem, with $\delta f_0 \neq 0$ and $\delta E_e = 0$, it is convenient to write (3.83) and (3.84) in a compact form,

$$\delta f(\mathbf{k},\mathbf{v}) = \int d\mathbf{v}' \, Q(\mathbf{k},\mathbf{v},\mathbf{v}') \, \delta f_0(\mathbf{k},\mathbf{v}')$$
$$\delta E(\mathbf{k}) = \int d\mathbf{v} \, P(\mathbf{k},\mathbf{v}) \, \delta f_0(\mathbf{k},\mathbf{v}) \tag{3.87}$$

where the functions Q and P, like δf and δE, depend not only on the indicated variables but also on ω or t,

$$Q(\mathbf{k},\mathbf{v},\mathbf{v}',\omega) = -i(\mathbf{k} \cdot \mathbf{v} - \omega)^{-1}\left[\delta(\mathbf{v},\mathbf{v}') - \frac{q}{km}\mathbf{k} \cdot \nabla_v f P(\mathbf{k},\mathbf{v}',\omega) \right]$$
$$P(\mathbf{k},\mathbf{v} \, \omega) = -\frac{4\pi nq}{k\epsilon(\mathbf{k},\omega)(\mathbf{k} \cdot \mathbf{v} - \omega)} \tag{3.88}$$

$$Q(\mathbf{k},\mathbf{v},\mathbf{v}',t) = (2\pi)^{-1}\int d\omega \, \exp\,(-i\omega t)Q(\mathbf{k},\mathbf{v},\mathbf{v}',\omega)$$
$$P(\mathbf{k},\mathbf{v},t) = (2\pi)^{-1}\int d\omega \, \exp\,(-i\omega t)P(\mathbf{k},\mathbf{v},\omega) \tag{3.89}$$

In carrying out the inverse Laplace transforms (3.89), the contour of integration in the ω plane must, as usual, be chosen above any singularities of the integrand. If we close the contour below, we are left with a sum of residues at the various poles of the integrand. Some of these poles are at the zeros of $\epsilon(\mathbf{k},\omega)$. If $f(\mathbf{v})$ is Maxwellian, or, more generally, if it is isotropic, i.e. a function of v^2, then all zeros of ϵ lie in the lower half of the ω plane. Any plasma whose ϵ has this property is called *stable*, and we restrict ourselves here to this case. Then, in the limit $t \to \infty$, the behavior of P and Q is determined only by the real poles at $\omega = \mathbf{k} \cdot \mathbf{v}$ or $\mathbf{k} \cdot \mathbf{v}'$,

$$P(\mathbf{k},\mathbf{v},t) \xrightarrow[t \to \infty]{} -\frac{4\pi nqi}{k\epsilon(\mathbf{k},\mathbf{k} \cdot \mathbf{v})} \exp\,(-i\mathbf{k} \cdot \mathbf{v}t)$$

$$Q(\mathbf{k},\mathbf{v},\mathbf{v}',t) \xrightarrow[t \to \infty]{} \delta(\mathbf{v},\mathbf{v}') \exp\,(-i\mathbf{k} \cdot \mathbf{v}t) \tag{3.90}$$

$$+ \frac{4\pi n'q'q}{k^2m}\frac{\mathbf{k} \cdot \nabla_v f}{\mathbf{k} \cdot (\mathbf{v} - \mathbf{v}')}\left[\frac{\exp\,(-i\mathbf{k} \cdot \mathbf{v}'t)}{\epsilon(\mathbf{k},\mathbf{k} \cdot \mathbf{v}')} - \frac{\exp\,(-i\mathbf{k} \cdot \mathbf{v}t)}{\epsilon(\mathbf{k},\mathbf{k} \cdot \mathbf{v})} \right]$$

since the residues at the zeros of ϵ are all damped. (This represents a phenomenon, peculiar to a plasma, known as *Landau damping*, which we shall discuss in more detail below.) Making use of the identities

$$\lim_{t \to \infty} \int_{-\infty}^{\infty} dx \, \frac{\exp (ixt)g(x)}{x \pm i\eta} = \begin{cases} 0 \\ 2\pi i g(0) \end{cases}$$

we can simplify the expression for Q,

$$Q(\mathbf{k},\mathbf{v},\mathbf{v}',t) \xrightarrow[t \to \infty]{} \exp (-i\mathbf{k} \cdot \mathbf{v}t) \left[\delta(\mathbf{v},\mathbf{v}') - \frac{4\pi n'qq'\mathbf{k} \cdot \nabla_v f}{mk^2(u + i\eta)\epsilon(\mathbf{k} \cdot \mathbf{v})} \right] \quad (3.91)$$

where $u = \mathbf{k} \cdot (\mathbf{v} - \mathbf{v}')$ and we have, for now, suppressed the $\mathbf{k}$ dependence of $\epsilon(\mathbf{k},\omega)$.

3.4.4. The Lenard-Balescu Kinetic Equation

We now return to the solution of (3.73) and (3.74) by the expansion in fluctuations, knowing some of the properties of the lowest-order approximation. Going to one higher order in the ε expansion, we must retain the $\langle \delta\mathfrak{F} \, \delta\mathfrak{F} \rangle$ term in (3.74). It is a straightforward matter to write an equation for $\langle \delta\mathfrak{F} \, \delta\mathfrak{F} \rangle$, or, equivalently, for the two-body function $f^{(2)}$. With the neglect of $\langle \delta\mathfrak{F} \, \delta\mathfrak{F} \, \delta\mathfrak{F} \rangle$, which of course occurs in the equation for $\langle \delta\mathfrak{F} \, \delta\mathfrak{F} \rangle$, the latter is soluble in principle, provided that one knows the one-body function $f^{(1)}$, which also occurs. In fact, a self-consistent solution of both the equation for $\langle \delta\mathfrak{F} \, \delta\mathfrak{F} \rangle$ and of (3.73) for $f^{(1)}$ is required, something which is currently a subject of many research investigations. However, for stable plasmas one can often simplify the problem by introducing an ansatz due to Bogolyubov: assume that, for given $f^{(1)}$, $\langle \delta\mathfrak{F} \, \delta\mathfrak{F} \rangle$ relaxes to its asymptotic form on a time scale short compared with that on which $f^{(1)}$ changes appreciably. This allows us to solve the equation for $\langle \delta\mathfrak{F} \, \delta\mathfrak{F} \rangle$, regarding $f^{(1)}$ as time-independent; take the $t \to \infty$ limit of the solution; and substitute the result, which will be a functional of $f^{(1)}$, into (3.74). This yields, then, a kinetic equation, i.e., one involving $f^{(1)}$ alone, from which the evolution of $f^{(1)}$ on its (slow) time scale can be determined.

The equation for $\langle \delta\mathfrak{F} \, \delta\mathfrak{F} \rangle$ was solved by Lenard[9] in the asymptotic limit. The full time-dependent solution was given by Guernsey,[10] who made elegant use of the techniques of singular integral equations to obtain an explicit solution in a closed, albeit algebraically complicated, form. We shall follow here the procedure of first solving for $\delta\mathfrak{F}$ and then taking the ensemble average of the product $\langle \delta\mathfrak{F} \, \delta\mathfrak{F} \rangle$; this leads to the same results, with a much simpler formalism.

The difference of (3.66) and (3.70) gives an (exact) equation for $\delta \mathcal{F}$,

$$\left[\frac{\partial}{\partial t} + \mathbf{v} \cdot \nabla + \frac{q}{m} \langle \mathbf{E} \rangle \cdot \nabla_v \right] \delta \mathcal{F} + \frac{q}{m} \delta \mathbf{E} \cdot \nabla_v f^{(1)}$$
$$= -q\nabla_p \cdot (\delta \mathbf{E} \, \delta \mathcal{F} - \langle \delta \mathbf{E} \, \delta \mathcal{F} \rangle) \quad (3.92)$$

If we derive from (3.92) an equation for $\langle \delta \mathcal{F} \, \delta \mathcal{F} \rangle$, the terms involving $\langle \delta \mathcal{F} \, \delta \mathcal{F} \, \delta \mathcal{F} \rangle$ will arise from the terms on the right side of (3.92). Neglect of third-order fluctuations is therefore tantamount to dropping the quadratic terms in (3.92), so that $\delta \mathcal{F}$ simply satisfies a *linearized* Vlasov equation like that studied in the previous section. Of course, this equation is not easy to solve if there is a nonvanishing average electric field, $\langle \mathbf{E} \rangle \neq 0$, or if $f^{(1)}$ is a function of $\mathbf{x}$ and t. Guernsey[10] and Dupree[8] have considered the case where the $(\mathbf{x},t)$ dependence of $f^{(1)}$ can be treated as a small perturbation, and Field and Fried[11] and Pearson[12] have studied the case $\langle \mathbf{E} \rangle \neq 0$. However, the Bogolyubov ansatz allows us to neglect the time dependence of $f^{(1)}$, and we shall here confine ourselves to the case where $f^{(1)}$ is also independent of $\mathbf{x}$, and $\langle \mathbf{E} \rangle = 0$.

For such a homogeneous plasma, bilinear expectation values, $\langle A(x)B(x') \rangle$, are functions only of $x - x'$; since

$$\langle A(x)B(x') \rangle = \int d\mathbf{k} \, d\mathbf{k}' \, \langle A(\mathbf{k})^* B(\mathbf{k}') \rangle \exp\left[i(\mathbf{k}' \cdot \mathbf{x}' - \mathbf{k} \cdot \mathbf{x})\right]$$

we have

$$\langle A^*(\mathbf{k})B(\mathbf{k}') \rangle = \langle\!\langle A^*(\mathbf{k})B(\mathbf{k}) \rangle\!\rangle \, \delta(\mathbf{k} - \mathbf{k}')$$
$$\langle\!\langle A^*(\mathbf{k})B(\mathbf{k}) \rangle\!\rangle \equiv (2\pi)^{-3} \int d\mathbf{x} \, \langle A(\mathbf{x})B(0) \rangle \exp i\mathbf{k} \cdot \mathbf{x}$$
and $\quad \langle A(\mathbf{x})B(\mathbf{x}) \rangle = \int d\mathbf{k} \, \langle\!\langle A^*(\mathbf{k})B(\mathbf{k}) \rangle\!\rangle$

Thus, the collisional term (3.74) is

$$\frac{\delta f^{(1)}}{\delta t} = -q \int d\mathbf{k} \, \langle\!\langle \delta \mathbf{E}^*(\mathbf{k},t) \cdot \nabla_p \, \delta \mathcal{F}(\mathbf{k},\mathbf{v},t) \rangle\!\rangle$$
$$= -q \int d\mathbf{k} \, \hat{\mathbf{k}} \cdot \nabla_p \langle \delta E^*(\mathbf{k},t) \, \delta \mathcal{F}(\mathbf{k},\mathbf{v},t) \rangle \quad (3.93)$$

Since $\delta \mathcal{F}$ satisfies the linear Vlasov equation obtained by dropping the right side of (3.92), we can use (3.87), (3.88), and (3.89) to write an explicit solution for $\delta \mathcal{F}$ and $\delta \mathbf{E}$ at time t in terms of the initial values, $\delta \mathcal{F}(t = 0)$. Substitution of this into (3.93) gives a time-dependent expression for $\delta f^{(1)}/\delta t$ equivalent to Guernsey's. Of more interest, however, in view of the Bogolyubov ansatz, is the asymptotic limit of $\delta \mathcal{F}$, $\delta \mathbf{E}$, and $\delta f^{(1)}/\delta t$. Specifically, we have from (3.87) and (3.93) an explicit expression for $\delta f^{(1)}/\delta t$ in terms of the initial fluctuations,

$$\frac{\delta f^{(1)}}{\delta t} = -q \int d\mathbf{k} \, \hat{\mathbf{k}} \cdot \nabla_p \oint d\mathbf{v}' \oint d\mathbf{v}'' P^*(\mathbf{k},\mathbf{v}',t)$$
$$Q(\mathbf{k},\mathbf{v},\mathbf{v}'',t) \langle\!\langle \delta \mathcal{F}_0(\mathbf{k},\mathbf{v}')^* \, \delta \mathcal{F}_0(\mathbf{k},\mathbf{v}'') \rangle\!\rangle \quad (3.94)$$

where $\hat{\mathbf{k}} = \mathbf{k}/k$ and

$$\delta\mathfrak{F}_0(\mathbf{k},\mathbf{v}) = \delta\mathfrak{F}(\mathbf{k},\mathbf{v},t)\Big|_{t=0}$$

It remains only to specify $\langle \delta\mathfrak{F}\ \delta\mathfrak{F}\rangle$ at $t = 0$.

The simplest choice is to assume that at $t = 0$ there are no correlations, i.e., that

$$\mathfrak{D}(\zeta_1,\zeta_2,\ \ldots\ ,\zeta_n) = \prod_{i=1}^{n} f(\zeta_i) \tag{3.95}$$

Then $\quad f^{(2)}(\zeta,\zeta') = f(\zeta)f(\zeta') = \langle\mathfrak{F}(\zeta)\rangle\langle\mathfrak{F}(\zeta')\rangle \tag{3.96}$

and (3.72) gives

$$\langle\delta\mathfrak{F}(\zeta)\ \delta\mathfrak{F}(\zeta')\rangle = n^{-1}\ \delta(\zeta,\zeta')f(\zeta) \tag{3.97}$$

Thus, at $t = 0$ we take

$$\langle\delta\mathfrak{F}_0{}^*(\mathbf{k},\mathbf{v})\ \delta\mathfrak{F}_0(\mathbf{k},\mathbf{v}')\rangle = \frac{1}{(2\pi)^3 n}\ f(\mathbf{v})\ \delta(\mathbf{v},\mathbf{v}') \tag{3.98}$$

which reduces (3.94) to

$$\frac{\delta f^{(1)}}{\delta t} = -\ q\int\frac{d\mathbf{k}}{(2\pi)^3}\ \hat{\mathbf{k}}\cdot\nabla_p\oint dv'\ \frac{f(\mathbf{v}')P^*(\mathbf{k},\mathbf{v}',t)Q(\mathbf{k},\mathbf{v},\mathbf{v}',t)}{n'} \tag{3.99}$$

Finally, using the asymptotic expressions (3.90) and (3.91) for P and Q, we find

$$\frac{\delta f^{(1)}}{\delta t}\ \xrightarrow[t\to\infty]{}\ -\ \frac{4\pi q^2}{(2\pi)^3}\int d\mathbf{k}\ \frac{\mathbf{k}\cdot\nabla_p}{k^2}\Bigg[\frac{if(\mathbf{v})}{\epsilon^*(\mathbf{k}\cdot\mathbf{v})}$$
$$-\ \frac{4\pi^2 n'(q')^2}{mk^2}\ \frac{\mathbf{k}\cdot\nabla_v f}{|\epsilon(\mathbf{k}\cdot\mathbf{v})|^2}\int dv'\ f(\mathbf{v}')\ \delta(u)\Bigg] \tag{3.100}$$

In the last term we have set

$$(u + i\eta)^{-1} = Pu^{-1} - \pi i\ \delta(u)$$

and have dropped the principal-value term because it vanishes by the Riemann-Lebesque lemma[13] for $t\to\infty$. It is conventional to rewrite the first term of (3.100), using the fact that $\mathrm{Re}\ \epsilon^{-1}$ is even in $\mathbf{k}$ and, being multiplied by $\mathbf{k}$, can be dropped; thus $i/\epsilon^*(\mathbf{k}\cdot\mathbf{v})$ can be replaced by

$$-\mathrm{Im}\ \frac{1}{\epsilon^*(\mathbf{k}\cdot\mathbf{v})} = \mathrm{Im}\ \frac{\epsilon}{|\epsilon|^2} = \oint dv'\ \frac{(\omega_p')^2\mathbf{k}\cdot\nabla_{v'}f(\mathbf{v}')\pi\ \delta(u)}{k^2[\epsilon(\mathbf{k}\cdot\mathbf{v})]^2}$$

and we have, finally,

$$\frac{\delta f^{(1)}}{\delta t} = -\nabla_p \cdot \mathbf{J} \tag{3.101}$$

$$\mathbf{J} = \oint d\mathbf{v}' \, [f(\mathbf{v})\nabla_{p'}f(\mathbf{v}') - f(\mathbf{v}')\nabla_p f(\mathbf{v})] \cdot \mathbf{K} \, (\mathbf{v},\mathbf{v}') \tag{3.102}$$

$$\mathbf{K}(\mathbf{v},\mathbf{v}') = 2q^2 q'^2 n' \int d\mathbf{k} \, \frac{\mathbf{kk} \, \delta[\mathbf{k} \cdot (\mathbf{v} - \mathbf{v}')]}{k^4 [\epsilon(\mathbf{k},\mathbf{k} \cdot \mathbf{v})]^2} \tag{3.103}$$

a result derived independently by Lenard and Balescu.

Before commenting on the various features of this result, we note that the derivation sketched above can serve as a model for calculation of other quantities of physical interest, e.g., the particle-particle correlations themselves, $\langle \delta \mathcal{F} \, \delta \mathcal{F} \rangle$, or other fluctuation quantities such as $\langle \delta E \, \delta E \rangle$, etc.

3.4.5. Some Properties of the Lenard-Balescu Kinetic Equation

1. Aside from the $|\epsilon|^2$ in the denominator [and a factor n' resulting from our convention of normalizing $\int f^{(1)} \, d\mathbf{v} = 1$ rather than $\int f \, d\mathbf{v} = n$ as in the development leading to (3.62)] we see that (3.103) agrees with the Landau-Boltzmann result (3.62). Moreover, the physical origin of the $|\epsilon|^2$ is clear; it corresponds to using for the interparticle interaction not the bare Coulomb field but rather one with the screening represented by the dielectric constant $\epsilon(\mathbf{k},\omega)$ of the correlationless approximation. The potential due to a particle having velocity $\mathbf{v}$ is

$$V(\mathbf{x} - \mathbf{v}t) = \int d\mathbf{k} \, \exp \, [i(\mathbf{k} \cdot \mathbf{x} - \mathbf{k} \cdot \mathbf{v}t)]V(\mathbf{k})$$
$$= \int d\mathbf{k} \, d\omega \, V(\mathbf{k}) \, \delta(\omega - \mathbf{k} \cdot \mathbf{v}) \exp \, [i(\mathbf{k} \cdot \mathbf{x} - \omega t)]$$

and so the screening results in an interaction

$$V_s = \int d\mathbf{k} \, d\omega \, \delta(\omega - \mathbf{k} \cdot \mathbf{v}) \exp \, [i(\mathbf{k} \cdot \mathbf{x} - \omega t)] \frac{V(\mathbf{k})}{\epsilon(\mathbf{k},\omega)}$$
$$= \int d\mathbf{k} \, \frac{V(\mathbf{k})}{\epsilon(\mathbf{k},\mathbf{k} \cdot \mathbf{v})} \exp \, [i\mathbf{k} \cdot (\mathbf{x} - \mathbf{v}t)] \tag{3.104}$$

Replacing the Coulomb potential, $V(\mathbf{k}) = q/k^2$, in (3.62) by $V(\mathbf{k})/\epsilon(\mathbf{k},\mathbf{k} \cdot \mathbf{v})$ results in the form (3.103).

2. The simplest way to see the effect of the screening represented by ϵ is just to use the static approximation

$$\epsilon(\mathbf{k},\omega) = \epsilon(\mathbf{k},0) = 1 + \sum_\alpha (k\lambda_D{}^\alpha)^{-2} \tag{3.105}$$

For example, if we neglect ion dynamics in a simple electron-ion plasma.

we have

$$\begin{aligned}
\mathsf{K} &= 2e^4 n \int d\mathbf{k} \frac{\mathbf{k}\mathbf{k}\,\delta(\mathbf{k}\cdot\mathbf{u})}{k^2(k^2 + k_D{}^2)} \\
&= \frac{2ne^4}{u}\left(\mathbf{I} - \frac{\mathbf{u}\mathbf{u}}{u^2}\right)\int dk_r \frac{k_r\,d\theta\,\sin^2\theta}{k_r{}^2 + k_D{}^2} \\
&= \frac{ne^4}{u}\,\pi\ln\left(\frac{k_{\max}{}^2}{k_D{}^2} + 1\right)\left(\mathbf{I} - \frac{\mathbf{u}\mathbf{u}}{u^2}\right)
\end{aligned} \tag{3.106}$$

As with (3.62), there is still a need for a large-k (or small-impact-parameter) cutoff. However, the divergence at large impact parameter is now taken care of automatically, the result being the same as if we had cutoff (3.62) at an impact parameter equal to λ_D. For large $\mathbf{k}$, $\epsilon(\mathbf{k},\mathbf{k}\cdot\mathbf{v}) \to 1$, so the divergence problem is the same as in the earlier calculations, and we again choose $k_{\max} = T/e^2$. Then

$$\mathsf{K} = \frac{ne^4\pi}{u}\,(\ln\Lambda)\left(\mathbf{I} - \frac{\mathbf{u}\mathbf{u}}{a^2}\right) \tag{3.107}$$

where $\ln\Lambda = \ln\dfrac{T^2}{k_D{}^2 e^4} = \ln\dfrac{T^3}{4\pi ne^6}$

is sometimes called the *Coulomb logarithm*.

3. As is clear from the derivation of (3.103), $\delta f/\delta t$ arises from a summation over the possible waves in the plasma. In addition to the contributions from the static limit, discussed above, there will be others, particularly if $\epsilon(\mathbf{k},\omega)$ has zeros near the real axis, corresponding to weakly damped waves, since the resulting small denominator in (3.103) will give a "resonant" contribution to K. Of course, the waves must not be too weakly damped, or else the approximations involved in deriving (3.90) from (3.89) will be invalid.

4. The Lenard-Balescu form of $\delta f/\delta t$ satisfies various canonical properties expected for a kinetic equation: f remains positive definite, conserves number of particles, momentum, and energy, and approaches a Maxwellian distribution.

5. Although the above derivation involved the special choice [(3.98)] of $\langle\delta\mathfrak{F}_0\,\delta\mathfrak{F}_0\rangle$ corresponding to a plasma without correlation, the result is a much more general one. In view of the oscillatory factors, $\exp(-i\mathbf{k}\cdot\mathbf{v}t)$, of Q and P for large t, the integrations over velocity in (3.94) would give zero, according to the Riemann-Lebesque lemma,[13] if $\langle\delta\mathfrak{F}_0\,\delta\mathfrak{F}_0\rangle$ were well behaved. That they do not is a consequence of the singular character of the $\langle\delta\mathfrak{F}_0\,\delta\mathfrak{F}_0\rangle$ we have chosen. However, we see that, for the large class of $\langle\delta\mathfrak{F}_0\,\delta\mathfrak{F}_0\rangle$ which differ from (3.98) by the addition of nonsingular integrable functions, the result of the integrations in (3.94) will differ from (3.100) only by terms which vanish as $t \to \infty$. By the same token, of course, it is possible to make special choices of $\langle\delta\mathfrak{F}_0\,\delta\mathfrak{F}_0\rangle$

which would lead to a different result, but these imply an initial preparation of the plasma which would be difficult or impossible to achieve physically.

The analysis of this section shows the basis for current kinetic equations of plasmas. We see that the Fokker-Planck or Landau-Boltzmann forms of $\delta f/\delta t$ are approximate versions of the Lenard-Balescu expression and that the latter involves a number of approximations and assumptions, not all of which can be fully justified, even a posteriori, the Bogolyubov ansatz concerning time scales, the expansion in fluctuations, and the homogeneity of $f^{(1)}$ being the principal ones. An actively pursued topic of current research involves attempts to solve the coupled equations for $f^{(1)}$ and $\langle \delta \mathfrak{F} \, \delta \mathfrak{F} \rangle$, together with higher-order terms in some cases, for problems where these approximations are invalid. In the "quasilinear" theory,[14] for example, one continues to linearize Eq. (3.92) for $\delta \mathfrak{F}$, but allows $f^{(1)}$ to have a (slow) time dependence determined, in a self-consistent fashion, from (3.73); the equation for $\delta \mathfrak{F}$ is typically solved in a WKB approximation. The nonlinear terms in (3.92) are, in this context, referred to as *mode-coupling effects*. If the wave amplitudes become large, then the situation has some resemblance to turbulence and there have been many studies, particularly in the Soviet literature, of the theory of a "turbulent" plasma. At the present time, a complete, well-defined theoretical treatment, buttressed by either rigorous arguments or experimental verification, has yet to be developed.

The application of the results of this section to relaxation and transport phenomena is considered in a subsequent chapter. In the following section we return to the lowest-order (Vlasov) formulation and discuss some of its consequences.

3.5 SOME PROPERTIES OF THE CORRELATIONLESS KINETIC EQUATION

3.5.1. Longitudinal Waves According to Landau

In a neutral gas, with its short-range interactions, nontrivial physics is obtained only when one takes account of correlations (or "collisions"). Because of the long-range Coulomb force, however, there is much to be learned about a plasma even from the correlationless approximation [(3.76) to (3.78)]. In the present section we shall summarize the principal features of this theory.

The essential features are seen most clearly by studying near-equilibrium situations and keeping only terms of first order in the deviation from equilibrium. To begin with, we shall consider only longitudinal waves ($\nabla \times \mathbf{E} = 0$). For a plasma consisting of electrons and ions, we have from (3.76) and (3.77), using lower-case and capital letters rather than an index to distinguish the species,

Electrons: $\dfrac{\partial f}{\partial t} + \mathbf{v} \cdot \nabla f - \dfrac{e}{m}\,\mathbf{E} \cdot \nabla_v f = 0$ (3.108)

Ions: $\dfrac{\partial F}{\partial t} + \mathbf{v} \cdot \nabla F + \dfrac{e}{M}\,\mathbf{E} \cdot \nabla_v F = 0$ (3.109)

$$\nabla \cdot \mathbf{E} = 4\pi n e \textstyle\int d\mathbf{v}\,(F - f)$$ (3.110)

Although we here neglect the effects of correlations ($\delta f/\delta t$, $\delta F/\delta t$), it is reasonable to suppose that they have established a nearly Maxwellian distribution and to study deviations from it.   Thus, we take

$$f(\mathbf{x},\mathbf{v},t) = f_0(\mathbf{v}) + f_1(\mathbf{x},\mathbf{v},t)$$ (3.111)

with $f_0(\mathbf{v}) = \exp \dfrac{[-(\mathbf{v} - \mathbf{v}_0)^2/a^2]}{(\sqrt{\pi}\,a)^3}$ (3.112)

and similar expressions for the ions ($a \to A$, $\mathbf{v}_0 \to \mathbf{V}_0$).   We shall linearize, i.e., neglect terms of second or higher order in f_1, and shall work chiefly with the Fourier-Laplace transform

$$f_1(\mathbf{k},\mathbf{v},\omega) = \int_0^\infty dt \int d\mathbf{x} \exp\,[-i(\mathbf{k} \cdot \mathbf{x} - \omega t)]f_1(\mathbf{x},\mathbf{v},t)$$ (3.113)

If g (or G) denotes the perturbation f_1 (or F_1) at time $t = 0$, then the transform of (3.108) to (3.110) gives

$$f_1(\mathbf{k},\mathbf{v},\omega) = i\,\frac{g(\mathbf{k},\mathbf{v}) + (e\mathbf{E}/m) \cdot \nabla_v f_0}{\omega - \mathbf{k} \cdot \mathbf{v}}$$

$$F_1(\mathbf{k},\mathbf{v},\omega) = i\,\frac{G(\mathbf{k},\mathbf{v}) - (e\mathbf{E}/M) \cdot \nabla_v F_0}{\omega - \mathbf{k} \cdot \mathbf{v}}$$ (3.114)

and $\mathbf{E}(\mathbf{k},\omega) = -4\pi n_0 e i\,\dfrac{\mathbf{k}}{k^2} \int d\mathbf{v}\,(F_1 - f_1)$ (3.115)

Substituting (3.114) into (3.115) gives

$$\epsilon(\mathbf{k},\omega)\mathbf{E}(\mathbf{k},\omega) = \frac{4\pi n_0 e}{k^2}\,\mathbf{k} \int d\mathbf{v}\,\frac{G - g}{\omega - \mathbf{k} \cdot \mathbf{v}} = N(\mathbf{k},\omega)\hat{\mathbf{k}}$$ (3.116)

where $\epsilon(\mathbf{k},\omega) = 1 - \dfrac{\omega_p{}^2}{k^2} \int d\mathbf{v}\,\mathbf{k} \cdot \dfrac{\nabla_v f_0 + (m/M)\nabla_v F_0}{\mathbf{k} \cdot \mathbf{v} - \omega}$ (3.117)

is the dielectric function for a (correlationless) two-component plasma. When we invert the Laplace transform

$$E(\mathbf{k},t) = (2\pi)^{-1} \int_{-\infty}^\infty d\omega\, E(\mathbf{k},\omega) \exp\,(-i\omega t)$$ (3.118)

the ω contour, as noted before, must pass above all singularities of $E(\mathbf{k},\omega)$, and since, for $t > 0$, it can be closed below, we obtain a sum of contributions, one from each singularity of $E(\mathbf{k},\omega)$. About those arising from singularities of $N(\mathbf{k},\omega)$ no general statement can be made, since their nature and location will depend upon the particular initial conditions. However, ϵ depends only upon the equilibrium distributions f_0 and F_0, and so the poles of $E(\mathbf{k},\omega)$ arising from zeros of $\epsilon(\mathbf{k},\omega)$ are "universal" in the sense that they are independent of the initial conditions.   The equation

$$\epsilon(\mathbf{k},\omega) = 0$$

is often called the *dispersion equation* of the plasma.   When we substitute Maxwellian distributions, with $\mathbf{v}_0$ parallel to $\mathbf{V}_0$, for f_0 and F_0 in (3.117), we can write this dispersion equation in the form

$$\epsilon = 1 - \frac{k_D{}^2}{2k^2}\left[Z'\left(\frac{\omega - kv_0}{ka}\right) + \frac{T_e}{T_i}\, Z'\left(\frac{\omega - kV_0}{kA}\right)\right] \tag{3.119}$$

where the function Z is defined by

$$Z(\zeta) = \pi^{-1/2} \int_{-\infty}^{\infty} dx\, \frac{\exp\,(-x^2)}{x - \zeta} \qquad \mathrm{Im}\,\zeta > 0 \tag{3.120}$$

and as the analytic continuation thereof for $\mathrm{Im}\,\zeta < 0$.   An alternative representation, valid for both signs of $\mathrm{Im}\,\zeta$, is†

$$Z(\zeta) = 2i \exp\,(-\zeta^2) \int_{-\infty}^{i\zeta} dt\, \exp\,(-t^2)$$
$$= \exp\,(-\zeta^2)\left[i\,\sqrt{\pi} - 2\int_0^{\zeta} \exp\,(s^2)\,ds\right] \tag{3.121}$$

The derivative is denoted by $Z'(\zeta) = dZ/d\zeta$.

The roots of (3.119) have been studied extensively.   We can here state only the results,[15] distinguishing three cases:

1.   There is no relative drift ($\mathbf{v}_0 = \mathbf{V}_0 = 0$), and ion dynamics are neglected ($m/M \rightarrow 0$).   All roots then lie in the lower half ω plane; i.e., the waves or oscillations are damped.   The least damped root is

$$\omega = \omega_R - i\omega_I$$

where, for $k \ll k_D = \sqrt{2}\,\omega_p/a$,

$$\omega_R = \sqrt{\omega_p{}^2 + \frac{3k^2a^2}{2}}$$
$$\text{and} \quad \omega_I = \sqrt{\pi}\,\omega_p \left(\frac{\omega_p}{ka}\right)^3 \exp\left(-\frac{3}{2}\right)\exp\left(-\frac{k_D{}^2}{2k^2}\right) \tag{3.122}$$

In the absence of the thermal spread (i.e., for $a = 0$) there is neither dispersion nor damping; the oscillations occur at the plasma frequency ω_p for all k, a result which follows readily from a continuum (i.e., hydro-

---

† Differentiating Eq. (3.120) with respect to ζ yields a linear first-order differential equation for Z which is easily solved (with an integrating factor) and yields (3.121).

dynamic) picture.   By including in such a picture the effects of (thermal) pressure, one gets a k dependence of ω_R such as that given by (3.131), but *never* any damping.   The damping can appear only in a treatment based on distribution functions, since it is due[16] to the resonant transfer of energy from waves to the small group of particles with velocity near the group velocity of the waves ($v \approx \omega/k$), and these are lost sight of in a formalism which involves only a few moments of f.   The term *Landau damping* is used for this phenomenon, since it was first pointed out in a classic paper by L. D. Landau.[17]

One can follow the root (3.122) to larger values of k, but this is of little physical interest since it becomes very heavily damped (that is, $|\omega_I| \gg |\omega_R|$) for larger k.   In the limit $k \gg k_D$, asymptotic evaluation of (3.122) gives

$$\omega_R = \pi \frac{ka}{2} \left(\ln \frac{k^2}{\sqrt{\pi}\, k_D{}^2} - \ln \sqrt{\ln \frac{k^2}{\sqrt{\pi}\, k_D{}^2}} + \cdots \right)^{-1/2} \tag{3.123}$$

$$\omega_I = ka \left(\ln \frac{k^2}{\sqrt{\pi}\, k_D{}^2} - \ln \sqrt{\ln \frac{k^2}{\sqrt{\pi}\, k_D{}^2}} + \cdots \right)^{-1/2}$$

2.   There is no relative drift ($v_0 = V_0 = 0$), but ion dynamics are included (m/M small but nonzero).   The *electron plasma oscillations* described in (1) are unaffected (except for corrections of order m/M), but a whole new set of roots of (3.121) appears, with frequencies smaller than the electron plasma oscillations by $\sqrt{m/M}$.   However, if the electron and ion temperatures are equal, $ma^2 = MA^2$, then even the least damped of these ion plasma oscillations has $|\omega_I| \gtrsim |\omega_R|$ for all k, including $k = 0$, in contrast to (3.122).   If, instead, the ions are cooler than the electrons, as is common in many discharge phenomena, then the damping decreases and in the limit $T_e/T_i \gg 1$ we have

$$\omega = ka \sqrt{\frac{m}{2M(1 + k^2/k_D{}^2)}} \left\{ 1 - \frac{i\sqrt{\pi m/M}}{[2(1 + k^2/k_D{}^2)]^{3/2}} \right\} \tag{3.124}$$

provided that $k/k_D \ll \sqrt{T_e/T_i}$, that is, for k smaller than the ion Debye wave number.

For $k \ll k_D$, these waves have a constant phase velocity,

$$\frac{\omega}{k} = a \left(\frac{m}{2M} \right)^{1/2}$$

This is of the same order of magnitude as one would expect for normal acoustic waves in a medium with $T_i = T_e$; so the term *ion acoustic wave* is frequently used in describing them.   Of course, the physical mechanism involved in their propagation is quite different from the collisional momentum transfer which gives rise to ordinary acoustic waves in a high-density neutral gas.   The restoring forces here are electrical, arising from a very slight electron-ion charge separation.   They have not only been observed but have also been studied quantitatively in experiments[18] with quiescent cesium plasmas in which both the phase velocity and Landau damping were observed and found to be in good agreement with the theoretical predictions.

3.   There is a relative drift of electrons and ions and hence a current, $j = ne(V_0 - v_0)$.   We find that the damping of the ion plasma waves decreases with increasing j until, at a critical value, the imaginary part of ω goes positive and we have growing waves.   If the ions and electrons have the same temperature, $T_e/T_i = 1$, then the minimum velocity for which there are growing waves is

$$V_0 - v_0 = 0.925 \left(1 + \sqrt{\frac{m}{M}}\right) a \tag{3.125}$$

The least stable waves are those with $k = 0$.   If, however, $T_e/T_i > 1$, we would expect to find the critical current to be smaller, since even in the absence of drift the waves are less damped than for $T_e/T_i = 1$.   This proves to be the case, and in the limit $T_e/T_i \gg 1$ the criterion for growth is just that

$$V_0 - v_0 \gtrsim 4A$$

corresponding to a critical velocity smaller by $\sqrt{(m/M)(T_i/T_e)}$.

The physical mechanism responsible for the growth is just the phenomenon of "bunching" familiar from klystrons and similar devices. A space-varying electric field causes velocity modulations of the particles, and these, through the conservation of particles ($nv = $ const), result in density modulations.   If the electric field resulting from these density variations is in phase with the initial field, then we have amplification, or growing waves.   When $T_e = T_i = 0$, we find that any nonzero drift velocity V_D will cause growth of waves, with

$$k \leq \frac{\omega_p}{V_D}\left(1 - \sqrt[3]{\frac{m}{M}}\right)^{3/2} \tag{3.126}$$

However, with finite temperatures the growth is opposed by Landau damping and exceeds it only for velocities above a critical threshold.

3.5.2. Normal Modes

A different approach to the problem of linearized waves was adopted by van Kampen[19] and subsequently extended by Case.[20]   Instead of solving an initial-value problem, as, following Landau, we have done here, he simply looked for "normal modes," i.e., for plane-wave solutions, $\exp[i(\mathbf{k} \cdot \mathbf{x} - \omega_0 t)]$, of (3.119) with both k and ω_0 real.   Consider the one-dimensional linearized version of (3.108) (which, for isotropic f_0, can be obtained by integration over the two components of velocity normal to **k**),

$$\frac{\partial f_1}{\partial t} + v \frac{\partial f_1}{\partial x} - \frac{e}{m} E \frac{\partial f_0}{\partial v} = 0 \tag{3.127}$$

The ansatz

$$f_1 = \psi(k,v,\omega_0) \exp [i(kx - \omega_0 t)]$$

gives

$$i(kv - \omega_0)\psi = \frac{e}{m} E \frac{\partial f_0}{\partial v} \tag{3.128}$$

whose general solution, if we allow singular functions or, in "fancier" language, Laurent-Schwartz "distributions," is

$$\psi(k,v,\omega_0) = - \frac{ieE}{mk} \left[\frac{1}{v - u_0} \frac{df_0}{dv} + \Lambda(k,\omega_0) \, \delta(v - u_0) \right] \tag{3.129}$$

where $u_0 = \dfrac{\omega_0}{k}$

and Λ is any constant. The singular terms in (3.129) are no cause for concern; since ψ is a distribution function, only velocity integrals of ψ have physical significance. To make (3.129) precise, we shall take the first term to be (when integrated) a Cauchy principal value. The value of Λ is then determined from (3.110),

$$ikE = + \frac{i\omega_p{}^2 E}{k} \left(P \int dv \, \frac{df_0/dv}{v - u_0} + \Lambda \right) \tag{3.130}$$

or $\Lambda(k,\omega_0) = \dfrac{k^2}{\omega_p{}^2} - P \int dv \, \dfrac{df_0/dv}{v - u_0}$

where the symbol P denotes the usual Cauchy principal-value integral. Thus, (3.127) is solved by

$$f_1 = - \exp [i(kx - \omega_0 t)] \frac{ieE}{km} \left[\frac{df_0}{dv} P \frac{1}{v - u_0} \right.$$
$$\left. + \left(\frac{k^2}{\omega_p{}^2} - P \int dv \, \frac{df_0/dv}{v - u_0} \right) \delta(v - u_0) \right] \tag{3.131}$$

Moreover, van Kampen and Case showed that such normal modes (suitably extended in the case of unstable plasmas) form a complete set, in terms of which any f (of a suitable class) can be expanded.

This result contrasts with our previous discussion based on the initial-value problem in that:

1.   The solutions (3.131) are valid for any k and for *any* ω_0—that is, there is no dispersion relation in the sense of a unique ω for each k.

2.   The solutions (3.131) are purely oscillatory and hence exhibit no damping.

In fact, however, the results of the two treatments are in complete agreement! On the one hand, van Kampen shows that if these normal modes are superposed in order to solve an initial-value problem the solution exhibits the usual Landau damping. Conversely, if one chooses (3.131), with $t = 0$, as the initial value of f_1 and then solves the initial-value problem, one finds that for this special initial condition there is no damping because the quantity $N(k,\omega)$ in (3.116) is proportional to $\epsilon(k,\omega)$ so that the zeros of ϵ do not produce singularities in $E(\mathbf{k},\omega)$. In fact, we have

$$N(k,\omega) = \frac{4\pi n_0 e}{k^2} \int dv \, \frac{g(v)}{v - u} \tag{3.132}$$

where, by hypothesis,

$$g(v) = f_1(k,v,t = 0) = -\frac{ieE}{km}\left[\frac{df_0}{dv} P \frac{1}{v - u_0}\right.$$
$$\left. + \left(\frac{k^2}{\omega_p{}^2} - P \int dv \, \frac{df_0/dv}{v - u_0}\right) \delta(v - u_0)\right] \tag{3.133}$$

Using the identity

$$\frac{1}{x - i\eta} = P\frac{1}{x} + \pi i \delta(x)$$

we can write

$$g(v) = -\frac{ieE}{km}\left[\frac{df_0}{dv}\frac{1}{v - u_0 - i\eta} + \frac{k^2}{\omega_p{}^2}\,\epsilon(k,\omega_0)\,\delta(v - u_0)\right] \tag{3.134}$$

so that

$$N(k,\omega) = \frac{i\omega_p{}^2 E}{k^3}\left\{\int dv\left[\frac{df_0/dv}{(v - u)(v - u_0 - i\eta)}\right]\right.$$
$$\left. + \frac{\epsilon(k,\omega_0)}{(u_0 - u)\omega_p{}^2}\,k^2\right\}$$
$$= \frac{iE}{\omega - \omega_0}\,\epsilon(k,\omega) \tag{3.135}$$

Thus, $E(k,\omega) = \dfrac{N(k,\omega)}{\epsilon(k,\omega)} = \dfrac{iE}{\omega - \omega_0}$ (3.136)

has no singularities at the zeros of ϵ, and

$$E(k,t) = \frac{1}{2\pi}\int d\omega \exp{(-i\omega t)}E(k,\omega) = E\exp{(-i\omega_0 t)} \tag{3.137}$$

simply oscillates at frequency ω_0.

3.5.3. Nonlinear Waves

The relation between the Landau and van Kampen treatments of the linear problem and the nature of the linearization approximation itself are further illustrated by the nonlinear solution of the correlationless equation

discovered by Bernstein, Greene, and Kruskal.[21]   They exploited the fact
that one can find exact solutions of the one-dimensional form of (3.108),

$$\frac{\partial f}{\partial t} + v \frac{\partial f}{\partial x} - \frac{e}{m} E \frac{\partial f}{\partial v} = 0 \tag{3.138}$$

in which f depends on x and t only through the combination $x - ut$, where
u is the (constant) velocity of a progressive wave.   By transforming to a
reference frame moving with velocity u, one obtains a time-independent
form of (3.138),

$$v \frac{\partial f}{\partial x} + \frac{e}{m} \frac{d\phi}{dx} \frac{\partial f}{\partial v} = 0 \tag{3.139}$$

where ϕ is the electrostatic potential,

$$E = - \frac{d\phi}{dx}$$

and we have chosen our signs so that (3.138) and (3.139) apply to electrons
(charge $-e$).   The general solution of (3.139) is just

$$f(x,v) = h\left(v^2 - \frac{2e\phi}{m}\right) \tag{3.140}$$

where h is an arbitrary function of $v^2 - 2e\phi/m \equiv \varepsilon$.   This result, easily
verified by differentiation, simply expresses the conservation of total
energy for each electron, i.e., the fact that in the xv plane f is constant
along curves of constant energy.   We shall consider here only the simplest
case, wherein ion dynamics are neglected $(m/M \to 0)$ but the entire treatment can be extended to the case of two species, with an equation analogous to (3.139) for the ion-distribution function, etc.   Combining (3.139)
with Poisson's equation

$$\phi'' = -4\pi e n_0 [1 - \int dv \, f(x,v)]$$

we have $\displaystyle \int_{-2e\phi/m}^{\infty} \frac{d\varepsilon \, h(\varepsilon)}{\sqrt{\varepsilon + 2e\phi/m}} = 1 + \frac{\phi''}{4\pi n_0 e}$ (3.141)

If ϕ is specified, this is an integral equation for h, whose solution can be
obtained in the sense that given f (as a function of v) at one value of x, say
$x = x_0$, we can find an explicit expression for $f(x,v)$ at all x and v.   Choose
the arbitrary constant in ϕ so that $\phi(x_0) = 0$.   Let x_1 be the minimum of
$-e\phi$ (that is, the maximum of ϕ) closest to x_0.   Then we see from (3.140)
that specification of $f(x_0,v)$ fixes $h(\varepsilon)$ for all $\varepsilon > 0$.   If ϕ happens to have
its absolute maximum value at x_0, then, for all other $x \neq x_0$, $f(x,v)$ is
determined, since the arguments of h are always nonnegative.   If ϕ has a
relative maximum at x_0, then by the same argument f is determined for
all x between x_0 and x_0', where x_0' is the nearest point for which $\phi(x_0') = \phi(x_0)$.
Therefore we need consider only the case where $\phi(x_0)$ is not a maximum.

Let x be the location of the relative maximum of ϕ closest to x_0. Then, for x between x_0 and x_1, f is not determined, since there we need $h(\varepsilon)$ for $0 > \varepsilon > -2e\phi/m$. The corresponding physical statement is just that electrons with energy below $-e\phi(x_0)$ cannot get to x_0 and hence are not counted in $f(x_0,v)$. Accordingly, we write (3.141) in the form

$$\int_{-2e\phi/m}^{0} \frac{h(\varepsilon)\,d\varepsilon}{\sqrt{\varepsilon + 2e\phi/m}} = g(\phi) \equiv 1 + \frac{\phi''}{4\pi n_0 e}$$

$$- \int_{0}^{\infty} \frac{h(\varepsilon)\,d\varepsilon}{\sqrt{\varepsilon + 2e\phi/m}} \qquad (3.142)$$

We may use ϕ as independent variable, since, in the interval between x_0 and x_1, ϕ is monotonic. The solution of (3.142), which is a form of Abel's integral equation,[22] is

$$h(\varepsilon) = +\frac{1}{\pi} \int_{0}^{-m\varepsilon/2e} \frac{d\phi\, g'(\phi)}{\sqrt{-2e\phi/m - \varepsilon}} \qquad \frac{-2e\phi(x_1)}{m} < \varepsilon < 0$$

$$(3.143)$$

Proceeding, piecewise, in this fashion, it is clear that we can find $h(\varepsilon)$ for all ε down to $-2e\phi_{max}/m$, where ϕ_{max} is the absolute maximum of ϕ, and hence that we can determine $f(x,v)$ for all x. (As a little thought will show, a solution exists only if, in the interval between two adjacent local minima, ϕ is symmetric about the included local maximum. This restriction, which disappears when the ion dynamics is included, will cause no difficulty here, since we shall later apply our results only to sinusoidal ϕ.)

It is particularly interesting to study the solution for potentials whose maximum variation is small, i.e., to consider (3.143) for small ε and to compare the result with the conventional small-amplitude (i.e., linearized) theory. Since (3.143), as it stands, is indeterminate in the limit $\varepsilon \to 0$, several integrations by parts and some algebraic manipulation are required.[21] The somewhat surprising result is

$$h(\varepsilon) = h(0) + a\sqrt{-\varepsilon} + b\varepsilon + \cdots \qquad (3.144)$$

where $a = \frac{2}{\pi}\left[\frac{d}{d\phi}\left(\frac{\phi''}{2\omega_p{}^2}\right)_{\phi=0} + \int_{-\infty}^{\infty} dv\, h'(v^2) \right]$ $\qquad (3.145)$

$b = h'(0)$

It is surprising, since we see that a power-series expansion for small-amplitude waves is not possible, the correct expansion beginning with a square root. As Bernstein, Greene, and Kruskal showed,[21] it is possible to find an expansion of f in integral powers of the wave amplitude by using, in place of the distribution function obtained from (3.141) and (3.144), one which is equivalent to it in the sense that the two give the same result for any velocity moment,

$$\langle v^n \rangle \equiv \int dv\, v^n f(v)$$

If $f_0(v)$ is the distribution function when $\phi = 0$, then to first order in the wave amplitude this "equivalent" distribution function is $f_0(v) + f_1(x,v)$, with

$$f_1 = -\frac{e\phi(x)}{m}\left\{P\frac{f_0'(v)}{v} - \delta(v)\left[P\int_{-\infty}^{\infty}dv\frac{f_0'(v)}{v}\right.\right.$$
$$\left.\left. + 2\frac{d}{d\phi}\left(\frac{\phi''}{\omega_p{}^2}\right)_{\phi=0}\right]\right\} \qquad (3.146)$$

(For the analyticity in the wave amplitude we pay the price of having a singular distribution, but since the singularities are integrable this is acceptable.) In particular, for a sinusoidal potential with wave number k we have $\phi'' = -k^2\phi$ and so

$$f_1 = -\frac{e\phi}{m}\left\{\left[\frac{k^2}{\omega_p{}^2} - P\int dv_1\frac{f_0'(v_1)}{v_1}\right]\delta(v) + P\frac{f_0'(v)}{v}\right\}$$
$$= -\frac{e\phi}{m}\lim_{\eta\to0}\left\{\left[\frac{k^2}{\omega_p{}^2} - \int dv_1\frac{f_0'(v_1)}{v_1 - i\eta}\right]\delta(v) + \frac{f_0'(v)}{v - i\eta}\right\} \qquad (3.147)$$

All this has been carried out for the rest frame of the wave. In a frame in which it moves with velocity $u_0 = \omega_0 k$, Galilean transformation of (3.147) gives

$$f_1 = -\frac{e\phi}{m}\left\{f_0'(v)P\frac{1}{v - u_0} + \delta(v - u_0)\left[\frac{k^2}{\omega_p{}^2}\right.\right.$$
$$\left.\left. - P\int\frac{dv_1 f_0'(v_1)}{v_1 - u_0}\right]\right\} \qquad (3.148)$$

This result is in complete agreement with the van Kampen solution (3.131). That the solution is undamped is a consequence of our restriction to solutions which depend only on $x - u_0 t$ with real u_0 and for which there consequently exists a frame where there is no time dependence, either oscillatory or damped. As in the case of the van Kampen solution, to which the nonlinear BGK result reduces in the small-amplitude limit, this can be achieved only by the choice of very special initial conditions.

3.5.4. Effect of an External Magnetic Field

So far we have discussed only longitudinal-wave solutions in the correlationless approximation. As we have seen, the interaction between the wave and the plasma is most important for those particles whose velocity is close to the phase velocity of the wave. [For example, these are the particles involved in singularities of (3.131).] In the case of transverse waves, the phase velocity is greater than or equal to the velocity of light, c, so that correct treatment of the resonant particles requires the relativistic generalization of our kinetic equation. Although offering no difficulty, this has led to no particularly significant results.

The situation becomes quite different, however, when (as is frequently the case) an external magnetic field is present.   The effect on the particle dynamics is such that a number of new wave motions become possible, many having phase velocities less than c.   A brief discussion of the many cases will be the subject of a later chapter.[23]   We simply note here that all the properties of these waves, and, indeed, all consequences of the linearized version of (3.76) to (3.78), can be summarized by giving the generalized "dielectric-constant," or "conductivity," tensors of the plasma, i.e., its response to an external electromagnetic field.

Let $\mathbf{B}_0$ be the constant external magnetic field, $\mathbf{E}_1$ and $\mathbf{B}_1$ the total space- and time-varying electric and magnetic fields.   For particles of charge q, mass m, we then have

$$\frac{\partial f_1}{\partial t} + \mathbf{v} \cdot \nabla f_1 + \mathbf{v} \times \mathbf{\Omega} \cdot \nabla_v f_1 = -\frac{q}{m}\left(\mathbf{E}_1 + \frac{\mathbf{v} \times \mathbf{B}_1}{c}\right) \cdot \nabla_v f_0 \quad (3.149)$$

$$\mathbf{\Omega} = \frac{q\mathbf{B}_0}{mc}$$

As in the analysis of longitudinal waves, it is convenient to take the Fourier transform on $\mathbf{x}$ and Laplace transform on t.   Since

$$\mathbf{v} \times \mathbf{\Omega} \cdot \nabla_v f_1 = -\Omega \frac{\partial f_1}{\partial \phi}$$

where ϕ is the azimuth angle in velocity space, with $\mathbf{B}_0$ as polar axis, (3.149) then reduces to an ordinary differential equation in ϕ, whose solution is readily obtained with an integrating factor,

$$f_1(\mathbf{k},\mathbf{v},\omega) = \frac{\exp\left[-i(\alpha\phi - \beta \sin \phi)\right]}{\omega_c} \int_\phi^\infty d\phi' \exp\left[i(\alpha\phi' - \beta \sin \phi')\right]$$
$$\left(\mathbf{E}_1 + \frac{\mathbf{v}' \times \mathbf{B}_1}{c}\right) \cdot \nabla_{v'} f_0(\mathbf{v}') \quad (3.150)$$

Here ϕ and ϕ' are, respectively, the azimuthal angles for $\mathbf{v}$ and $\mathbf{v}'$, and the plane containing $\mathbf{k}$ and $\mathbf{B}_0$ is taken to be the plane $\phi = 0$.   Also,

$$\alpha = \frac{\omega - \mathbf{k}_\parallel/v_\parallel}{\Omega}$$

$$\beta = \frac{k_\perp v_\perp}{\Omega}$$

where the subscripts $\parallel$ and $\perp$ denote components parallel or perpendicular to $\mathbf{B}_0$ and $\mathbf{E}_1$, $\mathbf{B}_1$ mean $\mathbf{E}_1(\mathbf{k},\omega)$ and $\mathbf{B}_1(\mathbf{k},\omega)$.   From (3.150) we compute the total current

$$\mathbf{j} = ne\int d\mathbf{v}\, \mathbf{v}f_1$$

which we see will be proportional to $\mathbf{E}_1$. In the term involving $\mathbf{B}_1$ we use Faraday's law,

$$-\dot{\mathbf{B}} = \nabla \times \mathbf{E}$$
or $\quad \omega \mathbf{B}_1 = \mathbf{k} \times \mathbf{E}_1$

Thus we can define a generalized "conductivity" tensor $\boldsymbol{\sigma}$,

$$\mathbf{j}(\mathbf{k},\omega) = \boldsymbol{\sigma}(\mathbf{k},\omega) \cdot \mathbf{E}_1(\mathbf{k},\omega)$$

and substitute this in Maxwell's equations,

$$\nabla \times (\nabla \times \mathbf{E}) = -\frac{\nabla \times \dot{\mathbf{B}}}{c} = -\frac{\partial}{\partial t}\left(\frac{4\pi\mathbf{j}}{c^2} - \frac{i\omega\mathbf{E}}{c^2}\right)$$

or $\quad \mathbf{k} \times (\mathbf{k} \times \mathbf{E}_1) + \left(\frac{\omega^2}{c^2}\right)\boldsymbol{\epsilon} \cdot \mathbf{E}_1 = 0$ $\qquad(3.151)$

with $\quad \boldsymbol{\epsilon} = \mathbf{I} + \dfrac{4\pi i\boldsymbol{\sigma}}{\omega}$ $\qquad(3.152)$

The integration of (3.150) over ϕ' and over $\mathbf{v}$ to find $\mathbf{j}$ and hence $\boldsymbol{\sigma}$ and $\boldsymbol{\epsilon}$ can be carried out explicitly if f_0 is Maxwellian,

$$f_0(\mathbf{v}) = \exp\left(-v^2/a^2\right)(\sqrt{\pi}\,a)^{-3} \qquad(3.153)$$

with the result

$$\boldsymbol{\sigma} = \frac{-4ine^2}{m\omega}\,\zeta_0 \int_0^\infty d\eta \exp\left(-\eta^2\right)\mathsf{T} \qquad(3.154)$$

where $\quad T_{xx} = \displaystyle\sum_{-\infty}^{\infty} \frac{n^2}{\lambda^2}\,\eta[J_n(\lambda\eta)]^2 Z_0(\zeta_n)$

$$T_{xy} = -T_{yx} = i\sum_{-\infty}^{\infty}\frac{n}{\lambda}\,\eta^2 J_n(\lambda\eta)J_n'(\lambda\eta)Z_0(\zeta_n)$$

$$T_{yy} = \sum_{-\infty}^{\infty}\eta^3[J_n'(\lambda\eta)]^2 Z_0(\zeta_n) \qquad(3.155)$$

$$T_{xz} = T_{zx} = \sum_{-\infty}^{\infty}\frac{n\eta}{\lambda}\,[J_n(\lambda\eta)]^2 Z_1(\zeta_n)$$

$$T_{zy} = -T_{yz} = i\sum_{-\infty}^{\infty}\eta^2 J_n(\lambda\eta)J_n'(\lambda\eta)Z_1(\zeta_n)$$

$$T_{zz} = \sum_{-\infty}^{\infty}\eta[J_n(\lambda\eta)]^2 Z_2(\zeta_n)$$

and $\quad \lambda = \dfrac{k_\perp a}{\omega_c \zeta_n} = \dfrac{\omega - n\Omega}{k_\parallel a}$ $\qquad(3.156)$

The labeling of the elements of T assumes that the magnetic field is along the z axis and $\mathbf{k}$ lies in the x-z plane. In terms of the function Z[24] given by

(3.120), we define

$$Z_0(\zeta) = Z(\zeta) \qquad Z_1(\zeta) = 1 + \zeta Z(\zeta) \qquad Z_2(\zeta) = \zeta Z_1(\zeta) \qquad (3.157)$$

Although we have written our equations for a single species, the results for two or more species are obtained by simply writing a sum of terms like (3.154), one for each species. Naturally, these equations are useful principally in those cases where it is a good approximation to keep only a few terms of the sum over n.

REFERENCES

1. Since 1961, when the lectures which led to this book were given, there has been a considerable development of the kinetic theory of plasmas. Rather than attempting to incorporate this newer material, we simply note that the reader wishing a more advanced or detailed treatment than that given here should consult:

    Montgomery, D., and D. Tidman: "Plasma Kinetic Theory," McGraw-Hill, New York, 1964.

    Balescu, R.: "Statistical Mechanics of Charged Particles," Interscience, New York, 1963.

    Thompson, W. B.: "An Introduction to Plasma Physics," Pergamon and Addison-Wesley, New York and Reading, Mass., 1962.

    Rosenbluth, M. N. (ed.): Advanced Plasma Theory, Course 25, *Proc. Intern. School Phys. Enrico Fermi*, 1964.

    Stix, T. H.: "The Theory of Plasma Waves," McGraw-Hill, New York, 1962.

2. Tolman, R. C.: "Principles of Statistical Mechanics," Oxford University Press, Fair Lawn, N. J., 1938.

3. Landau, L. D.: *Physik. Z. Sowjetunion*, **10**, 156 (1936).

4. Enoch, J.: *Phys. Fluids*, **3**, 353 (1960).

5. Chandrasekhar, S.: *Revs. Modern Phys.*, **15**, 3 (1943).

6. Rosenbluth, M. N., and N. Rostoker: *Phys. Fluids*, **3**, 1 (1960).

7. Dupree, T. H.: *Phys. Fluids*, **4**, 696 (1961).

8. Dupree, T. H.: *Phys. Fluids*, **6**, 1714 (1963). Also references cited there.

9. Lenard, A.: *Ann. Phys.*, **10**, 390 (1960).

10. Guernsey, R.: *Phys. Fluids*, **5**, 322 (1962).

11. Field, E. C., and B. D. Fried: *Phys. Fluids*, **7**, 1937 (1964).

12. Pearson, G. A.: "The Effect of Wave-Particle Interactions on the Stability of a Current Carrying Plasma," Dissertation, University of California, March, 1965.

13. Whittaker, E. T., and G. N. Watson: "Modern Analysis," sec. 9.41, Cambridge, New York, 1932.

14. Drummond, W. E., and D. Pines: *Nuclear Fusion Suppl.*, 1962, p. 1049.

15. Fried, B. D., and R. W. Gould: *Phys. Fluids*, **4**, 139 (1961).

16. Dawson, J.: *Phys. Fluids*, **4**, 869 (1961).

17. Landau, L. D.: *J. Phys. USSR*, **10**, 25 (1946).

18. Wong, A. Y., N. D'Angelo, and R. W. Motley: *Phys. Rev. Letters,* **9,** 415 (1962).
19. Van Kampen, N.: *Physica,* **21,** 949 (1955).
20. Case, K. M.: *Ann Phys. (N.Y.),* **7,** 349 (1959).
21. Bernstein, I., J. Greene, and M. Kruskal: *Phys. Rev.,* **108,** 546 (1957).
22. Courant, R., and D. Hilbert: "Methods of Mathematical Physics," vol. 1, p. 158, Interscience, New York, 1953.
23. See also T. H. Stix, Ref. 1.
24. Fried, B. D., and S. Conte: "The Plasma Dispersion Function," Academic, New York, 1961.

4

DISSIPATIVE EFFECTS

KAUFMAN

ALLAN N. KAUFMAN, *Associate Professor of Physics, University of California, Berkeley, California*

4

4.1 INTRODUCTION

To illustrate a dissipative process, let us consider a cylindrical plasma confined by an axial magnetic field (Fig. 4.1). In the course of time, the plasma diffuses radially across the field, thereby decreasing the degree of radial confinement. There are three ways to describe this process:

1. There is a *dissipation*, or loss, of order; by *order* we mean here the degree of confinement.
2. There is an *approach to equilibrium*. In thermal equilibrium the density is uniform throughout the available space. In the course of time, the nonuniformity of density decreases.
3. There is a *transport* of mass, to regions of lower density.

4.1.1. The Tendency toward Equilibrium

In the state of thermal equilibrium, a system has uniform properties—its density, temperature, and flow velocity are spatially uniform and constant in time. The several components, electrons and ions in the case of a fully ionized plasma,† are at the same temperature, and their flow velocities are

---

† The concept of a fully ionized plasma in thermal equilibrium is an idealization which is not realized in practice. At any temperature, there will be neutral atoms

equal. The distribution of particle velocities relative to the mean (the flow velocity) is Maxwellian, for each component, if the system is classical. The electromagnetic radiation has an energy spectrum characteristic of the temperature of the material system.

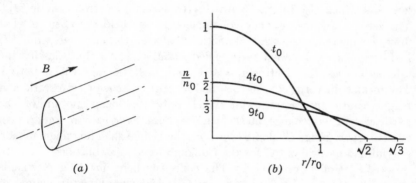

(a)   (b)

Fig. 4.1   (a) A cylindrical plasma confined (radially) by an axial magnetic field. (b) Density as a function of radial distance from the axis, for three successive times t_0, $4t_0$, and $9t_0$. These curves represent the similarity solution

$$\frac{n(r,t)}{n_0} = \left(\frac{t_0}{t}\right)^{1/2} \left[1 - \left(\frac{r}{r_0}\right)^2 \left(\frac{t_0}{t}\right)^{1/2} \right]$$

of the nonlinear diffusion equation

$$\frac{\partial n}{\partial t} = \nabla \cdot [D(n)\nabla n]$$

when the diffusivity $D(n)$ is proportional to the density n [see Eq. (4.14)]. (The constants n_0, r_0, and t_0 are related by $D(n_0) = r_0{}^2/8t_0$.) In contrast to linear diffusion, where the mean radius $\bar{r}$ varies as $t^{1/2}$, it varies here as $t^{1/4}$.

Many systems of interest, however, depart to some degree from a state of equilibrium. We may consider this departure from equilibrium as a measure of order. In the course of time, various mechanisms cause the system to approach equilibrium, and thus the order is dissipated. Associated with these mechanisms, there is often a spatial transport of some entity, such as mass, momentum, or energy, or a transfer of some entity among the components of the system. Thus one speaks of *transport* or *transfer* processes.

---

present, and ions in excited states. Further, the system must be contained in some way to be in equilibrium. However, we shall ignore interactions of the plasma with the walls of its container and shall suppose the plasma to be composed only of electrons and one species of singly charged ion without internal structure.

4.1.2. Dissipative Mechanisms

The mechanisms producing dissipation result from the interactions of the elements of the plasma. We may suppose the plasma to consist of charged particles, which interact *electrostatically* according to Coulomb's law, and which also interact with the (transverse) electromagnetic field. The zero- and low-frequency components of the field give rise to what may be called *magnetostatic* interactions between the particles, while the high-frequency components are called *radiation*. We thus classify the interactions as (1) electrostatic and magnetostatic and (2) radiative. The former, in turn, can be classified as (1a) *collective* electrostatic and magnetostatic interactions, for which only the mean motion of the particles is of importance, and which are characteristically of long range; and (1b) short-range Coulomb *collisions*, the Debye-shielding phenomenon providing a natural cutoff for the Coulomb force in a collision.

1a. Collective modes. The collective behavior of a plasma is studied by the use of the macroscopic equations (see the later sections of this chapter) or by the use of a kinetic equation (see Chap. 3). One investigates the response of the plasma to infinitesimal disturbances and finds stable oscillations (see Chap. 8) and possibly instabilities (see Chap. 5). As the amplitude of motion of an instability grows to finite size, the motion may become stably convective or turbulent. In either case, a major dissipative mechanism may result. The analysis of this dissipation is very difficult and is an active field of current research. We shall comment on it in Sec. 4.6.

1b. Collisions. Until recently, the study of plasma dissipative processes has been based almost solely on binary collisions as the dissipative mechanism. The resulting theory is called *classical*, since it is an outgrowth of the kinetic theory developed in the nineteenth century and perfected in the early decades of this century. Kinetic theory describes the behavior of a dilute gas of particles, interacting in collisions with short-range forces. The application of this theory to a plasma becomes possible through the introduction of the Debye cutoff on the long-range Coulomb interaction. This theory will form the basis of most of the discussion in this chapter. We must comment, however, on what is omitted in this approach. In the first place, there are microscopic interactions over distances larger than the Debye radius. These become important in stable plasmas departing appreciably from thermal equilibrium and may be studied by the more sophisticated kinetic equations of Chap. 3. In the second place, as pointed out in the preceding section, unstable plasmas produce dissipation through collective motion.

2. Radiative processes. The emission and absorption of elec-

tromagnetic radiation, by the particles of the plasma, are processes of major importance in astrophysical situations and in the energy balance and diagnostics of laboratory plasmas. We refer to Chap. 9 for details and ignore radiative processes in this chapter.

4.1.3. Departures from Equilibrium

NONUNIFORMITIES We may suppose that in any small region the plasma is essentially in thermal equilibrium, with definite density† n, temperature T, and flow velocity $\mathbf{u}$, but that these quantities are slowly varying functions of position and time.

Nonuniform Density A magnetic field can maintain an isothermal plasma in a state of *mechanical* equilibrium, with nonuniform density. The transverse pressure gradient is balanced by the Lorentz force:

$$\nabla_{\perp} p = \frac{1}{c}\mathbf{j} \times \mathbf{B} \tag{4.1}$$

While maintaining mechanical equilibrium, the approach to thermal equilibrium occurs via mass transport, with particle flux $n\mathbf{u}$ proportional to the density gradient (Fick's law):

$$n\mathbf{u}_D = -D_{\perp}\nabla_{\perp} n \tag{4.2}$$

The coefficient $D_{\perp}$ is called the *transverse diffusivity* and depends upon the local properties of the plasma. The mechanisms leading to this mass transport are Coulomb collisions and instabilities, leading, respectively, to laminar and turbulent diffusion.

Nonuniform Temperature Nonuniformity of temperature may arise from nonuniform heating of the plasma. There results a transport of energy, which may be laminar conduction (from Coulomb collisions) or turbulent conduction or laminar convection (from instabilities). The former case may be described by Fourier's law of heat flow:

$$\mathbf{Q} = -K\nabla(kT) \tag{4.3}$$

where $\mathbf{Q}$ = heat-flux density
  K = thermal conductivity
  k = Boltzmann's constant

† In a quiescent plasma the electron and ion densities are very nearly equal; the symbol n denotes the density of *either*, while p denotes the *total* pressure. Thus the perfect-gas law reads $p = 2nkT$.

Nonuniform Flow Velocity   The phenomenon of viscosity refers to the production of stress by a velocity gradient.   Momentum is transported in the direction of the gradient.   The mechanism may be collisions, leading to laminar viscosity, or Helmholtz instability, leading to turbulent viscosity.

DEPARTURES FROM LOCAL EQUILIBRIUM   In addition to spatial variation of the parameters characterizing local equilibrium, there may be further deviations from the equilibrium state.

Different Temperatures   Because of the small electron-ion mass ratio, it is often quite likely that a temperature difference may arise between the electron gas and the ion gas.   Collisions between electrons and ions transfer energy between the two components, tending toward equalization of their temperatures:

$$\frac{dT^e}{dt} = \frac{1}{\tau}(T^i - T^e) \tag{4.4}$$

The coefficient τ is called the *energy-exchange time*.

The temperature of electromagnetic radiation is commonly much lower than that of the particles in a laboratory plasma.   Radiative emission and absorption tend to equalize the temperatures (see Chap. 9).

Different Flow Velocities   If the electron and ion flow velocities differ locally, an electric current density

$$\mathbf{j} \equiv ne(\mathbf{u}^i - \mathbf{u}^e) \tag{4.5}$$

exists.   This relative velocity is opposed by friction between the electron and ion gases.   The source of friction may be electron-ion Coulomb collisions, or a collective instability (see Chap. 5).   The friction transfers momentum between the two gases, total momentum being conserved. The directed kinetic energy of relative flow is dissipated into thermal energy; this is called *Joule*, or *ohmic*, heating.

In the *absence* of a magnetostatic field, the relation between current and electric field is given by Ohm's law,

$$\mathbf{j} = \sigma \mathbf{E} \tag{4.6a}$$
$$\mathbf{E} = \eta \mathbf{j} \tag{4.6b}$$

where σ = (electrical) conductivity
    $\eta \equiv \sigma^{-1}$ = resistivity

Non-Maxwellian Velocity Distribution   Coulomb collisions will tend to restore the velocity distribution to the Maxwellian form, for either component.   In addition, instabilities may arise if the distribution is anisotropic (see Sec. 5.4.2).

4.1.4. Estimates of Dissipative Coefficients

In the absence of collisions, the motion of an electron transverse to a magnetostatic field **B** is circular, with angular gyrofrequency

$$\omega^e = \frac{eB}{m^e c} \tag{4.7a}$$

and gyroradius

$$a^e = \frac{v_\perp^{\ e}}{\omega^e} = \frac{m^e v_\perp^{\ e} c}{eB} \tag{4.7b}$$

This motion is considered to be interrupted by collisions, with a macroscopic cross section Σ given by the Rutherford formula with a Debye cutoff (see Chap. 1),

$$\Sigma \sim \frac{ne^4}{m^2 v^4} \ln \Lambda \sim \frac{ne^4}{(kT)^2} \ln \Lambda \tag{4.8}$$

The electron-electron or electron-ion collision frequency ν is

$$\nu \sim \Sigma \, \overline{v^e} \sim \frac{ne^4}{(kT^e)^{3/2}(m^e)^{1/2}} \ln \Lambda \tag{4.9}$$

The ion-ion collision frequency ν' is analogously

$$\nu' \sim \Sigma \, \overline{v_i} \sim \frac{ne^4}{(kT^i)^{3/2}(m^i)^{1/2}} \ln \Lambda \tag{4.10}$$

and is thus smaller.

The time for relaxing to a Maxwellian is thus $\sim \nu^{-1}$ for the electron gas and $\sim (\nu')^{-1}$ for the ion gas. Because the large mass ratio m^i/m^e makes energy transfer between electrons and ions slow, the energy-exchange time τ is still longer: $\tau \sim (m^i/m^e)\nu^{-1}$.

The conductivity σ (for **B** = 0) can be estimated by equating the rate of momentum gain by an electron from the field **E** to its rate of loss from electron-ion collisions,

$$-e\mathbf{E} \sim m^e(\mathbf{u}^e - \mathbf{u}^i)\nu$$

$$\therefore \quad \mathbf{j} = ne\,(\mathbf{u}^i - \mathbf{u}^e) \sim \frac{ne^2}{m^e \nu} \mathbf{E}$$

$$\therefore \quad \sigma \sim \frac{ne^2}{m^e \nu} \sim \frac{(kT^e)^{3/2}}{(m^e)^{1/2} e^2 \ln \Lambda} \tag{4.11}$$

Note that σ increases with T^e and is independent of density (except for the very weak implicit dependence through $\ln \Lambda$).

From dimensional reasoning ($[Q] = [n\,kT\,u]$), we estimate the thermal conductivity in the absence of **B** as

$$K(B = 0) \sim \frac{kT^e}{e^2}\sigma \tag{4.12}$$

and the viscosity μ (which will be precisely defined later) for **B** = 0 as

$$\mu(B = 0) \sim m^e K \tag{4.13}_e$$

Upon substitution of (4.12) and (4.11) into (4.13)$_e$, we see that the viscosity is proportional to $(m^e)^{1/2}$; hence the ionic contribution should predominate, and so we replace m^e by m^i and T^e by T^i,

$$\mu(B = 0) \sim \frac{(kT^i)^{5/2}(m^i)^{1/2}}{e^4 \ln \Lambda} \tag{4.13}_i$$

We note that K and μ are temperature-sensitive but density-independent.

We now return to the case **B** $\neq 0$. If $\nu \ll \omega^e$, the circular motion of an electron is interrupted occasionally by collisions, which produce a random transverse jump of the guiding center, by a distance $\sim a^e$. Thus there is a random walk of guiding centers, with a frequency ν and step a^e. From random-walk theory, we have the diffusivity

$$D_\perp \sim \nu(a^e)^2 \sim \frac{n}{B^2(kT^e)^{1/2}} e^2(m^e)^{1/2}c^2 \ln \Lambda \tag{4.14}$$

Note the dependence on density, temperature, and magnetic field. (We shall show later that only electron-ion collisions contribute to $D_\perp$ and that the ion and electron fluxes are equal.)

We estimate the transverse thermal conductivity by dimensional reasoning and find

$$K_\perp \sim nD_\perp$$

Here ion-ion collisions can contribute; so, noting the mass dependence of $D_\perp$, we expect a predominant ionic contribution,

$$K_\perp{}^i \sim \frac{n^2}{B^2(kT^i)^{1/2}} e^2(m^i)^{1/2}c^2 \ln \Lambda \tag{4.15}$$

Again we estimate the viscosity

$$\mu_\perp \sim m^i K_\perp{}^i \tag{4.16}$$

which is still more predominantly ionic.

4.2   LINEAR - TRANSPORT RELATIONS

If the departure from equilibrium (called a *force*) is sufficiently small, the response to it (called a *flux*) is *linear*. Curie's law states that the flux has the same vector character as the force driving it. Thus a temperature difference (a scalar) drives a temperature time derivative (a scalar) [see Eq. (4.4)]. A velocity gradient (a tensor) drives a viscous stress (a tensor). Any vector force—electric field, density gradient, temperature gradient— may drive any vector flux—current density, mass flux, heat flow. Further, the latter need not be in the same direction as the former if the unperturbed medium is anisotropic, as in the case of a plasma in a magnetostatic field.

The transport coefficients, which relate the fluxes to the forces, are not all independent. For example, the thermoelectric coefficient relates both the heat flow to the electric field and the electric current to the temperature gradient.

We shall study in detail the connection between the electrical resistivity and the transverse diffusivity, (1) first from the macroscopic point of view, (2) then from microscopic considerations, in order to gain an understanding of plasma behavior.

4.2.1. Macroscopic Treatment

Ohm's law (4.6a) can be applied to a plasma in a magnetic field if $\mathbf{E}$ is replaced by $\mathbf{E}'$, the electric field referred to a frame translating with the plasma velocity $\mathbf{u}$,

$$\mathbf{E}' = \mathbf{E} + \frac{1}{c}\mathbf{u} \times \mathbf{B} \tag{4.17}$$

$$\mathbf{j} = \sigma\mathbf{E}' \tag{4.18}$$

(This statement will be justified and refined in Sec. 4.5.) We solve for the transverse velocity $\mathbf{u}_\perp$,

$$\mathbf{u}_\perp = \frac{\mathbf{E} \times \mathbf{B}}{B^2} c - \frac{\eta c}{B^2}\mathbf{j} \times \mathbf{B} \tag{4.19}$$

and use the statement of mechanical equilibrium (4.1) to obtain

$$\mathbf{u}_\perp = \frac{\mathbf{E} \times \mathbf{B}}{B^2} c - \frac{\eta c^2}{B^2}\nabla_\perp p \tag{4.20}$$

We recognize the first term on the right of (4.20) as the $\mathbf{E} \times \mathbf{B}$ drift $\mathbf{u}_E$,

$$\mathbf{u}_E \equiv \frac{\mathbf{E} \times \mathbf{B}}{B^2} c \tag{4.21}$$

It may be interpreted as the velocity of the magnetic field lines. The other term is then the relative velocity of the plasma and the field lines, i.e., the *diffusion* velocity $\mathbf{u}_D$,

$$\mathbf{u}_D \equiv \mathbf{u}_\perp - \mathbf{u}_E = -\frac{\eta c^2}{B^2}\nabla_\perp p \tag{4.22}$$

We use the perfect-gas law to eliminate p,

$$p = n(kT^e + kT^i) \tag{4.23}$$

and ignore a possible temperature gradient:

$$\mathbf{u}_D = -\frac{\eta c^2}{B^2}(kT^e + kT^i)\nabla_\perp n \tag{4.24a}$$

$$n\mathbf{u}_D = -\frac{\eta c^2 p}{B^2}\nabla_\perp n \tag{4.24b}$$

Comparing this with (4.2), we see that the diffusivity $D_\perp$ is

$$D_\perp = \frac{\eta c^2 p}{B^2} \tag{4.25}$$

This agrees with our qualitative considerations of the preceding section; from (4.11) we have

$$\nu \sim \frac{ne^2}{m^e}\eta \tag{4.26}$$

and then from (4.14) and (4.7b)

$$D_\perp \sim \frac{ne^2}{m^e}\eta\left(\frac{m^e v^e c}{eB}\right)^2 \sim \frac{\eta c^2 p}{B^2}$$

Thus the resistivity and diffusivity are intimately related; since only ion-electron collisions contribute to η, the same is true for $D_\perp$.

4.2.2. Microscopic Treatment

Now we study the same problem[†] from the microscopic point of view, using the guiding-center theory of Chap. 2. An electron charge $q^e = -e$, acted upon by a transverse force $\mathbf{F}^e$ (in *addition* to the Lorentz force), undergoes a guiding-center drift $\mathbf{v}_d$,

$$\mathbf{v}_d{}^e = \frac{\mathbf{F}^e \times \mathbf{B}}{q^e B^2}c \tag{4.27}$$

The force $\mathbf{F}^e$ consists of an electric part

$$\mathbf{F}_E{}^e = q^e\mathbf{E} \tag{4.28}$$

[†] In contrast to the preceding discussion, where we *postulated* Ohm's law and *derived* Fick's law, we here derive *both* laws from a microscopic analysis.

and a frictional part

$$\mathbf{F}_f{}^e = -m^e\nu(\mathbf{v}^e - \mathbf{u}^i) \tag{4.29}$$

proportional to its velocity relative to the ion gas and due to electron-ion collisions. (Here ν is again the collision frequency. We assume ν to be independent of $\mathbf{v}$, for simplicity; hence our analysis is only semi-quantitative.) Averaging over all the electrons in a macroscopically small region, and using $\langle \mathbf{v}^e \rangle \equiv \mathbf{u}^e$, we obtain

$$\langle \mathbf{F}_f{}^e \rangle = -m^e\nu(\mathbf{u}^e - \mathbf{u}^i) \tag{4.30}$$

as the mean frictional force on an electron. (Replacing $\mathbf{u}^i$ by $\mathbf{u}^e$ shows that electron-electron collisions do not contribute to the mean friction.) From (4.5), we can write this as

$$\langle \mathbf{F}_f{}^e \rangle = \frac{m^e\nu}{ne}\,\mathbf{j} \tag{4.31}$$

The law of conservation of momentum in Coulomb collisions (or Newton's third law) shows that the mean friction on an ion must be equal and opposite,

$$\langle \mathbf{F}_f{}^i \rangle = -\langle \mathbf{F}_f{}^e \rangle \tag{4.32}$$

Hence the mean frictional drifts, which we identify as the diffusion velocity $\mathbf{u}_D$, are the *same* for both species,

$$\mathbf{u}_D \equiv \langle \mathbf{v}_{df} \rangle = \frac{\langle \mathbf{F}_f \rangle \times \mathbf{B}}{qB^2}\,c \tag{4.33a}$$

$$= -\frac{m^e\nu c}{ne^2 B^2}\,\mathbf{j} \times \mathbf{B} \tag{4.33b}$$

Let us here quantitatively *define* the resistivity and conductivity in terms of ν,

$$\eta \equiv \sigma^{-1} \equiv \frac{m^e\nu}{ne^2} \tag{4.34}$$

and use (4.1) to obtain

$$\mathbf{u}_D = -\frac{\eta c^2}{B^2}\,\nabla_\perp p \tag{4.35}$$

in agreement with (4.22).

The electric drift, from (4.28), is of course just $\mathbf{u}_E$ [(4.21)]. The total drift is thus

$$\mathbf{u}_E + \mathbf{u}_D = \frac{(\mathbf{E} - \eta\mathbf{j}) \times \mathbf{B}}{B^2}\,c \tag{4.36}$$

Solving for $\mathbf{j}_\perp$, we have Ohm's law,

$$\mathbf{j}_\perp = \sigma \left[\mathbf{E}_\perp + \frac{1}{c}(\mathbf{u}_E + \mathbf{u}_D) \times \mathbf{B} \right] \tag{4.37}$$

Thus both Fick's law and Ohm's law arise from the same guiding-center motion.

      Let us summarize the diffusion mechanism. A transverse pressure gradient implies the existence of a current density, or relative ion-electron mean velocity. Collisions between the particles of the two gases produce friction, which in turn causes guiding-center drift relative to the magnetic field.

4.3 EQUATIONS OF CHANGE

We wish to formulate a quantitative set of macroscopic equations which may be used to describe the macroscopic behavior of a plasma under suitable conditions. In this section we introduce the so-called equations of change, which express conservation of mass, charge, momentum, and energy. Let us first define the macroscopic quantities appearing in these equations. They are defined in terms of $f^a(\mathbf{r},\mathbf{v},t)$, the Boltzmann distribution function for each species a (ions or electrons).

Particle density:	$n^a(\mathbf{r},t) \equiv \int d^3v\, f^a(\mathbf{r},\mathbf{v},t)$
Mass density:	$\rho^a(\mathbf{r},t) \equiv m^a n^a(\mathbf{r},t)$
Charge density:	$\epsilon^a(\mathbf{r},t) \equiv q^a n^a(\mathbf{r},t)$
Momentum density:	$\mathbf{g}^a(\mathbf{r},t) \equiv \int d^3v\, m^a \mathbf{v} f^a(\mathbf{r},\mathbf{v},t)$
Flow velocity:	$\mathbf{u}^a(\mathbf{r},t) \equiv \dfrac{\mathbf{g}^a(\mathbf{r},t)}{\rho^a(\mathbf{r},t)}$
Current density:	$\mathbf{j}^a(\mathbf{r},t) \equiv \epsilon^a(\mathbf{r},t)\mathbf{u}^a(\mathbf{r},t)$
Shear-flow tensor:	$\mathsf{U}^a(\mathbf{r},t) \equiv \tfrac{1}{2}(\nabla \mathbf{u}^a + \mathbf{u}^a \nabla) - \tfrac{1}{3}\mathbf{I}\nabla \cdot \mathbf{u}^a$
Pressure tensor:	$\mathbf{p}^a(\mathbf{r},t) \equiv \int d^3v\, m^a(\mathbf{v} - \mathbf{u}^a)(\mathbf{v} - \mathbf{u}^a)f^a$
Scalar pressure:	$p^a(\mathbf{r},t) \equiv \int d^3v\, \tfrac{1}{3}m^a(\mathbf{v} - \mathbf{u}^a)^2 f^a$
Shear-stress tensor:	$\boldsymbol{\pi}^a(\mathbf{r},t) \equiv \mathbf{p}^a - p^a\mathbf{I}$
Thermal energy density:	$\mathcal{E}^a(\mathbf{r},t) \equiv \int d^3v\, \tfrac{1}{2}m^a(\mathbf{v} - \mathbf{u}^a)^2 f^a = \tfrac{3}{2}p^a$
Heat flow:	$\mathbf{Q}^a(\mathbf{r},t) \equiv \int d^3v \tfrac{1}{2}m^a(\mathbf{v} - \mathbf{u}^a)^2(\mathbf{v} - \mathbf{u}^a)f^a$
Momentum-transfer rate density to species a from species b:	$\mathbf{P}^{ab}(\mathbf{r},t) \equiv \int d^3v\, \langle \Delta m^a\, \mathbf{v}\rangle^{ab} f^a$
Energy-transfer rate density:	$R^{ab}(\mathbf{r},t) \equiv \int d^3v\, \langle \Delta \tfrac{1}{2}m^a v^2\rangle^{ab} f^a$

The quantity $\langle \Delta m^a \mathbf{v}\rangle^{ab}$ is the mean rate of momentum change for a particle of species a, with velocity $\mathbf{v}$, due to collisions with all the particles of species b.   It is therefore a function of $\mathbf{v}$ and f^b.   The quantity $\langle \Delta \frac{1}{2} m^a v^2 \rangle^{ab}$ is analogously defined.

4.3.1. Conservation of Mass and Charge

Consider a volume V fixed in space.   Conservation of particles states that

$$\frac{d}{dt} \int_V d^3r \, n^a(\mathbf{r},t) = - \oint_V d\mathbf{\sigma} \cdot n^a \mathbf{u}^a \tag{4.38}$$

That is, the number of particles in V is reduced by particle flux through the surface.   With the divergence theorem, we may write this as

$$\int_V d^3r \left[\frac{\partial n^a}{\partial t} + \nabla \cdot (n^a \mathbf{u}^a) = 0 \right] \tag{4.39}$$

or since V is arbitrary,

$$\frac{\partial n^a}{\partial t} = - \nabla \cdot (n^a \mathbf{u}^a) \tag{4.40}$$

This is called the *continuity equation*.   An alternative form utilizes the convective derivative,

$$\frac{D^a}{Dt} \equiv \frac{\partial}{\partial t} + \mathbf{u}^a \cdot \nabla \tag{4.41}$$

Thus $\quad \dfrac{D^a n^a}{Dt} = -n^a \nabla \cdot \mathbf{u}^a \tag{4.42}$

Still other forms follow from the second to the sixth of the preceding definitions,

$$\frac{\partial \rho^a}{\partial t} = - \nabla \cdot \mathbf{g}^a \tag{4.43}$$

$$\frac{\partial \epsilon^a}{\partial t} = - \nabla \cdot \mathbf{j}^a \tag{4.44}$$

the continuity equations for mass and charge.   We add the equations for the two species, to obtain

$$\frac{\partial \rho}{\partial t} = - \nabla \cdot \mathbf{g} \tag{4.45}$$

$$\frac{\partial \epsilon}{\partial t} = - \nabla \cdot \mathbf{j} \tag{4.46}$$

[The absence of a superscript indicates a *total* quantity (*sum* over species, for example, $\rho \equiv \rho^e + \rho^i$), with the exception of $n \equiv n^i \approx n^e$.]

4.3.2. Conservation of Momentum

From conservation of momentum, we have

$$\frac{\partial \mathbf{g}^a}{\partial t} = -\nabla \cdot (\mathbf{p}^a + \rho^a \mathbf{u}^a \mathbf{u}^a) + \epsilon^a \left(\mathbf{E} + \frac{1}{c} \mathbf{u}^a \times \mathbf{B} \right) + \mathbf{P}^{ab} \qquad (4.47)$$

The terms on the right represent momentum change due to microscopic momentum flow ($\mathbf{p}^a$), macroscopic momentum flow ($\rho^a \mathbf{u}^a \mathbf{u}^a$), interaction with the macroscopic (self-consistent) fields $\mathbf{E}$ and $\mathbf{B}$, and interspecies collisions. Summing over species, and using $\mathbf{P}^{ei} = -\mathbf{P}^{ie}$, we have

$$\frac{\partial \mathbf{g}}{\partial t} = -\nabla \cdot \left(\mathbf{p} + \sum_a \rho^a \mathbf{u}^a \mathbf{u}^a \right) + \epsilon \mathbf{E} + \frac{1}{c} \mathbf{j} \times \mathbf{B} \qquad (4.48)$$

Use of (4.43) and of the convective derivative (4.41) converts (4.47) to

$$\rho^a \frac{D^a \mathbf{u}^a}{Dt} = -\nabla \cdot \mathbf{p}^a + \epsilon^a \left(\mathbf{E} + \frac{1}{c} \mathbf{u}^a \times \mathbf{B} \right) + \mathbf{P}^{ab} \qquad (4.49)$$

This is called the *equation of motion for species a*. Let us formally solve (4.49) for $\mathbf{u}_\perp{}^a$, appearing in the third term on the right,

$$\mathbf{u}_\perp{}^a = \frac{\mathbf{E} \times \mathbf{B}}{B^2} c + \frac{1}{\omega^a} \hat{\mathbf{n}} \times \frac{D^a \mathbf{u}^a}{Dt} + \frac{c}{\epsilon^a B} \hat{\mathbf{n}} \times (\nabla \cdot \mathbf{p}^a)$$

$$- \frac{c}{\epsilon^a B} \hat{\mathbf{n}} \times \mathbf{P}^{ab} \quad (4.50)$$

where $\quad \hat{\mathbf{n}} \equiv \dfrac{\mathbf{B}}{B}$ \hfill (4.51)

and† $\quad \omega^a \equiv \dfrac{q^a B}{m^a c}$ \hfill (4.52)

Each term on the right has a physical interpretation. The first is $\mathbf{u}_E$, as we know; the second,

$$\mathbf{u}_A{}^a \equiv \frac{1}{\omega^a} \hat{\mathbf{n}} \times \frac{D^a \mathbf{u}^a}{Dt} \qquad (4.53)$$

is the accelerational drift (see Chap. 2); the third is the flow (not drift) due to pressure gradient (see Chap. 2),

$$\mathbf{u}_p{}^a \equiv \frac{c}{\epsilon^a B} \hat{\mathbf{n}} \times (\nabla \cdot \mathbf{p}^a) \qquad (4.54)$$

the fourth is just the frictional drift of (4.33a),

$$\mathbf{u}_D \equiv -\frac{c}{\epsilon^a B} \hat{\mathbf{n}} \times \mathbf{P}^{ab} \qquad (4.55)$$

† From here on ω^e is a *negative* number.

Thus we write

$$\mathbf{u}_\perp{}^a \equiv \mathbf{u}_E + \mathbf{u}_A{}^a + \mathbf{u}_p{}^a + \mathbf{u}_D \tag{4.56}$$

We note that $\mathbf{u}_E$ and $\mathbf{u}_D$ are the *same* for ions and electrons, while $\mathbf{u}_A$ and $\mathbf{u}_p$ are *different*. Adding the former velocities only, we obtain an equation

$$\frac{\mathbf{P}^{ei}}{ne} = \mathbf{E} + \frac{1}{c}\,(\mathbf{u}_E + \mathbf{u}_D) \times \mathbf{B} \tag{4.57}$$

which will resemble Ohm's law after $\mathbf{P}^{ei}$ is expressed in terms of $\mathbf{j}$.

4.3.3. Conservation of Energy

Lastly, we have the energy-conservation equation

$$\frac{\partial}{\partial t}\,[\mathcal{E}^a + \tfrac{1}{2}\,\rho^a(\mathbf{u}^a)^2] = -\nabla \cdot \{[\mathcal{E}^a + \tfrac{1}{2}\rho^a(\mathbf{u}^a)^2]\mathbf{u}^a + \mathbf{p}^a \cdot \mathbf{u}^a + \mathbf{Q}^a\}$$
$$+ \mathbf{j}^a \cdot \mathbf{E} + R^{ab} \tag{4.58}$$

the terms on the right representing kinetic-energy change due to convection, work by the pressure, heat flow, work by the electric field, and collisions. Algebraic manipulation, with the use of (4.43) and (4.47), yields the simpler form

$$\frac{D^a \mathcal{E}^a}{Dt} = -\tfrac{5}{3}\mathcal{E}^a \nabla \cdot \mathbf{u}^a - \nabla \cdot \mathbf{Q}^a - \boldsymbol{\pi}^a \cdot \mathbf{U}^a + R^{ab} - \mathbf{P}^{ab} \cdot \mathbf{u}^a \tag{4.59}$$

where the terms on the right are interpreted as adiabatic work, heat conduction, viscous heating, energy transfer, and work performed by momentum transfer.

In summary, we have (for *each* species) three independent equations [say (4.43), (4.47), (4.58); or (4.42), (4.49), (4.59)] in the 14 independent unknowns ρ^a, $\mathbf{u}^a$, $\mathcal{E}^a$, $\boldsymbol{\pi}^a$, $\mathbf{Q}^a$, $\mathbf{P}^{ei} = -\mathbf{P}^{ie}$, $R^{ei} = -R^{ie}$, $\mathbf{E}$, $\mathbf{B}$; we may add the Maxwell equations

$$\nabla \times \mathbf{B} = \frac{4\pi}{c}\,\mathbf{j} + \frac{1}{c}\,\frac{\partial \mathbf{E}}{\partial t} \tag{4.60}$$

$$\nabla \times \mathbf{E} = -\frac{1}{c}\,\frac{\partial \mathbf{B}}{\partial t} \tag{4.61}$$

Thus we have altogether eight equations for 14 unknowns. The six missing equations, discussed earlier and to be found later, will be the thermal-conduction and viscosity laws for each species, Ohm's law (or its equivalent, Fick's law), and the law of energy exchange (see Sec. 4.1.3).

4.3.4. Alternative Definitions

Certain definitions introduced at the beginning of this section are unconventional but are used because of the possibility of simple physical interpretations of the results. We now relate our definitions to the conventional ones. One defines the fluid velocity $\mathbf{u}$ by

$$\mathbf{u} \equiv \frac{\mathbf{g}}{\rho} = \sum_a \frac{\mathbf{u}^a \rho^a}{\rho} \tag{4.62}$$

Since $\rho^i \gg \rho^e$, $\mathbf{u}$ is essentially the *ion*-flow velocity. Then, wherever $\mathbf{v} - \mathbf{u}^a$ appears in the original definitions, one uses instead $\mathbf{v} - \mathbf{u}$. We shall denote the corresponding quantities by a prime, e.g.,

$$\mathbf{p}^{a'} \equiv \int d^3v \, m^a (\mathbf{v} - \mathbf{u})(\mathbf{v} - \mathbf{u}) f^a$$

The primed and unprimed quantities are related as follows:

$$\mathbf{p}^{a'} - \mathbf{p}^a = \rho^a (\mathbf{u}^a - \mathbf{u})(\mathbf{u}^a - \mathbf{u}) \tag{4.63}$$

$$\mathbf{p}' + \rho \mathbf{u}\mathbf{u} = \mathbf{p} + \sum_a \rho^a \mathbf{u}^a \mathbf{u}^a \tag{4.64}$$

$$\mathcal{E}^{a'} - \mathcal{E}^a = \tfrac{1}{2}\rho^a (\mathbf{u}^a - \mathbf{u})^2 \tag{4.65}$$

$$\mathcal{E}' + \tfrac{1}{2}\rho u^2 = \mathcal{E} + \sum_a \tfrac{1}{2}\rho^a (\mathbf{u}^a)^2 \tag{4.66}$$

$$\mathbf{Q}^{a'} - \mathbf{Q}^a = \mathbf{p}^a \cdot (\mathbf{u}^a - \mathbf{u}) + \mathcal{E}^{a'}(\mathbf{u}^a - \mathbf{u}) \tag{4.67}$$

$$(\mathcal{E}' + \tfrac{1}{2}\rho u^2)\mathbf{u} + \mathbf{p}' \cdot \mathbf{u} + \mathbf{Q}'$$
$$= \sum_a \{[\mathcal{E}^a + \tfrac{1}{2}\rho^a (\mathbf{u}^a)^2]\mathbf{u}^a + \mathbf{p}^a \cdot \mathbf{u}^a + \mathbf{Q}^a\} \tag{4.68}$$

With (4.64), we can write (4.48) as

$$\frac{\partial \mathbf{g}}{\partial t} = -\nabla \cdot (\mathbf{p}' + \rho \mathbf{u}\mathbf{u}) + \epsilon \mathbf{E} + \frac{1}{c}\mathbf{j} \times \mathbf{B} \tag{4.69}$$

or $$\rho \frac{D\mathbf{u}}{Dt} = -\nabla \cdot \mathbf{p}' + \epsilon \mathbf{E} + \frac{1}{c}\mathbf{j} \times \mathbf{B} \tag{4.70}$$

With (4.66) and (4.68) we can write the sum over species of (4.58) as

$$\frac{\partial}{\partial t}(\mathcal{E}' + \tfrac{1}{2}\rho u^2) = -\nabla \cdot [(\mathcal{E}' + \tfrac{1}{2}\rho u^2)\mathbf{u} + \mathbf{p}' \cdot \mathbf{u} + \mathbf{Q}'] + \mathbf{j} \cdot \mathbf{E} \tag{4.71}$$

or $$\frac{D\mathcal{E}'}{Dt} = -\tfrac{5}{3}\mathcal{E}'\nabla \cdot \mathbf{u} - \nabla \cdot \mathbf{Q}' - \boldsymbol{\pi}' : \mathbf{U} + (\mathbf{j} - \epsilon \mathbf{u}) \cdot (\mathbf{E} + \frac{1}{c}\mathbf{u} \times \mathbf{B}) \tag{4.72}$$

Equations (4.70) and (4.72) are the conventional total momentum and internal-energy equations; the last term of (4.72) is the work done by $\mathbf{E}'$ in the fluid rest frame. We see that the conventional definitions allow a more concise presentation for the sum-over-species equations, although not for the single-species equations.

4.4 LOCAL EQUILIBRIUM

The equations of the previous section require no assumptions about the spatial or temporal scale of the macroscopic phenomenon studied, except that the macroscopic variation be sufficiently slow so that the concept of collision be meaningful.   Further progress is made by assuming that the characteristic macroscopic time be long compared with the mean time between collisions and that the characteristic macroscopic distance be large.† Under such conditions, we may expect the plasma to be locally in thermal equilibrium, at least to a first approximation.   By the phrase *local equilibrium*, we mean that the Boltzmann function $f^a(\mathbf{r},\mathbf{v},t)$ for each component is Maxwellian about its temperature $T^a(\mathbf{r},t)$ and flow velocity $\mathbf{u}^a(\mathbf{r},t)$, with a common density $n(\mathbf{r},t)$,

$$f^a(\mathbf{r},\mathbf{v},t) = n(\mathbf{r},t)\left[\frac{m^a}{2\pi k T^a(\mathbf{r},t)}\right]^{3/2} \exp\left\{-\frac{\tfrac{1}{2}m^a[\mathbf{v}-\mathbf{u}^a(\mathbf{r},t)]^2}{kT^a(\mathbf{r},t)}\right\} \quad (4.73)$$

We must realize that such a form is only a first approximation; in the next section we study the corrections to it.   Here we note that, because of collisions, we may expect the differences $\mathbf{u}^e - \mathbf{u}^i$ and $T^e - T^i$ to be small, in some sense.

If we use (4.73) in the definitions of Sec. 4.3, we find that

$$p^a(\mathbf{r},t) = n(\mathbf{r},t)kT^a(\mathbf{r},t) \tag{4.74}$$
and $\quad \boldsymbol{\pi}^a = \mathbf{Q}^a = 0 \tag{4.75}$

The corrections to (4.73) will allow us to calculate nonzero values for $\boldsymbol{\pi}^a$ and $\mathbf{Q}^a$.

4.4.1. Entropy Production

We may now construct a quasithermodynamic theory.   Let us define the entropy per particle for component a by

$$s^a(\mathbf{r},t) \equiv -\ln n(\mathbf{r},t) + \tfrac{3}{2}\ln kT^a(\mathbf{r},t) \tag{4.76}$$

(The conventional entropy S equals $k \sum_a \int d^3r\, ns^a$ to within an additive constant, for fixed number of particles and $\mathbf{j} = 0$.)   We calculate its convective derivative and find, from Eqs. (4.42) and (4.59),

$$\frac{D^a s^a}{Dt} = \frac{1}{p^a}\left(-\nabla\cdot\mathbf{Q}^a - \boldsymbol{\pi}^a{:}\mathbf{U}^a + R^{ab} - \mathbf{P}^{ab}\cdot\mathbf{u}^a\right) \tag{4.77}$$

† Along B, it must be large compared with the mean free path, while, in the transverse plane, it must be large compared with the gyroradii.

Next we introduce the entropy density,

$$\mathcal{S}(\mathbf{r},t) \equiv \sum_a n(\mathbf{r},t) s^a(\mathbf{r},t) \tag{4.78}$$

and find

$$\frac{\partial \mathcal{S}}{\partial t} = -\nabla \cdot \sum_a \left(n\mathbf{u}^a s^a + \frac{\mathbf{Q}^a}{kT^a} \right) + \sum_a [\mathbf{Q}^a \cdot \nabla(kT^a)^{-1}$$
$$- (kT^a)^{-1}\boldsymbol{\pi}^a : \mathbf{U}^a + (kT^a)^{-1}(R^{ab} - \mathbf{P}^{ab} \cdot \mathbf{u}^a)] \tag{4.79}$$

The divergence term represents entropy flux density, while the bracket is the irreversible entropy production density. It is a quantitative measure of dissipation. Its three terms correspond to dissipation from heat conduction, viscosity, and energy and momentum transfer.

4.4.2. Energy Transfer between Species

Let us study the latter in more detail. The quantity $R^{ab} - \mathbf{P}^{ab} \cdot \mathbf{u}^a$, appearing in (4.79) and in (4.59), is invariant under a change of reference frame to one moving with any velocity $\mathbf{u}'$. Let us choose $\mathbf{u}' \equiv \mathbf{u}_E + \mathbf{u} \cdot \hat{\mathbf{n}}\hat{\mathbf{n}}$, with $\mathbf{u}$ defined by (4.62). Relative to $\mathbf{u}'$, we evaluate R^{ei} by studying energy exchange in collisions, neglecting $\mathbf{u}^a - \mathbf{u}'$. We find

$$R^{ei} = 3n\nu \frac{m^e}{m^i} (kT^i - kT^e) \tag{4.80}$$

in agreement with the qualitative considerations of Sec. 4.1.3. The collision frequency ν is now quantitatively defined as

$$\nu \equiv \frac{4}{3} \sqrt{2\pi}\, n \left(\frac{e^2}{kT^e} \right)^2 \left(\frac{kT^e}{m^e} \right)^{1/2} \ln \Lambda \tag{4.81}$$

in agreement with (4.9). The rate of entropy production from R^{ab} is thus

$$\sum_a (kT^a)^{-1} R^{ab} \equiv R^{ei}[(kT^e)^{-1} - (kT^i)^{-1}]$$
$$= 3n\nu \frac{m^e}{m^i} \frac{(kT^i - kT^e)^2}{(kT^i)(kT^e)} \tag{4.82}$$

4.4.3. Momentum Transfer between Species

Relative to the frame $\mathbf{u}'$, the momentum-transfer term in Eq. (4.59) reads

$$-P_\parallel^{ab} \cdot (\mathbf{u}^a - \mathbf{u})_\parallel - P_\perp^{ab} \cdot (\mathbf{u}_\perp^a - \mathbf{u}_E) \tag{4.83}$$

The longitudinal momentum transfer can be expressed in terms of $\mathbf{E}$ and ∇T by means of the longitudinal equation

$$0 = -\nabla_\parallel(nkT^a) + nq^a\mathbf{E}_\parallel + \mathbf{P}_\parallel^{ab} \tag{4.84}$$

obtained from (4.49) by dropping the acceleration term (as a simplification) and neglecting $\boldsymbol{\pi}^a$. Using (4.84) for both electrons and ions, we obtain

$$\mathbf{P}_{\parallel}^{ei} = ne\mathbf{E}_{\parallel} + n\,\frac{(kT^i)\nabla_{\parallel}(kT^e) - (kT^e)\nabla_{\parallel}(kT^i)}{kT^i + kT^e} \equiv ne\mathbf{E}'' \tag{4.85}$$

(We note that, if $T^e = T^i$, then $\mathbf{E}''$ reduces to $\mathbf{E}_{\parallel}$.) The first term of (4.83) now reads (for electrons)

$$-\mathbf{P}_{\parallel}^{ei} \cdot (\mathbf{u}^e - \mathbf{u})_{\parallel} = -\mathbf{E}'' \cdot ne(\mathbf{u}^e - \mathbf{u})_{\parallel} = +\mathbf{E}'' \cdot (\mathbf{j}^e - \epsilon^e\mathbf{u})_{\parallel} \tag{4.86a}$$

and (for ions)

$$-\mathbf{P}_{\parallel}^{ie} \cdot (\mathbf{u}^i - \mathbf{u})_{\parallel} = \mathbf{E}'' \cdot (\mathbf{j}^i - \epsilon^i\mathbf{u})_{\parallel} \tag{4.86b}$$

In terms of the *conduction* current

$$\mathbf{j}_c^a \equiv \mathbf{j}^a - \epsilon^a\mathbf{u} \tag{4.87}$$

the terms are

$$\mathbf{E}'' \cdot \mathbf{j}_{c\parallel}^a \tag{4.88}$$

which is recognizable as Joule heating.

For the transverse term of (4.83), we use Eqs. (4.55) and (4.56) to write

$$-\mathbf{P}_{\perp}^{ab} \cdot (\mathbf{u}_{\perp}^a - \mathbf{u}_E) = -\mathbf{P}_{\perp}^{ab} \cdot (\mathbf{u}_p^a + \mathbf{u}_A^a) \tag{4.89}$$

Now we may obtain two alternative forms, by eliminating $\mathbf{P}_{\perp}^{ab}$ either by (4.57) or by (4.55). In the former case we have

$$\left[\mathbf{E}_{\perp} + \frac{1}{c}(\mathbf{u}_E + \mathbf{u}_D) \times \mathbf{B} \right] \cdot [\mathbf{j}_p^a + \mathbf{j}_A^a]_{\perp} \tag{4.90}$$

which is analogous to (4.88). In the latter case, we have

$$-\mathbf{u}_D \cdot \left(\nabla \cdot \mathbf{p}^a - \rho^a \frac{D^a\mathbf{u}^a}{Dt} \right)_{\perp} \tag{4.91}$$

which relates it to the diffusion process.

In summary, the results (4.80), (4.88), and (4.90) or (4.91) may be substituted into Eqs. (4.59) and (4.79) to express the energy and momentum-transfer contributions in terms of macroscopic quantities. For example, Eq. (4.59) may now read

$$\frac{D^a\mathcal{E}^a}{Dt} = -\tfrac{5}{3}\mathcal{E}^a\nabla \cdot \mathbf{u}^a - \nabla \cdot \mathbf{Q}^a - \boldsymbol{\pi}^a{:}\mathbf{U}^a + 3n\nu\,\frac{m^e}{m^i}(kT^b - kT^a)$$

$$+ \mathbf{E}'' \cdot \mathbf{j}_{c\parallel}^a - \mathbf{u}_D \cdot \left(\nabla \cdot \mathbf{p}^a - \rho^a \frac{D^a\mathbf{u}^a}{Dt} \right)_{\perp} \tag{4.92}$$

4.5   CHAPMAN - ENSKOG EXPANSION

In the preceding section, we presented the energy-transfer equation (4.80), and now we must find the five transport equations mentioned at the end of Sec. 4.3.3.   This requires the use of a kinetic equation; from the discussion in Chap. 3, we have the Fokker-Planck equation

$$\frac{\partial f^a}{\partial t}\,(\mathbf{r},\mathbf{v},t) + \mathbf{v}\cdot\frac{\partial f^a}{\partial \mathbf{r}} + \frac{q^a}{m^a}\left(\mathbf{E}+\frac{1}{c}\,\mathbf{v}\times\mathbf{B}\right)\cdot\frac{\partial}{\partial \mathbf{v}}\,f^a$$
$$= -\frac{\partial}{\partial \mathbf{v}}\cdot\left[\langle\Delta\mathbf{v}\rangle^a f^a - \frac{1}{2}\frac{\partial}{\partial \mathbf{v}}\cdot(\langle\Delta\mathbf{v}\,\Delta\mathbf{v}\rangle^a f^a)\right] \equiv C^a \quad (4.93)$$

The terms in $\mathbf{E}$ and $\mathbf{B}$ represent the effect of the self-consistent field, while the right side represents the effect of collisions.   The conservation equations of Sec. 4.3 can be obtained by taking velocity moments of (4.93).

Let us change variables from $\mathbf{r}$, $\mathbf{v}$, t to $\mathbf{r}$, $\mathbf{w}$, t, where

$$\mathbf{w} \equiv \mathbf{v} - \mathbf{u}^a(\mathbf{r},t) \tag{4.94}$$

is the peculiar velocity of a particle, appearing already in the definitions of Sec. 4.3.   In terms of the new variables, the Fokker-Planck equation becomes

$$\frac{D^a f^a}{Dt} + \mathbf{w}\cdot\frac{\partial}{\partial \mathbf{r}}\,f^a + \left[-\mathbf{w}\cdot\nabla\mathbf{u}^a + \frac{1}{\rho^a}\,(\nabla\cdot\mathbf{p}^a - \mathbf{P}^{ab})\right]\cdot\frac{\partial}{\partial \mathbf{w}}\,f^a$$
$$= \omega^a\hat{\mathbf{n}}\cdot\mathbf{w}\times\frac{\partial}{\partial \mathbf{w}}\,f^a + C^a \quad (4.95)$$

[We have used Eq. (4.49) to transform the expression in brackets.]

We shall solve Eq. (4.95) by successive approximations, utilizing a formal expansion of f^a in an implicit small parameter δ,

$$f^a = f^{a(0)} + f^{a(1)} + \cdots \tag{4.96}$$

where $f^{a(0)}$ is given by the local Maxwellian (4.73), which we write as

$$f^{a(0)}(\mathbf{r},\mathbf{w},t) = \left(\frac{m^a}{2\pi}\right)^{3/2}\exp\left[-s^a(\mathbf{r},t) - \frac{\frac{1}{2}m^a w^2}{kT^a(\mathbf{r},t)}\right] \tag{4.97}$$

using Eqs. (4.76) and (4.94).   For this to be a zero-order solution, and for our ordering scheme to be consistent, we make the following assumptions:

1.   The operator $\nabla(\equiv \partial/\partial\mathbf{r})$ is first-order in δ.
2.   The operator D/Dt contains first and higher orders of δ.
3.   The relative flow $\mathbf{u}^a - \mathbf{u}^b$ is first-order, and therefore so is $\mathbf{P}^{ab}$.
4.   The mass ratio m^e/m^i is second-order, so that R^{ab} is a second-order dissipative effect, like the other dissipative terms.

It follows that the left side of Eq. (4.95) vanishes to zero order, while (4.97) is a solution of the right side to zero order.

To find $f^{a(1)}$, we evaluate the terms of Eq. (4.95) to first order. The term $D^a f^a / Dt$ is evaluated by replacing f^a by $f^{a(0)}$,

$$
\begin{aligned}
\left(\frac{D^a f^a}{Dt}\right)^{(1)} &= \left(\frac{D}{Dt} f^{a(0)}\right)^{(1)} = f^{a(0)} \left(\frac{D}{Dt} \ln f^{a(0)}\right)^{(1)} \\
&= f^{a(0)} \left[-\frac{D}{Dt} s^a - \tfrac{1}{2} m^a w^2 \frac{D}{Dt} (kT^a)^{-1}\right]^{(1)}
\end{aligned}
\tag{4.98}
$$

From Eq. (4.77), we see that the dissipation is *second*-order, so

$$
\left(\frac{D}{Dt} s^a\right)^{(1)} = 0
$$

$$
\left[\frac{D}{Dt} (kT^a)^{-1}\right]^{(1)} = \frac{2}{3} (kT^a)^{-1} \nabla \cdot \mathbf{u}^{(0)}
\tag{4.99}
$$

[using Eq. (4.76)].   Thus we have

$$
\left(\frac{D^a f^a}{Dt}\right)^{(1)} = -\tfrac{1}{3} m^a w^2 (kT^a)^{-1} (\nabla \cdot \mathbf{u}^{(0)}) f^{a(0)}
\tag{4.100}
$$

Analogously, we evaluate the rest of Eq. (4.95) to first order and find

$$
\omega^a \hat{\mathbf{n}} \cdot \mathbf{w} \times \frac{\partial}{\partial \mathbf{w}} f^{a(1)} = -C^{a(1)} + f^{a(0)} \left[\frac{\mathbf{w} \cdot \mathbf{P}^{ab(1)}}{p^a} \right.
$$
$$
\left. + \frac{m^a}{kT^a} \mathbf{ww} : \mathbf{U}^{(1)} + \frac{\tfrac{1}{2} m^a w^2 - \tfrac{5}{2} kT^a}{(kT^a)^2} \mathbf{w} \cdot \nabla kT^a \right]
\tag{4.101}
$$

This is an integrodifferential equation for $f^{a(1)}$, since $C^{a(1)}$ contains $f^{a(1)}$ implicitly.   We shall suppose that the Fokker-Planck coefficients $\langle \Delta \mathbf{v} \rangle^a$ and $\langle \Delta \mathbf{v} \, \Delta \mathbf{v} \rangle^a$, which appear in C^a, are independent of magnetic-field strength;[†] this is a good approximation if the Debye length is small compared with the mean gyroradii.   Then, upon averaging over the directions in velocity space transverse to $\hat{\mathbf{n}}$, we obtain an equation (to be called the *longitudinal equation*) involving only the longitudinal components of the forces.   Subtracting that equation from (4.101), we obtain the remainder (to be called the *transverse equation*), which involves only the transverse components.

An exposition of the methods of solution of these equations would be too lengthy.   Hence in the remainder of this section we limit ourselves to presenting the results.

4.5.1. The Limit of Strong Magnetic Field

The transverse equation can be solved analytically if the gyrofrequency exceeds the collision frequency by a large factor.   Referring to Eqs.

[†] See Refs. 2 and 3.

(4.35) and (4.34), we here obtain the generalization† of Fick's law,

$$\mathbf{u}_D = -\frac{\eta_\perp c^2}{B^2}\left[\nabla p + \rho\,\frac{D\mathbf{u}}{Dt} - \tfrac{3}{2}n\nabla(kT^e)\right]_\perp \qquad (4.102)$$

where $\eta_\perp \equiv \sigma_\perp^{-1} \equiv \dfrac{m^e\nu}{ne^2}$ $\qquad\qquad$ (4.103)

and ν is given by Eq. (4.81). [The term in $\rho\,D\mathbf{u}/Dt$ could have been obtained already in (4.35) if mechanical equilibrium had not been assumed in Eq. (4.1).] The term in $\nabla_\perp(kT^e)$ represents *thermal diffusion*. Crossing Eq. (4.102) with **B**, and adding the electric drift $\mathbf{u}_E$, we find the generalization of (4.37), *Ohm's law*,

$$\mathbf{j}_\perp = \sigma_\perp\left[\mathbf{E}_\perp + \frac{1}{c}\,(\mathbf{u}_E + \mathbf{u}_D)\times\mathbf{B}\right] + \frac{3}{2}\frac{nc}{B}\,\hat{\mathbf{n}}\times\nabla(kT^e) \qquad (4.104)$$

The last term here, of course, comes from the $\nabla_\perp(kT^e)$ term of (4.102); but now it is called the *thermoelectric effect*.

In agreement with Eqs. (4.3) and (4.15), we find an ionic thermal conduction,

$$\mathbf{Q}_\perp^i = -K_\perp^i\nabla_\perp(kT^i) \qquad\qquad (4.105)$$

$$K_\perp^i = \frac{8}{3}\sqrt{\pi}\,\frac{n^2e^2c^2(m^i)^{1/2}}{B^2(kT^i)^{1/2}}\ln\Lambda \qquad (4.106)$$

The electron heat flow includes the thermoelectric effect,

$$\mathbf{Q}_\perp^e = -K_\perp^e\nabla_\perp(kT^e) + \frac{3}{2}\frac{\eta_\perp c^2}{B^2}\,p^e\left(\nabla p + \rho\,\frac{D\mathbf{u}}{Dt}\right)_\perp \qquad (4.107)$$

$$\frac{K_\perp^e}{K_\perp^i} = \left(1 + \frac{13}{8}\sqrt{2}\right)\left(\frac{m^e}{m^i}\right)^{1/2}\left(\frac{T^i}{T^e}\right)^{1/2} \qquad (4.108)$$

Finally, we discuss the shear stress $\boldsymbol{\pi}^a$. To zero order in the collision frequency, one finds terms independent of collision frequency, and valid as well for a collisionless plasma (we take the z axis along $\hat{\mathbf{n}}$),

$$\begin{aligned} \pi_{xx}^a &= -\pi_{yy}^a = -\frac{p^a}{\omega^a}\,U_{xy} \\ \pi_{xz}^a &= -\frac{2p^a}{\omega^a}\,U_{xz} \\ \pi_{yz}^a &= +\frac{2p^a}{\omega^a}\,U_{yz} \\ \pi_{xy}^a &= \frac{p^a}{2\omega^a}\,(U_{xx} - U_{yy}) \end{aligned} \qquad (4.109)$$

† An additional ion flow of the same order as (4.102) was found by J. B. Taylor, *Phys. Fluids* **4**, 1142 (1961). If $T^e \neq T^i$, temperature relaxation may occur [Eqs. (4.4) and (4.80)], causing the flow $\mathbf{u}_p^i$ [Eq. (4.54)] to be time-dependent; accelerational drift [Eq. (4.53)] then results.

These terms should be included in the momentum-conservation equation [(4.48) or (4.50)], but they do not contribute to the viscous dissipation (4.77). The viscous contributions are of first order in the collision frequency,

$$
\begin{aligned}
\pi_{xx}^a &= -\pi_{yy}^a = -\mu_\perp{}^a(U_{xx} - U_{yy}) \\
\pi_{zz}^a &= -2\mu_{\perp\parallel}{}^a U_{zz} \\
\pi_{yz}^a &= -2\mu_{\perp\parallel}{}^a U_{yz} \\
\pi_{xy}^a &= -2\mu_\perp{}^a U_{xy}
\end{aligned}
\tag{4.110}
$$

where
$$
\mu_\perp{}^a = \tfrac{1}{4}\mu_{\perp\parallel}^a
\tag{4.111}
$$

$$
\mu_{\perp\parallel}^i = \tfrac{3}{5}m^i K_\perp{}^i
\tag{4.112}
$$

$$
\mu_{\perp\parallel}^e = \left[1 + \left(1 + \frac{2}{3}\frac{T^i}{T^e}\right)\sqrt{2}\right]\left(\frac{m^e}{m^i}\right)^{3/2}\left(\frac{T^i}{T^e}\right)^{1/2}\mu_{\perp\parallel}^i
\tag{4.113}
$$

The viscous (4.110) and nonviscous (4.109) contributions are additive. Because of the mass dependence, the electron viscosity is usually negligible in its effect.

4.5.2. Longitudinal Coefficients

The longitudinal equation must be solved numerically. The generalization of Ohm's law (4.6a) is found to be

$$
\mathbf{j}_\parallel = \sigma_\parallel \mathbf{E}'' + \alpha \nabla_\parallel(kT^e)
\tag{4.114}
$$

where $\mathbf{E}''$ is defined in (4.85),

$$
\sigma_\parallel = 1.98\sigma_\perp
\tag{4.115}
$$

and
$$
\alpha = \frac{0.70\sigma_\parallel}{e}
\tag{4.116}
$$

The electron heat flow is

$$
\mathbf{Q}_\parallel^e = -K_\parallel^e \nabla_\parallel(kT^e) - \alpha(kT^e)\mathbf{E}''
\tag{4.117}
$$

with
$$
K_\parallel^e = \frac{2.10(kT^e)\sigma_\parallel}{e^2}
\tag{4.118}
$$

in agreement with (4.12), and with the same α. The ion heat flow is

$$
\mathbf{Q}_\parallel^i = -K_\parallel^i \nabla_\parallel(kT^i)
\tag{4.119}
$$

with
$$
K_\parallel^i = 1.37\left(\frac{m^e}{m^i}\right)^{1/2}\left(\frac{T^i}{T^e}\right)^{5/2}K_\parallel^e
\tag{4.120}
$$

and thus is much less except when T^i/T^e is comparable with or larger than $(m^i/m^e)^{1/5}$.

The longitudinal ion viscosity [see $(4.13)_i$] is

$$\mu_\parallel^i = \frac{(m^i)^{1/2}(kT^i)^{5/2}}{(2\pi)^{1/2}e^4 \ln \Lambda} \tag{4.121}$$

and contributes to

$$\pi_{zz}^i = -(\pi_{xx}^i + \pi_{yy}^i) = -2\mu_\parallel^i U_{zz} \tag{4.122}$$

The electron viscosity is usually negligible; it presumably has a form analogous to (4.121).

4.5.3. Transverse Coefficients for Arbitrary Magnetic Field

In Ref. 7, the transverse transport coefficients are evaluated by a variational technique for arbitrary values of ν/ω^a, the ratio of collision- to gyrofrequency. We present the results, following Ref. 8 with slightly different notation:†

$$\mathbf{j}_\perp = \sigma_1 \mathbf{E}' - \sigma_2 \mathbf{E}' \times \hat{\mathbf{n}} + \alpha_1 \nabla_\perp (kT) - \alpha_2 \nabla(kT) \times \hat{\mathbf{n}} \tag{4.123}$$

$$\mathbf{Q}_\perp{}^{e'} = -K_1{}^e \nabla_\perp(kT) + K_2{}^e \nabla(kT) \times \hat{\mathbf{n}}$$
$$- \beta_1 \mathbf{E}' + \beta_2 \mathbf{E}' \times \hat{\mathbf{n}} \tag{4.124}$$

$$\mathbf{Q}_\perp{}^{i'} = -K_1{}^i \nabla_\perp(kT) + K_2{}^i \nabla(kT) \times \hat{\mathbf{n}} \tag{4.125}$$

where $\quad \mathbf{E}' \equiv \mathbf{E}_\perp + \dfrac{1}{c}\mathbf{u}^i \times \mathbf{B} + \dfrac{kT}{ne}\nabla_\perp n \tag{4.126}$

and $\quad \sigma_1 = \dfrac{ne^2}{m^e}\dfrac{\nu g_\sigma}{(\nu g_\sigma)^2 + (\omega^e h_\sigma)^2} \qquad \sigma_2 = \dfrac{ne^2}{m^e}\dfrac{|\omega^e|h_\sigma}{(\nu g_\sigma)^2 + (\omega^e h_\sigma)^2}$

$$\tag{4.127}$$

$$\alpha_1 = \dfrac{ne}{m^e}\dfrac{\nu g_\tau}{(\nu g_\tau)^2 + (\omega^e h_\tau)^2} \qquad \alpha_2 = \dfrac{ne}{m^e}\dfrac{|\omega^e|h_\tau}{(\nu g_\tau)^2 + (\omega^e h_\tau)^2}$$

$$\tag{4.128}$$

$$K_1{}^e = \dfrac{5nkT}{m^e}\dfrac{\nu g_K}{(\nu g_K)^2 + (\omega^e h_K)^2} \qquad K_2{}^e = \dfrac{5nkT}{m^e}\dfrac{|\omega^e|h_K}{(\nu g_K)^2 + (\omega^e h_K)^2}$$

$$\tag{4.129}$$

$$\beta_1 = \dfrac{5}{2}\dfrac{nekT}{m^e}\dfrac{\nu g_\mu}{(\nu g_\mu)^2 + (\omega^e h_\mu)^2} \qquad \beta_2 = \dfrac{5}{2}\dfrac{nekT}{m^e}\dfrac{|\omega^e|h_\mu}{(\nu g_\mu)^2 + (\omega^e h_\mu)^2}$$

$$\tag{4.130}$$

$$K_1{}^i = \dfrac{5nkT}{m^i}\dfrac{\nu' g_{K_I}}{(\nu' g_{K_I})^2 + (\omega^i h_{K_I})^2} \qquad K_2{}^i = \dfrac{5nkT}{m^i}\dfrac{\omega^i h_{K_I}}{(\nu' g_{K_I})^2 + (\omega^i h_{K_I})^2}$$

$$\tag{4.131}$$

The symbols $\hat{\mathbf{n}}$, ω^a, and ν have been defined in Eqs. (4.51), (4.52), and

† In Ref. 7, it was assumed that $T^e = T^i \equiv T$.   Note that $\mathbf{E}'$ is newly defined here.

(4.81).   In accordance with Eq. (4.10),

$$\nu' \equiv \nu \left(\frac{m^e}{m^i}\right)^{1/2} \tag{4.132}$$

The quantities g and h (with various subscripts) are slowly varying numerical functions (near unity) of ω^e/ν (or ω^i/ν'); they are tabulated and graphed in Ref. 8.   (For example, g_σ varies from 0.564 at $B = 0$ to 1 at $B = \infty$.)

To compare (4.123) with the strong-field limit (4.102), we solve for $\mathbf{u}_D$,

$$\mathbf{u}_D = -g_\sigma \frac{\eta_\perp c^2}{B^2}\left(\nabla p + \rho \frac{D\mathbf{u}}{Dt}\right)_\perp + \frac{ne^2}{(m^e)^2}\eta_\perp \frac{h_\tau g_\sigma - h_\sigma g_\tau}{(\nu g_\tau)^2 + (\omega^e h_\tau)^2}\nabla_\perp(kT)$$
$$+ \frac{h_\sigma^{-1}}{ne}\mathbf{j}_\perp + \frac{1}{m^e|\omega^e|}\frac{\nu^2 g_\tau(g_\sigma - g_\tau) + (\omega^e)^2 h_\tau(h_\sigma - h_\tau)}{(\nu g_\tau)^2 + (\omega^e h_\tau)^2}\nabla(kT) \times \mathbf{\hat{n}} \tag{4.133}$$

Comparing with (4.102), we see that the direct diffusion (the first term) has a simple factor g_σ, that the thermal-diffusion term (the second) has a more complicated coefficient, and that there are two additional terms which vanish in the strong-field limit.

For the viscosity results, we consult Ref. 8 directly.

4.6. ANOMALOUS EFFECTS

In the preceding discussion, it was assumed that the plasma was locally in thermal equilibrium (to lowest approximation) and that the plasma was stable.   If these conditions are violated, the dissipative processes may be quite different in character; one calls them *anomalous* as opposed to *classical*.

Suppose that the plasma is stable, but non-Maxwellian; for example, the velocity profile may be qualitatively like that leading to the two-stream instability.   For such a "nearly unstable" plasma, the thermal fluctuations of nearly unstable oscillations will be anomalously large. The wave-particle interactions (called *Cerenkov emission and absorption*), which are negligible for a Maxwellian plasma, may dominate the Coulomb interactions and enhance the effective collision frequency.   Thus one expects anomalously small longitudinal thermal and electrical conductivity and anomalously large transverse diffusivity and thermal conductivity.

If the plasma is weakly unstable, one may apply the quasilinear instability theory to study the dissipation process.   For example, the "universal" instability is due to a small transverse density or temperature gradient and would lead to transverse mass and energy flux.

Discussions of the anomalous diffusion often refer to the Bohm diffusion formula.   For a heuristic derivation, we may use the random-walk formula (4.14), but with ν replaced by its maximum, ω (the gyrofrequency), for which a step of a (the gyroradius) could be taken.   Since

$$a^2 \sim \frac{v^2}{\omega^2} \sim \frac{kT}{m\omega^2}$$

we have

$$D \sim \frac{ckT}{eB} \tag{4.134}$$

This Bohm diffusivity is independent of density and is *directly* proportional to temperature and inversely proportional to only the *first* power of magnetic field strength; also, it is independent of species mass.   More rigorous attempts to derive (4.134) generally lead to a coefficient dependent on the unstable property of the plasma.

If the plasma is strongly unstable, a convective or turbulent means of transport will arise.   Theoretical treatment of this regime is generally only semiquantitative, owing to the lack of a basic theory of turbulence.

REFERENCES

Section 4.4

The material in the latter part of this section is taken from:
1.   Kaufman, A.: Plasma Transport Theory, in C. de Witt and J. F. Detouef (eds.), "The Theory of Neutral and Ionized Gases," Wiley, New York, 1960.

Section 4.5

The effect of a magnetostatic field on the Fokker-Planck coefficients is calculated by:
2.   Rostoker, N., and M. Rosenbluth: *Phys. Fluids*, **3**, 1 (1960).
3.   Rostoker, N.: *Phys. Fluids*, **3**, 922 (1960).

Subsection 4.5.1

This material is derived in Ref. 1.   The viscosity coefficients are discussed in terms of guiding-center motion (as in Sec. 4.2) in:
4.   Kaufman, A.: *Phys. Fluids*, **3**, 610 (1960).
In addition to providing viscous drag and heating, the shear stress is responsible for finite-ion-gyroradius stabilization:
5.   Roberts, K., and J. B. Taylor: *Phys. Rev. Letters*, **8**, 197 (1962).
For higher-order diffusion due to ion-ion collisions:
6.   Simon, A.: *Phys. Rev.*, **100**, 1557 (1955); A. Kaufman, *Phys. Fluids*, **1**, 252 (1958).

Subsections 4.5.2 and 4.5.3

This material is taken from:

7. Robinson, B., and I. Bernstein: *Ann. Phys. (N.Y.)*, **18,** 110 (1962).
8. Shkarofsky, I., I. Bernstein, and B. Robinson: *Phys. Fluids*, **6,** 40 (1963).
The variational techniques for solving the Chapman-Enskog equations are fully discussed in:
9. Bond, J. W., Jr., K. M. Watson, and J. A. Welch, Jr.: "Atomic Theory of Gas Dynamics," chap. 7, Addison-Wesley, Reading, Mass., 1965; F. Reif: "Fundamentals of Statistical and Thermal Physics," chap. 14, McGraw-Hill, New York, 1965.

Section 4.6

Anomalous transport in a non-Maxwellian stable plasma is discussed by:
10. Rostoker, N.: *Nuclear Fusion*, **1,** 101 (1960).
11. Silin, V.: *Nuclear Fusion*, **2,** 125 (1962).   Translation from Russian available from editor.
Anomalous transport arising from a microinstability is treated by:
12. Drummond, W., and M. Rosenbluth: *Phys. Fluids*, **5,** 1507 (1962).
A derivation of the Bohm diffusion from Brownian-motion theory is presented by:
13. Taylor, J. B.: *Phys. Rev. Letters*, **6,** 262 (1961); *Nuclear Fusion Suppl.*, pt 2, p. 477, 1962.
A thorough coverage of the present state of turbulence theory is to be found in:
14. Kadomtsev, B. B.: "Plasma Turbulence," Academic Press, New York, 1965.

5

PLASMA STABILITY

ROSTOKER

NORMAN ROSTOKER, *General Atomic Division of General Dynamics Corporation, John Jay Hopkins Laboratory for Pure and Applied Science, San Diego, California*

5

5.1 INTRODUCTION

We are interested in the question of how long a hot plasma can be confined in a magnetic bottle without touching material walls. An upper bound can be obtained by considering ordinary diffusion due to Coulomb collisions. For a plasma in the temperature range 10 to 100 kev, the upper bound is several seconds with reasonable dimensions and magnetic fields. This is adequate for the thermonuclear problem. The plasma may, however, escape confinement in a much shorter time, of the order of 1 to 100 μsec, owing to instabilities of an almost endless variety. To produce a confined plasma for interesting times we must eliminate intolerable instabilities, tolerate instabilities that cannot be eliminated, and know which is which.

To illustrate the nature of the questions and the techniques for answering them, consider the relatively simple example of a fluid of density ρ supported by a fluid of density $\rho_0 < \rho$ in a gravitational field g (cf Fig. 5.1). The first question is: Does there exist a lower energy state? Since the potential energy per unit area is $U = g \int \rho z \, dz$ we know that the state in which the light fluid is on top of the heavy fluid is lower

This work was carried out under a joint General Atomic–Texas Atomic Energy Research Foundation program on controlled thermonuclear reactions.

in energy by

$$\Delta U = g(\rho - \rho_0)ab \qquad (5.1)$$

The next question is: Can the system reach this state? Consider a possible small displacement of the fluid such as the boundary perturbation $\Delta z_B = \varepsilon \sin 2n\pi x/L$. The corresponding potential-energy change is

$$\delta U = \tfrac{1}{4}g(\rho_0 - \rho)\varepsilon^2 \qquad (5.2)$$

This is negative when $\rho > \rho_0$ so that the system is unstable. The total energy of the system $W = U + T$ must remain constant so that if the boundary perturbation takes place the fluid must acquire the kinetic energy $\delta T = -\delta U$. It is apparent that the perturbation will grow, and

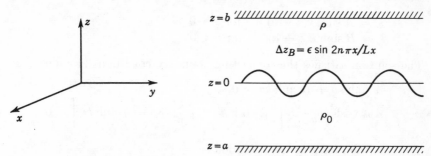

Fig. 5.1 Interchange instability: heavy fluid supported by light fluid.

to determine the growth rate, we must make use of the hydrodynamic equations

$$\frac{\partial \rho}{\partial t} + \nabla \cdot \rho \mathbf{V} = 0 \qquad (5.3)$$

$$\rho \frac{d\mathbf{V}}{dt} = -\nabla P - \rho g \mathbf{e}_z \qquad (5.4)$$

We assume incompressible fluids so that $d\rho/dt = \partial\rho/\partial t + \mathbf{V} \cdot \nabla\rho = 0$ and the initial state $\mathbf{V} = 0$, $-\partial P/\partial z = \rho g$. The procedure is to assume small perturbations such as $\mathbf{V}' = \mathbf{V} + \delta\mathbf{V}$, etc., and linearize the equations of motion. For convenience we introduce the displacement vector $\delta\boldsymbol{\xi} = \int_0^t \delta\mathbf{V}(t')\,dt'$. The linearized equations of motion are

$$\delta\rho + \delta\boldsymbol{\xi} \cdot \nabla\rho = -\rho\nabla \cdot \delta\boldsymbol{\xi} = 0$$

$$\rho \frac{\partial^2 \delta\boldsymbol{\xi}}{\partial t^2} = -\nabla\,\delta P - \delta\rho g \mathbf{e}_z \qquad (5.5)$$

An initial-value problem may be posed; i.e., assume some initial value for $\delta\boldsymbol{\xi}(\mathbf{x},0)$, and find $\delta\boldsymbol{\xi}(\mathbf{x},t)$. For example, consider normal modes of the form

$$\delta\boldsymbol{\xi}(\mathbf{x},t) = \boldsymbol{\xi}(z) \exp\left[pt + i(k_x x + k_y y)\right] \qquad (5.6)$$

After substituting this expression into Eq. (5.5) and eliminating $\delta\xi_x$, $\delta\xi_y$, $\delta\rho$, and δP in favor of $\delta\xi_z$, a differential equation is obtained for $\xi_z(z)$.

$$\frac{d}{dz} \rho \frac{d\xi_z}{dz} - \frac{k^2}{p^2}\left(\rho p^2 - g\frac{d\rho}{dz}\right)\xi_z = 0 \tag{5.7}$$

where $k^2 = k_x{}^2 + k_y{}^2$. Since $d\rho/dz = (\rho - \rho_0)\,\delta(z)$, ξ_z obeys the differential equation $\xi_z'' - k^2\xi_z = 0$ except at $z = 0$ and the boundary conditions connecting the solutions at $z = 0$ are $\xi_z(-0) = \xi_z(+0)$,

$$\rho\xi_z'(+0) - \rho_0\xi_z'(-0) = (gk^2/p^2)\xi_z(0)(\rho_0 - \rho)$$

The solution that satisfies the boundary conditions $\xi_z(b) = \xi_z(-a) = 0$ is

$$\begin{aligned}\xi_z &= A \sinh k(z - b) \qquad \text{for } z > 0\\ \xi_z &= B \sinh k(z + a) \qquad \text{for } z < 0\end{aligned}$$

The solution satisfies the connecting boundary conditions provided that

$$A \sinh kb + B \sinh ka = 0$$

$$A\rho k \cosh kb - B\left[\rho_0 k \cosh ka + g\frac{k^2}{p^2}(\rho_0 - \rho)\sinh ka\right] = 0$$

The condition for a nonzero solution is

$$\begin{aligned}p^2 &= \frac{kg(\rho_0 - \rho)}{\rho \coth kb + \rho_0 \coth ka}\\ p &\cong \sqrt{kg} \qquad \text{for } \frac{\rho}{\rho_0},\ ka,\ kb \gg 1\end{aligned} \tag{5.8}$$

Since we have linearized the hydrodynamic equations, the above growth rate will be incorrect when the amplitude of the perturbation approaches the wavelength. Experimentally[1] it has been found that the growth of a perturbation changes from exponential to linear when $\delta\xi \sim 2.5/k$. Then the light fluid rises with a velocity $\sim \sqrt{g/k}$, and the heavy fluid approaches free fall.

A plasma in a magnetic bottle is an object of considerably greater complexity, but the questions that concern us and the techniques employed to find answers are similar.

5.2 PLASMA EQUATIONS

To investigate the stability of a hot plasma, the Vlasov, or collisionless Boltzmann, equation is an adequate description (cf. Chap. 3),

$$\frac{\partial f_i}{\partial t} + \mathbf{v}\cdot\frac{\partial f_i}{\partial \mathbf{x}} + \frac{e_i}{m_i}\left(\mathbf{E} + \frac{1}{c}\mathbf{v}\times\mathbf{B}\right)\cdot\frac{\partial f_i}{\partial \mathbf{v}} = 0 \tag{5.9}$$

The fields $\mathbf{E}$ and $\mathbf{B}$ are determined by Maxwell's equations with

Charge density: $\qquad \rho_c = \sum_i e_i \int f_i \, d\mathbf{v}$

Current density: $\qquad \mathbf{j} = \sum_i e_i \int \mathbf{v} f_i \, d\mathbf{v}$

To obtain a more tractable description, we take moments and obtain (cf. Chap. 4)

$$\frac{\partial \rho}{\partial t} + \nabla \cdot \rho \mathbf{V} = 0 \tag{5.10}$$

$$\rho \frac{d\mathbf{V}}{dt} = -\nabla \cdot \mathbf{P} + \rho_e \mathbf{E} + \frac{1}{c} \mathbf{j} \times \mathbf{B} \tag{5.11}$$

$$\rho \frac{dU}{dt} = \mathbf{V} \cdot (\nabla \cdot \mathbf{P}) - \nabla \cdot (\mathbf{P} \cdot \mathbf{V}) + \mathbf{j} \cdot \left(\mathbf{E} + \frac{1}{c} \mathbf{V} \times \mathbf{B} \right)$$
$$- \rho_e (\mathbf{E} \cdot \mathbf{V}) - \nabla \cdot \mathbf{Q} \tag{5.12}$$

etc., where $\quad \rho = \sum_i m_i \int f_i \, d\mathbf{v} \qquad\qquad \rho \mathbf{V} = \sum_i m_i \int \mathbf{v} f_i \, d\mathbf{v}$

$$\mathbf{P} = \sum_i m_i \int (\mathbf{v} - \mathbf{V})(\mathbf{v} - \mathbf{V}) f_i \, d\mathbf{v} \qquad \rho U = \tfrac{1}{2} \operatorname{Tr} \mathbf{P}$$

$$\mathbf{Q} = \sum_i \frac{m_i}{2} \int |\mathbf{v} - \mathbf{V}|^2 (\mathbf{v} - \mathbf{V}) f_i \, d\mathbf{v}$$

This chain of equations continues indefinitely, and to make any progress, it must be terminated. Two ad hoc terminations have been employed called the *magnetohydrodynamic* (MHD) *approximation* and the *Chew-Goldberger-Low* (CGL)[2] *approximation*. Both assume $\mathbf{Q} = 0$ and $\mathbf{E} + (1/c)\mathbf{V} \times \mathbf{B} = 0$. The MHD approximation assumes in addition that $\mathbf{P} = P\mathbf{I}$ so that Eq. (5.12) reduces to $(d/dt)P\rho^{-5/3} = 0$. In the CGL approximation $\mathbf{P} = P_\perp \mathbf{I} + (P_\parallel - P_\perp)\mathbf{nn}$, where $\mathbf{n}$ is a unit vector parallel to $\mathbf{B}$. Instead of Eq. (5.12) we can obtain two equations for $\rho(dP_\parallel/dt)$ and $\rho(dP_\perp/dt)$ which reduce to $(d/dt)(P_\parallel B^2/\rho^3) = 0$ and $(d/dt)(P_\perp/\rho B) = 0$. Since we are concerned with the application to stability, we consider a linearized problem about an initial state characterized by $\mathbf{V} = 0$, $\mathbf{B}(\mathbf{x})$, and $P(\mathbf{x})$ or $P_\parallel(\mathbf{x})$ and $P_\perp(\mathbf{x})$. In terms of the displacement vector $\delta\boldsymbol{\xi} = \int_0^t \delta\mathbf{V}(t') \, dt' = \boldsymbol{\xi} e^{pt}$, the differential equation is of the form

$$A_{\alpha\beta} \xi_\beta = \rho p^2 \xi_\alpha \tag{5.13}$$

where $A_{\alpha\beta} = \dfrac{1}{4\pi} \dfrac{\partial}{\partial x_a} M_{\alpha\beta;\,ab} \dfrac{\partial}{\partial x_b}$

For the MHD approximation,

$$M_{\alpha\beta;\,ab} = \frac{B^2}{2}\left(\delta_{a\alpha}\delta_{b\beta} + \delta_{\alpha\beta}\delta_{b\alpha}\right) + B_a B_b \delta_{\alpha\beta} - \left(B_\beta B_b \delta_{a\alpha} + B_a B_\alpha \delta_{b\beta}\right)$$
$$+ 4\pi P \delta_{\alpha\beta}\delta_{b\alpha} + 4\pi(\gamma - 1)P\delta_{\alpha\alpha}\delta_{b\beta}$$

For the CGL approximation the terms involving P are replaced by

$$4\pi P_\perp (\delta_{\alpha\alpha}\delta_{bb} + \delta_{\alpha\beta}\delta_{b\alpha} - n_\alpha n_a \delta_{bb} - n_\beta n_b \delta_{\alpha\alpha})$$
$$- 4\pi(P_\parallel - P_\perp)n_a n_b \delta_{\alpha\beta} - 4\pi(P_\perp - 4P_\parallel)n_a n_b n_\alpha n_\beta$$

If the boundaries, for example, are conducting walls, $\xi_n = B_n = 0$ and we can show that

$$\int \eta_\alpha A_{\alpha\beta}\xi_\beta \, d\tau = \int \xi_\alpha A_{\alpha\beta}\eta_\beta \, d\tau$$

for any two vectors ξ_α, η_β that satisfy the boundary conditions; that is, $A_{\alpha\beta}$ is a self-adjoint operator. This follows from integrating by parts, use being made of the symmetry property $M_{\alpha\beta;\,ab} = M_{\beta\alpha;\,ba}$ and the boundary conditions. If we multiply Eq. (5.13) by ξ_α^* and sum, the complex conjugate of Eq. (5.13) by ξ_α and sum, then subtract the two results

$$[p^2 - (p^2)^*]\int \rho \xi_\alpha^* \xi_\alpha \, d\tau = \int (\xi_\alpha^* A_{\alpha\beta}\xi_\beta - \xi_\alpha A_{\alpha\beta}\xi_\beta^*) \, d\tau = 0$$

it follows that p^2 is real. We can always take linear combinations of ξ_α, ξ_α^* so that the eigenfunctions corresponding to p^2 are real.

The energy change corresponding to a perturbation is

$$W = \int \left[\int (\delta\mathbf{E} \cdot \delta\mathbf{j}) \, dt + \frac{\delta\mathbf{B}^2}{8\pi} \right] d\tau \tag{5.14}$$
$$= T + U$$

where
$$T = \frac{p^2}{2}\int \rho_m \xi_\alpha \xi_\alpha \, d\tau$$

$$U = -\frac{1}{2}\int \xi_\alpha A_{\alpha\beta}\xi_\beta \, d\tau$$

$W = 0$ and the first variation $\delta W = 0$ for perturbations that satisfy Eq. (5.13). The necessary and sufficient condition for stability is $U > 0$ for all possible perturbations, that is, if $U > 0$, all solutions with $p^2 > 0$ result in $W > 0$ so that there are no unstable solutions; if there are only stable solutions, $p^2 < 0$ and since $T < 0$, $W = 0$, it follows that $U > 0$.

It can be shown that $U_{CGL} > U_{MHD}$ so that the CGL approximation is always more optimistic about stability. The relevance of either approximation to the problem of plasma stability has been clarified by expanding the Vlasov equations in terms of the parameters $\lambda = a_i/L$, p/ω_p, p/ω_i, where L is a characteristic length over which unperturbed quantities vary, $a_i = v_\perp/\omega_i$ is the Larmor radius for particles of species i, $\omega_p = \sqrt{4\pi n e^2/m}$ is the plasma frequency, and $\omega_i = q_i B/m_i c$ is the Larmor frequency for particles of species i. Note that $\lambda \propto m/e$ so that the expansion applies to particles of very large charge and small mass. It can of course apply only to the low-frequency behavior of the plasma and is limited to slow spatial variations. This expansion terminates the moment equations naturally. In lowest order it yields a set of integro-differential equations in which e, m do not appear explicitly as in the MHD or CGL approximation. Of particular interest are the properties of the potential energy U. It has been proved that[3]

$$U_{\text{MHD}} < U < U_{\text{CGL}} \tag{5.15}$$

This provides a sound justification for using the MHD approximation and not using the CGL approximation for stability analyses. In addition it has recently been proved by F. Low[4] that, if a mode is stable to lowest order in $\lambda \propto m/e$, it is then stable to all orders in λ. This does not, however, relieve us of the necessity of investigating beyond the lowest order, because there are nonexpandable trapping instabilities and additional modes that did not appear in lowest order about which Low's theorem says nothing.

In view of the fact that configurations of a plasma in a magnetic bottle have been found that have complete stability according to the MHD approximation, relatively little effort has been devoted to the time evolution of instabilities. The present discussion will concentrate mainly on the question of stability and the application of the energy principle.

For the high-frequency behavior of a plasma where $p > \omega_p$, ω_i, no satisfactory method of terminating the moment equations has been found, and no useful energy principle exists. The Vlasov equations are employed in their original form. It has been possible to carry out the required analysis only for an infinite plasma. The question of avoiding instabilities is much more difficult so that a considerable effort has been devoted to calculating growth rates and more recently to the nonlinear evolution to a turbulent state. It should be noted that almost no experiments have been devoted to this aspect of plasma behavior.

When the plasma is relatively cold and resistive, there are some additional instabilities that are of interest in connection with positive-column experiments and some devices presently under study in the thermonuclear program.

5.3 LOW-FREQUENCY INSTABILITIES

5.3.1. Plasma Supported in a Gravitational Field

Assume a plasma density $\rho(z)$ and magnetic fields $B_x(z)$, $B_y(z)$ (cf. Fig. 5.2) in the initial state. On the assumption that $\mathbf{V} = 0$, the equilibrium state satisfies

$$\nabla P = -\frac{1}{8\pi}\nabla B^2 + \frac{1}{4\pi}\mathbf{B}\cdot\nabla\mathbf{B} - \rho g \mathbf{e}_z$$

or $$\frac{\partial}{\partial z}\left(P + \frac{B^2}{8\pi}\right) = -\rho g$$

On assuming perturbations of the form of Eq. (5.6), the linearized MHD equations for an incompressible plasma reduce to a single equation similar to Eq. (5.7), that is,

$$\frac{d}{dz}\left[\rho + \frac{(\mathbf{k}\cdot\mathbf{B})^2}{4\pi p^2}\right]\frac{d\xi_z}{dz} - \frac{k^2}{p^2}\left[\rho p^2 + \frac{(\mathbf{k}\cdot\mathbf{B})^2}{4\pi} - g\frac{\partial\rho}{\partial z}\right]\xi_z = 0 \quad (5.16)$$

We note that if the magnetic field has no shear, or does not change direction as a function of z, perturbations for which $\mathbf{k}\perp\mathbf{B}$ are unaffected by

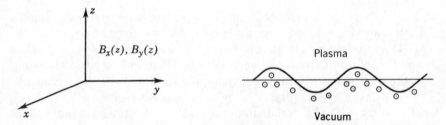

Fig. 5.2   Interchange instability: plasma supported by magnetic field.

the magnetic field.   The sharp transition case with $B_y = 0$ leads then to the same growth rate as predicted in Eq. (5.8), $p \cong \sqrt{k_y g}$ for the mode illustrated above, where $k_x = 0$.   This is because the deformation leads to no change in magnetic energy.   When there is shear, any deformation distorts the field lines and changes the magnetic energy.

To illustrate how shear provides some stabilization, consider the sharp-transition case,

$$B_x(z) = \begin{cases} 0 & z > 0 \\ B_0 & z < 0 \end{cases}$$
$$B_y(z) = \begin{cases} \alpha_P B_0 & z > 0 \\ \alpha_V B_0 & z < 0 \end{cases} \quad (5.17)$$

Since p^2 must be real, we go through $p^2 = 0$ on passing from stability to instability.   If we wish only a stability criterion, it is sufficient to solve the $p = 0$ problem,

$$\frac{d}{dz}\frac{(\mathbf{k}\cdot\mathbf{B})^2}{4\pi}\frac{d\xi_z}{dz} - k^2\left[\frac{(\mathbf{k}\cdot\mathbf{B})^2}{4\pi} - g\frac{\partial\rho}{\partial z}\right]\xi_z = 0 \qquad (5.18)$$

For $z \neq 0$, $\xi_z'' - k^2\xi_z = 0$ so that $\xi_z = Ae^{-kz}$ for $z > 0$ and $\xi_z = Ce^{kz}$ for $z < 0$. The boundary conditions at $z = 0$ follow from Eq. (5.18),

$$\xi_z(+0) = \xi_z(-0)$$
$$\frac{B_0{}^2}{4\pi}[(k_y\alpha_p)^2\xi_z'(+0) - (k_x + \alpha_V k_y)^2\xi_z'(-0)] = -k^2\rho g\xi_z(0)$$

The condition for marginal stability is therefore

$$\frac{B_0{}^2}{4\pi\rho}[(k_x + \alpha_V k_y)^2 + (k_y\alpha_p)^2] = kg \qquad (5.19)$$

For stability the left-hand side must be greater than the right-hand side. Consider, for example, the application to a rail-gun plasma accelerator.[5] Let the dimensions of the plasma be L in the x, y directions and d in the z direction, which is the direction of acceleration. The acceleration is obtained from $\rho L^2\,dg = (B_0{}^2/8\pi)L^2$ on assuming that $\alpha_p \sim \alpha_V$. The longest significant wavelength is $k \sim 2\pi/L$ so that stability should be obtained if $\alpha^2 > (1/8\pi)(L/d)$.

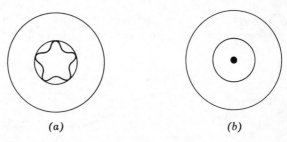

(a)   *(b)*

Fig. 5.3 End-on view of linear pinch. *(a)* Showing flutes; *(b)* flutes suppressed by B_θ generated in "hard core."

Direct observations of the flute instability have been obtained with a linear pinch tube by photographing through a perforated anode with a Kerr cell camera.[6] Deceleration of the inner surface of the plasma is obtained due to compression of the B_z field, and flutes appear as in Fig. 5.3*a*. When a B_θ is provided by a central conductor such that $B_\theta \gtrsim B_z/8$, the flutes disappear as in Fig. 5.3*b*.

For subsequent applications it is instructive to examine the instability from a particle point of view (cf. Fig. 5.4). The guiding centers of

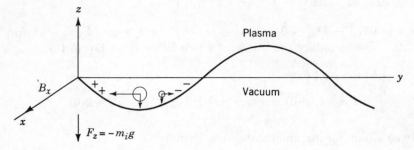

Fig. 5.4 Interchange instability derived from guiding-center motion.

particles experience a drift $\mathbf{V}_d = (c/e_i B)(\mathbf{F}_i \times \mathbf{B})$, where $\mathbf{F}_i = -m_i g \mathbf{e}_z$ (cf. Chap. 2). Electrons and ions drift in opposite directions, producing a charge density as indicated and a corresponding electric field (E_y, E_z). The drift due to $\mathbf{F}_i = e_i E_y \mathbf{e}_y$ is in the same direction for electrons and ions, which is the direction that makes the perturbation grow. An elementary calculation[7] shows that the amplitude of the perturbation grows at the rate $\sqrt{gk}$.

5.3.2. Mirror Machines

Consider Fig. 5.5. A particle spiraling along one of the field lines experiences a drift due to the centripetal force $M\mathbf{R}v_\parallel^2/R^2$, where $\mathbf{R}$ is the local

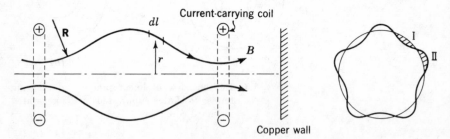

Fig. 5.5 Gravity equivalent and flutes in mirror machine.

radius of curvature of the line. In addition, owing to the gradient of B perpendicular to $\mathbf{B}$, there is a drift due to an equivalent force $M\mathbf{R}v_\perp^2/2R^2$. The physical effect is the same as gravity, with $\mathbf{g} = (\mathbf{R}/R^2)(v_\parallel^2 + v_\perp^2/2)$. If $\mathbf{R}$ points from the plasma to vacuum, we should expect flute instabilities [perturbations proportional to exp $(pt + im\theta)$] to grow with the rate $p \sim \sqrt{gk}$, where $k = m/r_0$ and r_0 is the radius of the plasma surface.

From this discussion we can see that convex plasma surfaces lead to instability and concave surfaces to stability. There are many field configurations[8] that are stable, such as the cusp (cf. Fig. 5.6); however, the

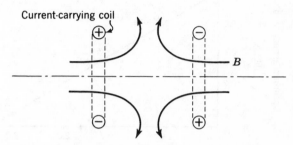

Current-carrying coil

Fig. 5.6 Cusp configuration.

price for stability is a larger leak rate than for a mirror machine unless the two configurations are combined in some fashion, as mentioned in Sec. 5.5 and as described more fully in Chap. 13.

The instability of a mirror machine is also apparent from a simple application of the energy principle.[7] Consider a case where the particle energy is small compared with the magnetic energy; i.e., the magnetic field is nearly the vacuum field which has the minimum field energy. The worst displacement is one which leaves the magnetic field unchanged. Such a displacement is the interchange of particles and flux of two tubes of equal flux I and II. Since the magnetic energy is unchanged, $\delta U = \delta \int \frac{p/\rho}{\gamma - 1} \, d\tau$. An elementary[7] calculation shows that for $\delta U > 0$ we must have

$$\int \frac{dl}{RrB^2} > 0 \tag{5.20}$$

where R is positive at the ends and negative in the middle. Since B is larger at the ends, the middle region dominates and the mirror should be unstable. In fact the mirror machine is often observed to be stable. Two explanations have been proposed to explain this:

1. Conducting end walls. If the end walls are good conductors, the magnetic field lines are tied at the ends and the interchange discussed above is not a possible displacement. In the limit $\beta = 8\pi P/B^2 \to 0$, where the particles do not alter the fields, the plasma should be stable. From the previous discussion of the gravity instability with the particle point of view, the effect of the end walls is to short-circuit the field lines

and prevent E_y from developing.   However, if the change in magnetic field is considered for finite β, it is found that stability obtains up to some critical β,[9] as indicated in Fig. 5.7.

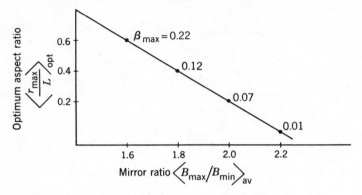

Fig. 5.7   Critical β in mirror machines with "tied" lines.

2.   Finite Larmor radius effect.   In the previous discussion of the gravity instability, it was assumed that the Larmor radius of electrons and ions was negligibly small.   If we reconsider and include the fact that the ion Larmor radius (which is much larger than the electron Larmor radius) is finite, it is apparent that the ions take a different sampling of the field E_y and should drift in the z direction more slowly than the electrons. Thus a current in the z direction is produced which competes with the current in the y direction and significantly alters the charge distribution. The result is that the perturbation does not simply grow but oscillates as well, and if g is sufficiently small, it can be a stable oscillation.   For perturbations $\delta\xi \propto \exp[i(\omega t + m\theta)]$ of a cylindrical plasma of radius r_0[10]

$$\omega = -\left(\frac{a_I}{r_0}\right)^2 \omega_I(1-m) \pm \left\{\left[\left(\frac{a_I}{r_0}\right)^2 \omega_I(1-m)\right]^2 - \frac{m}{r_0}g\right\}^{1/2}$$

$$(5.21)$$

where a_I is the mean ion Larmor radius and ω_I is the ion cyclotron frequency.   If $a_I = 0$, we recover $\omega = \pm \sqrt{kg}$, where $k = m/r_0$.   As a_I is "switched on," the oscillation changes to an overstable growing one and finally becomes stable when

$$\frac{a_I}{r_0} > \frac{\sqrt{m}}{m-1} \sqrt{\frac{r_0}{R}}$$

$$(5.22)$$

We have assumed $g = v_\perp^2/R$ for a mirror machine and $a_I = v_I/\omega_I$.   Note that no stabilization takes place for $m = 1$.   This is because the electric-field perturbation has no spatial dependence so that the difference between electron and ion Larmor radii is of no consequence.

Recent experiments with high-temperature plasmas in mirror machines[11] have indeed given support to the notion that both finite Larmor radius effects and line-tying at conducting walls can have a stabilizing influence.

5.3.3. Θ PINCHES

The theta-pinch configuration and behavior are sketched in Fig. 5.8. The experiment usually consists of a single-turn coil that produces mainly a B_z field, but sometimes with mirrors at the ends. For short coils the flute instability characteristic of mirror fields[12] has been observed. $m = 1$ and $m = 2$ instabilities predominate as in Fig. 5.8b and c. As the length

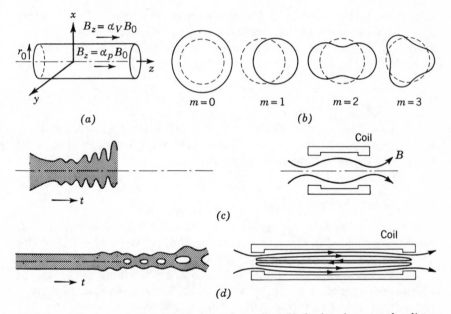

Fig. 5.8   The theta pinch.   (*a*) Basic configuration; (*b*) the four lowest-order distortions; (*c*) smear camera picture as seen through transverse slit, indicating $m = 1$ oscillation, typical for behavior in short coils with end mirrors; (*d*) smear camera picture showing $m = 2$ "bifurcation," typical behavior for long pinches with internally trapped reversed magnetic field.

of the coil increases, the growth rate is reduced and complete elimination of the instability has been observed.   This is consistent with the previous discussion of the finite Larmor radius effect.

For very long coils[13] the observations are quite different.   With no internal reversed B_z field, flutes are not observed. With an internal reversed field as illustrated in Fig. 5.8d, the plasma is observed to be stable for about 5 μsec, after which an $m = 2$ instability is observed to develop to the point where the plasma appears to have fissioned into two parts. There is an apparent rotation of frequency $\Omega \sim \omega_{\mathrm{I}}/10$.

A j_θ current must flow in the plasma to produce the internal reversed field. Diffusional mixing of the fields is expected to occur. After the internal reversed field and the corresponding j_θ disappear, the plasma should be hotter. In addition, if part of the j_θ is carried by ions, in order to conserve angular momentum a radial electric field E_r should develop, producing a guiding-center drift in the θ direction with frequency $\Omega = -cE_r/rB_z$. According to the MHD approximation, modes proportional to exp $[i(\omega t + m\theta)]$ obey the dispersion relation

$$\omega + m\Omega = \Omega(1 \pm \sqrt{1 - m}) \tag{5.23}$$

so that modes for which $m > 1$ are increasingly unstable. If we consider finite Larmor radius effects consistently (the predicted $\Omega = O[(a_i/r_0)^2\omega_{\mathrm{I}}]$) there is a stabilization effect similar to that discussed for mirrors and the result is that the plasma is stable for

$$\frac{1 - \sqrt{m}}{2} < \frac{\Omega}{\Omega_0} < \frac{1 + \sqrt{m}}{2} \tag{5.24}$$

where $\Omega_0 = 2(a_{\mathrm{I}}/r_0)^2\omega_{\mathrm{I}}$. For $m = 2$, $-0.207 < \Omega/\Omega_0 < 1.207$, and for $m = 3$, $-0.366 < \Omega/\Omega_0 < 1.366$. The observed value is $\Omega/\Omega_0 \cong -0.3$ when only the $m = 2$ mode is apparent.

5.3.4. Θ-Z and Hard-core Pinches

Θ-Z PINCH   The field configuration in the so-called stabilized, or Θ-Z, pinch[14] is indicated in Fig. 5.9a. It can be shown that, if $m = 0, 1$ modes are stable, all modes are stable so that it is sufficient to discuss these modes. If $\alpha_V = \alpha_p = 0$, $m = 0$ and $m = 1$ modes are unstable owing to the obvious changes of magnetic pressure accompanying the distributions of the B_θ field indicated in Fig. 5.9b. The B_z field and the conducting wall at $R = r$ produce stabilization. The internal B_z field has a pressure which inhibits the $m = 0$ deformation by creating internal pressure changes that oppose the external changes of the B_θ field. This stabilizes $m = 0$ for sufficiently large $\alpha_p{}^2 \gtrsim 0.5$. The external conductor does not help this mode. The tension of the internal B_z field inhibits the $m = 1$ deformations and eliminates large k instabilities by itself. This is not a large enough effect for the long-wavelength instabilities for which the external conductor is essential. Pinches designed to be in the stable range do not show stability. The reason is that the idealization of surface currents is never realized in the laboratory. If a finite-thickness current layer is considered corresponding to $B_\theta(r)$, $B_z(r)$, there is always some wavelength $(k, m/r)$ for which $\mathbf{k} \cdot \mathbf{B} = kB_z + (m/r)B_\theta = 0$. This resonance produces an additional instability whose growth rate is proportional to the thickness of the current layer and is consistent[15] with observed field structures and stability times of the order of a few microseconds observed with magnetic probes.

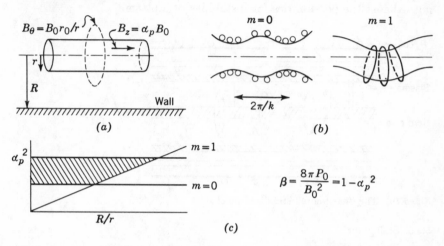

Fig. 5.9 The "stabilized" Θ-Z pinch. (a) Field configuration; (b) sausage and kink modes; (c) stability domain.

HARD-CORE PINCH   For general fields $B_\theta(r)$, $B_z(r)$ it is possible to see a simple sufficient condition for stability making use of the energy principle. We first vary U with respect to ξ_θ, ξ_z to obtain the minimum corresponding to the most dangerous displacements,

$$U_{\min} = \int r \, dr \left\{ \left[(\mathbf{k} \cdot \mathbf{B})^2 - \frac{2B_\theta}{r^2} \frac{d}{dr} (rB_\theta) \right] \xi_r^2 \right.$$
$$\left. + \frac{[(kB_z - mB_\theta/r)\xi_r + r(\mathbf{k} \cdot \mathbf{B}) \xi_r']^2}{k^2 r^2 + m^2} \right\} \quad (5.25)$$

It is apparent that there is only one term that is not positive definite and it will be positive definite if $(d/dr)(rB_\theta)^2 < 0$, for example, if $B_\theta = 1/r^n$, where $n > 1$. This condition is satisfied in the hard-core pinch for which a typical case is illustrated in Fig. 5.10.

For $0 < r < a$: $B_\theta = \dfrac{2I}{cr}$

For $a < r < b$: $B_\theta = \dfrac{2}{cr} (I - I_p)$

where $I_p = \displaystyle\int_a^r |j_z(r)| 2\pi r \, dr$

For $b < r < c$: $B_\theta = 0$

It is thus clear that the sufficient condition is satisfied.   Extensive experimental studies[16] have been carried out with linear hard-core pinches as well as with a toroidal configuration called *Levitron*.   Magnetic-probe measurements indicate much greater stability than with the ordinary pinch, but complete stability does not always obtain when the MHD criterion is satisfied.   It is thus clear that the MHD approximation does not exhaust the possibilities for instability of a plasma.

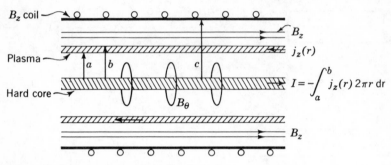

Fig. 5.10   The "hard-core" stabilized pinch.

5.3.5. Stellarator

A simple torus where the field lines close on themselves is unstable because there is a $\mathbf{B} \times \nabla B$ drift that produces charge separation and an electric field.   The resultant $\mathbf{E} \times \mathbf{B}$ drift causes both signs of charges to drift to the walls.   The simple torus is a degenerate case.   In general only one line of force has the property of closing on itself.   This is called the *magnetic axis*.   Other lines start at P_1 and after going once around the torus appear at P_2 at an angle ι with respect to P_1 (see Fig. 5.11*a*).   The angle ι is called the *rotational transform*, and the operation that carries P_1 to P_2 is called a *magnetic transform*.   If $\iota = 2\pi/n$, closure will be achieved after going n times around the torus.   In general, as we follow a line around the torus, it will generate a closed curve and all such closed curves lie in a magnetic surface.   A magnetic field with a rotational transform can be obtained by twisting a torus into a figure 8.   In this case ι is independent of the distance r from the magnetic axis.   An alternative procedure is to employ helical cusp fields from $2n$ helical conductors with pitch $2\pi n/K$ and currents flowing oppositely in adjacent conductors.   The case $l = 3$ is illustrated in Fig. 5.11*b* for which $\iota \propto r^2$.   The point of such fields is that particles must follow the field lines so that the $\mathbf{B} \times \nabla B$ drifts tend to cancel.   Alternatively, charge separations are inhibited because they cannot be maintained along field lines, but only between field lines.

If ohmic heating is employed which involves currents parallel to the $\mathbf{B}$ lines, the rotational transform will be destroyed for some critical current.   For example, a cylindrical shell of plasma carrying a current I_z produces $B_\theta = 2I_z/cr$.   Adding this to an externally maintained B_z field produces a spiral field of pitch B_θ/rB_z.   That is, in a distance $2\pi R$, or once around a torus, $\Delta\iota = \dfrac{2\pi R}{r}\dfrac{B_\theta}{B_z}$ is the angular transform of the field line. Therefore when $\dfrac{4\pi}{c}\dfrac{R}{r^2}\dfrac{I_z}{B_z} = \iota_0 + 2\pi n$, the initial transform will be unwound if $n = 0$, and for any integer n it will also be destroyed.   This is

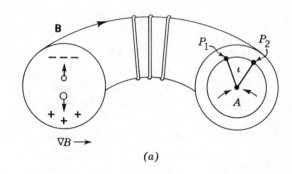

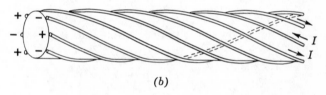

Fig. 5.11   (a) Section of torus showing charge separation caused by ∇B drift and explaining rotational transform ι;
(b) helical windings for hexapole ($l = 3$) cusp fields.

called the *Kruskal limit*,[17,18] for which kink instabilities have been predicted and observed.

On the basis of the MHD approximation stable stellarator designs have been demonstrated for values of $\beta = 4\pi P/B^2$ up to 0.1 to 0.2.   Under these circumstances the plasma is lost to the walls in times of the order of hundreds of microseconds rather than of microseconds when MHD stability is violated.   This phenomenon, which has been called *pump-out*, is further evidence that the MHD description is incomplete.

5.4   MICROINSTABILITIES

A large class of instabilities depend on the details of the velocity-space distribution function.   Although some of them may be treated with terminated moment equations for electrons and ions treated separately, it is in general necessary to employ the Vlasov equations. This aspect is mathematically complex so that we shall attempt to indicate only the essential features.

5.4.1. Instabilities with $\mathbf{B} = 0$

Consider the linearized Vlasov equations

$$\frac{\partial}{\partial t} \delta f_i + \mathbf{v} \cdot \frac{\partial}{\partial \mathbf{x}} \delta f_i + \frac{e_i}{m_i} \frac{\partial \delta \phi}{\partial \mathbf{x}} \cdot \frac{\partial f_i}{\partial \mathbf{v}}$$

$$\nabla^2 \delta \phi = -4\pi n \sum_i e_i \int \delta f_i \, dv \qquad (5.26)$$

where we include only electrostatic coupling of particles. Suppose that, at $t = 0$, the distribution is perturbed by $\delta f_i^{(0)}(x,v)$ and we wish to find the subsequent evolution of $\delta f_i(x,v,t)$, $\delta \phi(x,t)$. This problem can be solved with Fourier and Laplace transforms. Thus

$$\delta \bar{\phi}(\mathbf{k},p) = \frac{4\pi \rho_0(\mathbf{k},p)}{k^2 \epsilon(\mathbf{k},p)}$$

where $\quad \rho_0(\mathbf{k},p) = \sum_i e_i \int \frac{\delta f_i^{(0)}(k,v)}{p + ik \cdot v} \, dv$

is determined by the initial condition and

$$\epsilon(\mathbf{k},p) = 1 - \sum_i \frac{\omega_{pi}^2}{k^2} \int \frac{i\mathbf{k} \cdot \partial f_i/\partial \mathbf{v}}{p + i\mathbf{k} \cdot \mathbf{v}} \, dv \qquad (5.27)$$

is the dielectric coefficient which is characteristic of the unperturbed state $f_i(\mathbf{v})$. The Laplace inversion produces terms like $e^{p_n t}$, where p_n are solutions of $\epsilon(\mathbf{k},p) = 0$. If there are any solutions where p has a positive real part, the initial state $f_i(\mathbf{v})$ is unstable. In general p will be complex; there is no useful energy principle and no alternative to hard labor.

PLASMA-WAVE INSTABILITY   Suppose that $f_i = \delta(\mathbf{v})$ or that the particles are cold. Then $\epsilon(\mathbf{k},p) = 1 + \omega_p^2/p^2$, where $\omega_p^2 = \sum_i \omega_{pi}^2$. The normal modes are $p = \pm i\omega_p$. For well-behaved distribution functions[19]

$$p = \pm i\omega_p + \frac{\pi}{2} \frac{\omega_p^2}{k^2} \frac{\partial F_e}{\partial u} \bigg|_{u = \omega_p/k}$$

where $\quad F_e(u) = \int f_e(v) \delta\left(\frac{u - \mathbf{k} \cdot \mathbf{v}}{k}\right) dv.$ $\qquad (5.28)$

The growth term comes from the pole at $\mathbf{k} \cdot \mathbf{v}/k = ip \cong \omega_p$ in Eq. (5.27). Thus, if $\partial F_e/\partial u < 0$ for all u as in the case of a Maxwell distribution, the plasma waves are stable. The plasma will be unstable if it contains a small bump due, for example, to runaway electrons from ohmic heating (cf. Fig. 5.12a). The instability has a simple physical interpretation (see Fig. 5.12b). Plasma waves have phase velocity ω_p/k. Particles moving

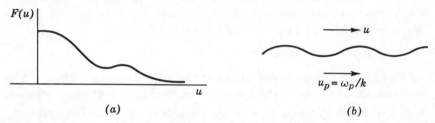

(a) (b)

Fig. 5.12   (a) Velocity-distribution function $F(u)$ with a "bump"; (b) particle and phase velocities that can give rise to trapping instabilities.

parallel to the wave with $u \cong \omega_p/k$ couple with the wave and produce a resonant transfer of energy. The wave grows if the particles which go slightly faster than ω_p/k outnumber those which go slower. Such instabilities have been called *trapping instabilities*. A large family of them exists which can be classified in terms of the normal modes of a cold plasma. It should be noted that they cannot be obtained by expanding in powers of the temperature Θ_e. For example, in the case of a Maxwell distribution of electrons the damping is

$$\sqrt{\frac{\pi}{8}} \frac{\omega_p{}^4}{(k\bar{v})^3} \exp\left[-\frac{\omega_p{}^2}{2(k\bar{v})^2} \right] \qquad \text{where } m_e\bar{v}^2 = \Theta_e$$

TWO-STREAM INSTABILITY   Consider the singular distribution

$$f(\mathbf{v}) = \tfrac{1}{2}[\delta(\mathbf{v} - \mathbf{V}) + \delta(\mathbf{v} + \mathbf{V})] \tag{5.29}$$

Substituting into Eq. (5.27), we find the dispersion relation

$$\frac{\omega_p{}^2}{(p + i\mathbf{k} \cdot \mathbf{V})^2} + \frac{\omega_p{}^2}{(p - i\mathbf{k} \cdot \mathbf{V})^2} = -2$$

The solution is

$$p^2 = -\left[(\mathbf{k} \cdot \mathbf{V})^2 + \frac{\omega_p{}^2}{2} \right]$$
$$\pm \sqrt{\left[(\mathbf{k} \cdot \mathbf{V})^2 + \frac{\omega_p{}^2}{2} \right]^2 + (\mathbf{k} \cdot \mathbf{V})^2[\omega_p{}^2 - (\mathbf{k} \cdot \mathbf{V})^2]} \tag{5.30}$$

There will be an unstable mode whenever $\omega_p > \mathbf{k} \cdot \mathbf{V}$. For small $\mathbf{k} \cdot \mathbf{V}$, there are two branches, the stable one with $p = \pm i(\omega_p/2)$ and the unstable one with $p = \pm \mathbf{k} \cdot \mathbf{V}$. The growth rate is independent of plasma properties. This is clearly quite different from the trapping instability. It is more like the hydrodynamic instability of coupled fluids, such as wind blowing over water and producing whitecaps. As long as there is some coupling, the nature of it is unimportant.

ION WAVES   We assume that electrons are much hotter than ions. Ions in slow motion are accompanied by their shield clouds. Consider a displacement of the ions $\delta\xi \sim e^{pt+ikx}$, which changes the ion density by $\delta n_I = -ikn_I\,\delta\xi$. The equation of motion for the ions is

$$n_I m_I p^2\,\delta\xi = -n_I eik\,\delta\Phi$$

where $\delta\Phi$ is the change in potential due to the displacement of ions. The electrons shield the ions, which can be described by a dielectric coefficient $\epsilon_e = 1 + 1/(kL_e)^2$, where $L_e = \sqrt{\Theta_e/4\pi n_e e^2}$ is the electron Debye length. That is, $\delta\Phi = 4\pi e\,\delta n_I/k^2\epsilon_e$. Upon combining these equations, the result is

$$p^2 = -\frac{\omega_{pI}^2}{\epsilon_e(k)} = -\omega_L{}^2 \tag{5.31}$$

These oscillations are strongly damped unless $\Theta_e \gg \Theta_I$, which is apparent if they are obtained from Eq. (5.27). If hot electrons have a drift relative to cold ions of velocity $V \ll v_e$, $\epsilon_e(k)$ must be replaced by

$$\epsilon_e(k, -ik \cdot V) = 1 + \frac{1}{(kL_e)^2}\left(1 + \sqrt{\frac{\pi}{2}}\,i\,\frac{V}{v_e}\right) \tag{5.32}$$

so that ion oscillations grow with the rate $\operatorname{Re} p \sim \dfrac{\omega_L}{1 + (kL_e)^2}\,\dfrac{V}{v_e}$, where $\operatorname{Im} p = \omega_L$, the Langmuir frequency. For ions of finite temperature $\Theta_I = m_I v_I{}^2$ it can be shown by solving Eq. (5.27) that the threshold of instability is $V \sim v_I$. Unstable ion oscillations have recently been observed in the study of plasma diodes.[20]

TRANSVERSE-WAVE INSTABILITY[21]   The dispersion relation analogous to Eq. (5.27) for transverse waves is

$$1 + \frac{\omega_p^2 + k^2 c^2}{p^2} + \sum_i \frac{\omega_{pi}^2}{p^2}\int d\mathbf{v}\,\frac{v_\perp{}^2}{2}\,\frac{i\mathbf{k}\cdot\partial f_i/\partial \mathbf{v}}{p + i\mathbf{k}\cdot\mathbf{v}} = 0 \tag{5.33}$$

For $\quad f = \dfrac{m}{2\pi\Theta_\perp}\,\delta(v_z)\exp\left(-\dfrac{mv_\perp{}^2}{2\Theta_\perp}\right)$

$p \cong \omega_p k_z L_\perp$, where $L_\perp{}^2 = \Theta_\perp/4\pi n e^2$ and we have assumed $\mathbf{k} = (0,0,k_z)$, $kL_\perp < 1$. This mode can be described [22] as the spontaneous formation of pinches with current parallel and antiparallel to the axis of maximum thermal velocity. As long as electron and ion temperatures are finite, any anisotropy leads to instability. However, this mode can be stabilized with a $\mathbf{B}$ field perpendicular to the low-temperature direction.

It should be noted that there are no trapping instabilities for transverse waves because there are no particles going with the phase velocity $\sqrt{(\omega_p^2 + k^2 c^2)/k^2} > c$.

5.4.2. Instabilities with $B \neq 0$

FIRE-HOSE INSTABILITY   Consider an Alfvén wave propagating parallel to $\mathbf{B}$ when $P_\parallel > P_\perp$. Particles following the field lines experience a centripetal force $mv_\parallel^2/R$ so that there is a reaction force $F_c \cong P_\parallel/R$ which makes the perturbation grow (see Fig. 5.13). Opposing this are the

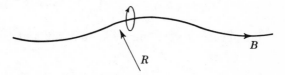

Fig. 5.13   Curved field lines giving rise to the fire-hose instability.

restoring forces $F_\mu \sim \nabla n\mu \cdot \mathbf{B} \sim P_\perp/R$ and $F_T \sim \nabla(B^2/8\pi) \sim (B^2/8\pi R)$. The perturbation will grow if

$$P_\parallel - P_\perp > \frac{B^2}{8\pi}$$

MIRROR INSTABILITY   Consider a hydromagnetic wave propagating perpendicular to $\mathbf{B}$ when $P_\perp > P_\parallel$. The perturbation shown in Fig. 5.14

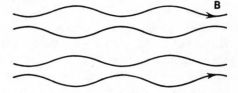

Fig. 5.14   Field lines in the mirror instability.

produces magnetic mirrors that concentrate particles where the field is weakest. The perturbation will grow if

$$\frac{P_\perp^2}{P_\parallel} > P_\perp + \frac{B^2}{8\pi} \tag{5.34}$$

This instability has been deliberately induced and observed in the mirror machine.[23]

TRAPPING INSTABILITIES   In the absence of a magnetic field, resonant transfer of energy takes place for particles such that $\omega - \mathbf{k} \cdot \mathbf{v} = 0$, where ω is the frequency of a normal mode such as a plasma wave. In the presence of a magnetic field in the z direction, the corresponding condition is $\omega - k_z v_z - n\Omega = 0$, where Ω is the electron or ion cyclotron frequency and ω corresponds to plasma, Alfvén, electromagnetic waves, etc. Since this subject is mathematically complex, the present treatment will be limited to mentioning a few results.

For $T_\perp/T_\parallel = \infty$ plasma oscillations with frequencies equal approximately to integral multiples of Ω_I are unstable.[24] The onset of instability[25] is at $T_\perp/T_\parallel \cong 8$. There are also instabilities at multiples of Ω_e for which $T_\perp/T_\parallel > 2$ is required.

Alfvén waves are unstable for any anisotropy, but the growth rate is

$$\mathrm{Re}\; p \sim \sqrt{\frac{\pi}{8}}\, \frac{T_\perp}{T_\parallel}\, \Omega_I \exp\left[-\frac{1}{(1 - T_\perp/T_\parallel)^2} \right] \tag{5.35}$$

which becomes very small for small differences in $T_\perp$, $T_\parallel$.

5.4.3. Instabilities of a Slightly Inhomogeneous Plasma

Since the study of velocity-space instabilities is too complex for a finite plasma, considerable effort has gone into the study of an infinite plasma, where the field varies slowly, for example $B\,(x) = B_0(1 + \varepsilon x)$ where $\varepsilon \ll 1$. The objective was to see whether or not there are universal instabilities that always obtain in an inhomogeneous field and therefore always rule out the possibility of a confined plasma. The first result was that $p(k) = p_0(k) + \varepsilon^2 p_1(k)$,[26] where $p_0(k)$ is the normal mode of the plasma for $\varepsilon = 0$. If p_0 has no positive real part, this is not altered up to terms of order ε^2. This was a precursor to Low's[4] more general theorem that stability is unaltered to all orders in ε. This, however, excludse trapping instabilities[27] which are not expandable and new modes[28] which do not appear to lowest order in ε. The new physical feature of a drift in the unperturbed orbits implied by $B(x)$ suggests that there should be new modes. New modes in the limit $\beta = 8\pi P/B^2 \to 0$ are listed in the following table. V_D is the drift velocity, V_I is the mean ion velocity, and $p = i\omega + \gamma$.

Type	Characteristic	Stability	Growth or Damping Rate, γ
Ion cyclotron	Longitudinal $\omega \sim \Omega_I \sim kV_D$	Unstable	$\dfrac{m_e}{m_I}\Omega_I$
Thermal convection	Longitudinal $\omega = 0$	$\dfrac{d\,(\ln P)}{d\,(\ln T)} > 2$	$+kV_D$
Drift waves	Longitudinal $\omega \sim kV_D$	Stable	$-(kR_L)^2$
Ion convection waves	Longitudinal $\omega \sim kV_I$	Unstable	$kV_D \left(\dfrac{m_e}{m_I}\right)^2 \dfrac{e^{-2/\beta}}{\beta^3}$
Transverse	$\mathbf{E} \parallel B_0,\; ka_e \sim 1$	Unstable	$kV_D \left(\dfrac{m_e}{m_I}\right)^{3/2}$

5.4.4. Quasilinear Theory

A veritable jungle of microinstabilities is predicted by the Vlasov equations. It seems unlikely that they could all be avoided, although there is very little direct evidence for them in the laboratory. It is particularly important to investigate the time evolution of an instability to determine how serious the consequences are. Considerable[29,30] progress has recently been made in this connection. Let $F(x,vt) = F_0(v) + f(x,v,t)$,

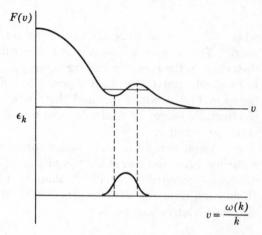

Fig. 5.15   Velocity-distribution function $F(v)$ with two maxima, and corresponding dielectric function ϵ_k.

where $F_0(v)$ is the initial distribution function as illustrated in Fig. 5.15. The Fourier component of $f(x,v,t)$ satisfies

$$\frac{\partial f_k}{\partial t} + i\mathbf{k} \cdot \mathbf{v} f_k = \frac{e}{m} \mathbf{E}_k \cdot \nabla_v F_0 + \frac{e}{m} \sum_{k'} \mathbf{E}_{k-k'} \cdot \nabla_v f_{k'} \qquad (5.36)$$

If the last term on the right-hand side is omitted, the linearized solutions are $\exp[p(k)t + i\mathbf{k} \cdot \mathbf{x}]$, where $p(k) = i\omega(k) + \gamma(k)$,

$$\gamma(k) = \frac{\pi}{2} \frac{\omega^2(k)}{k^2} \frac{\partial F_0}{\partial v}\bigg|_{v=\omega/k}$$

and the electrostatic energy $\varepsilon_k = |E_k|^2/8\pi$ satisfies $d\varepsilon_k/dt = 2\gamma_k\varepsilon_k$. If coupling between modes of different k is omitted except for $k = 0$, Eq. (5.36) reduces to

$$\frac{\partial f_k}{\partial t} + i\mathbf{k} \cdot \mathbf{v} f_k = \frac{e}{m} \mathbf{E}_k \cdot \nabla_v(F_0 + f_0) \qquad \text{for } k \neq 0 \qquad (5.37)$$

and $$\frac{\partial}{\partial t} f_0 = \frac{e}{m} \sum_{k'} \mathbf{E}_{-k'} \cdot \nabla_v f_{k'} \qquad \text{for } k = 0 \qquad (5.38)$$

This approximation has been justified[29] for $\gamma(k)/\omega(k) \ll 1$.   Equations (5.37) and (5.38) can be solved by a WKB approximation, with the result that the spatially homogeneous part $g = F_0 + f_0$ satisfies a diffusion equation

$$\frac{\partial g}{\partial t} = \frac{\partial}{\partial \mathbf{v}} \left[8\pi^2 \frac{e^2}{m^2} \sum_k \varepsilon_k \delta(\omega_k - \mathbf{k} \cdot \mathbf{v}) \right] \frac{\partial g}{\partial \mathbf{v}} \tag{5.39}$$

and $$\frac{\partial \varepsilon_k}{\partial t} = 2\gamma_k \varepsilon_k \tag{5.40}$$

where $\gamma_k(t)$ now corresponds to the instantaneous distribution function $g(v,t)$.   The average effect of the fast oscillations is to produce a slow distortion in the spatially homogeneous part of the distribution.   Equations (5.39) and (5.40) then describe a diffusion process in which the bump in $F(v)$ is wiped out and there is a corresponding growth in the electrostatic energy of unstable modes until a quasistationary turbulent state is reached.

Anomalous diffusion in such a turbulent state arising from electrostatic ion cyclotron waves has recently been calculated.   The anomalous diffusion predicted is smaller[31] than that observed in the stellarator (Bohm diffusion) by a factor $(v_D/v_e)^6(\Theta_e/\Theta_I)^2$, where v_D is the drift of electrons relative to ions

5.5 RESISTIVE INSTABILITIES

If the assumption of perfect conductivity is relaxed, plasma can slip across magnetic field lines and now instabilities obtain in magnetohydrodynamically stable situations.

Consider, for example, a stabilized pinch.   If a kink mode develops without carrying the B_z field with it (see Fig. 5.16a), the original j_z current now has a θ component and the $j_\theta \times B_z$ force will make the perturbation grow.   In the case of a sausage instability the resultant $j_r \times B_z$ force will cause the plasma to spin, which is destabilizing (Fig. 5.16b).

In the case of a magnetic field which has shear or which changes direction as illustrated in Fig. 5.16c, the magnetic energy could be reduced by allowing the fields to mix and annihilate.   This is prevented by perfectly conducting plasma.   However, for finite conductivity an instability of the type shown in Fig. 5.16d will develop.[32]

Many cherished ideas about stability that depend on perfect conductivity must be abandoned for a cold plasma.   For example, finite electric fields can now exist along magnetic field lines so that terminating the field lines of a mirror machine in conducting plates does not necessarily

eliminate flutes for a low-β plasma.   The addition of simple shear does not
help because of the resistive instability previously discussed.   Ioffe[33] and
Yushmanov have observed flutes in a mirror machine and succeeded in
eliminating them by adding cusp fields with external longitudinal conductors having opposite currents in adjacent conductors.   This configuration has no shear stabilization and produces stability only because the
field increases away from the axis.   It is a *minimum-B* configuration.

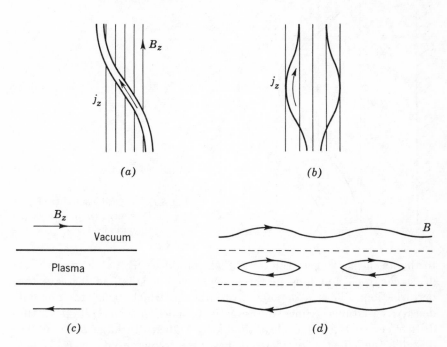

Fig. 5.16   Resistive instabilities.   (a) Stabilized pinch with "kink" in $\mathbf{j}$ $(m = 1)$;
(b) stabilized pinch with "sausage" in $\mathbf{j}$ $(m = 0)$; (c) reversed-field layer;
(d) resistive "tearing" instability of reversed-field layer.

The importance of resistive instabilities has been recognized only
recently so that this subject is still in its infancy.

Another type of instability in cold plasma has been observed and
explained in considerable detail.   It is a helical instability observed in
a positive column[34] and in low-pressure discharge tubes with streaming[35]
in the axial direction.   The observation is that for axial magnetic fields
$B < B_c$ diffusion to the walls is in accordance with classical theory.   For
$B \geq B_c$ the diffusion rate is anomalously high, and a spinning motion is
apparent from probe measurements.   A quantitative theory of the observations has been given,[36] with the following physical interpretation:[37]

Consider helical perturbations of electron and ion density as depicted in Fig. 5.17. The E_z field tends to lift the electron helix relative

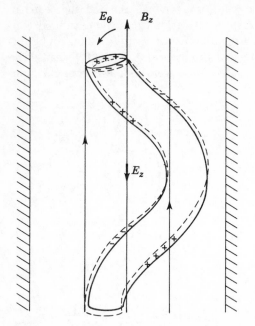

Fig. 5.17 Helical instability of discharge column in magnetic field.

to the ion helix. The resultant E_θ field produces $c\mathbf{E} \times \mathbf{B}/\mathbf{B}^2$ drifts that drive electrons and ions in the same direction so as to increase the perturbation. This is opposed by diffusion, which tends to iron out density variations. Since diffusion is proportional to $1/B_z{}^2$ and the drift effect to $1/B_z$, it is clear that for a sufficiently large $B_z > B_c$ the destabilizing effect triumphs. It has been shown that an E_z is not necessary, but streaming[38] of electrons and ions at different rates produces the same effect of separating the helices in the z direction. This is consistent with the low-pressure experiments.[35] This instability does not obtain in a perfectly conducting plasma but vanishes like $1/\sigma$. Although it may be unimportant in a thermonuclear plasma, it may be the source of many current difficulties such as pump-out in the stellarator.

REFERENCES

1. Lewis, D. J.: *Proc. Roy. Soc. (London)*, **A202,** 81 (1950).
2. Chew, G. F., M. L. Goldberger, and F. E. Low: *Proc. Roy. Soc. (London)*, **A236,** 112 (1956).
3. Rosenbluth, M. N., and N. Rostoker: *Phys. Fluids*, **2,** 23 (1959).

4. Low, F. E.: *Phys. Fluids*, **4**, 842 (1961).
5. Krall, N. A.: Stability of Generalized Rail Guns, *Gen. Atomic Internal Rept.* GAMD-1595, Aug. 16, 1960.
6. Albares, D. J., N. A. Krall, and C. L. Oxley: *Phys. Fluids*, **4**, 1033 (1961).
7. Rosenbluth, M. N., and C. L. Longmire: *Ann. Phys.*, **1**, 120 (1957).
8   Berkowitz, J., K. O. Friedrichs, H. Goertzel, H. Grad, J. Killeen, and E. Rubin: *Proc. Second Intern. Conf.*, Geneva, 1958, vol. 31, p. 171.
9. Berkowitz, J., H. Grad, and H. Rubin: *Proc. Second Intern. Conf.*, Geneva, 1958, vol. 31, p. 177.
10. Rosenbluth, M. N., N. A. Krall, and N. Rostoker: *Nuclear Fusion Suppl.*, **1**, 143 (1962).
11. Kuo, L. G., et al.: *Phys. Fluids*, **7**, 988 (1964); F. H. Coensgen: *Phys. Fluids*, **9**, 187 (1966); F. R. Scott et al.: Flute Characteristics of and Microwave Emission from a Plasma in a Mirror in "Plasma Physics and Controlled Nuclear Fusion Research," Intern. Atomic Energy Agency, Vienna, 1966 (*Proc. Second Intern. Conf. on Controlled Fusion, Culham, England, 1965*), to be published.
12. Little, E. M., W. E. Quinn, F. L. Ribe, and G. A. Sawyer: *Nuclear Fusion Suppl.*, **2**, 497 (1962).
13. Rostoker, N., and A. C. Kolb: *Phys. Rev.*, **125**, 2208 (1962).
14. Rosenbluth, M. N.: Stability of the Pinch, *Los Alamos Sci. Lab. Rept.* LA-2030, 1956.
15. Rosenbluth, M. N.: *Proc. Second Intern. Conf.*, Geneva, 1958, vol. 31, p. 85.
16. Colgate, S. A., and H. P. Furth: *Phys. Fluids*, **3**, 982 (1960); K. L. Aitken et al.: *Nuclear Fusion* **5**, 330 (1965); H. P. Furth: Experiments in Toroidal Plasma Confinement, in "Plasma Physics," pp. 411–420, International Atomic Energy Agency, Vienna, 1965.
17. Kruskal, M. D., J. L. Johnson, M. B. Gottlieb, and L. M. Goldman: *Phys. Fluids*, **1**, 421 (1958).
18. Ware, A. A., H. K. Forsen, and A. A. Schupp: *Phys. Rev.*, **125**, 417 (1962).
19. Landau, L. D., *J. Phys. (USSR)*, **10**, 25 (1946).
20. Garvin, H. L., W. B. Teutsch, and R. W. Pidd: *J. Appl. Phys.*, **31**, 1508 (1960); N. D'Angelo: *Phys. Fluids*, **4**, 1054 (1961).
21. Weibel, E. S.: *Phys. Rev. Letters*, **2**, 83 (1959).
22. Furth, H. P.: *Phys. Fluids*, **6**, 48 (1963).
23. Post, R. F., and W. A. Perkins: *Phys. Rev. Letters*, **6**, 85 (1961).
24. Harris, E. G.: *J. Nuclear Energy*, **C2**, 138 (1961).
25. Drummond, W. E., and M. N. Rosenbluth: *Bull Am. Phys. Soc.* (2), **6**, 185 (1961).
26. Krall, N. A., and M. N. Rosenbluth: *Phys. Fluids*, **4**, 163 (1961).
27. Krall, N. A., and M. N. Rosenbluth: *Phys. Fluids*, **5**, 1435 (1962).
28. Rosenbluth, M. N., and N. A. Krall: *Phys. Fluids*, **6**, 254 (1963).
29. Drummond, W. E., and D. Pines, *Nuclear Fusion Suppl.*, **3**, 1049 (1962).
30. Vedenov, A. A., E. P. Velikov, and R. Z. Sagdeev: *Nuclear Fusion Suppl.*, **2**, 465 (1962); A. A. Vedenov: *J. Nuclear Energy*, **C5**, 196 (1963).
31. Drummond, W. E., and M. N. Rosenbluth: *Phys. Fluids*, **5**, 1507 (1962).
32. Furth, H. P., M. N. Rosenbluth, and J. Killeen: *Phys. Fluids*, **6**, 459 (1963).

33. Ioffe, M. S., and E. E. Yushmanov: *Nuclear Fusion Suppl.*, **1,** 177 (1962); Yu. T. Baiborodov, M. S. Ioffe, V. M. Petrov, and R. I. Sobolev: *J. Nuclear Energy*, **C5,** 409 (1963).

34. Lehnert, B.: *Proc. Second Intern. Conf.*, Geneva, 1958, vol. 32, p. 349; G. A. Paulikas and R. V. Pyle: *Phys. Fluids*, **5,** 348 (1962).

35. Neidigh, R. V., and C. H. Weaver: *Proc. Second Intern. Conf.*, Geneva, 1958, vol. 32, p. 315.

36. Kadomtsev, B. B., and A. V. Nedospasov: *J. Nuclear Energy*, **C1,** 230 (1961).

37. Hoh, F. C., and B. Lehnert: *Phys. Rev. Letters*, **7,** 75 (1961).

38. Guest, G., and A. Simon: *Phys. Fluids*, **5,** 503 (1962).

6

MHD CHARACTERISTICS AND SHOCK WAVES

KANTROWITZ and PETSCHEK

ARTHUR KANTROWITZ, *Vice-president and Director, Avco Corporation, and Director, Avco-Everett Research Laboratory Everett, Massachusetts*

HARRY E. PETSCHEK, *Principal Research Scientist, Avco-Everett Reseach Laboratory, Everett, Massachusetts*

6

6.1 INTRODUCTION

In ordinary gas dynamics the theory of nonlinear flows which either are time-dependent or involve supersonic velocities is well developed. Such flows tend to contain planes across which the significant changes in gas conditions occur almost discontinuously. These discontinuities, which are called *shock waves*, become so thin (of the order of a few mean-free paths) that the dissipation rates within them become extremely large. One of the prime reasons for interest in shock waves is that they provide a mechanism for converting flow kinetic energy ahead of the wave into thermal energy behind the wave and thus provide a controlled means of producing high-temperature gases. Our ability to deal with flows containing shock waves is greatly facilitated by two facts. First, the changes in flow properties across a shock are independent of the detailed structure of the shock wave. Thus, the conservation laws yield algebraic relations which define conditions behind a shock in terms of its velocity and the conditions ahead. Second, the shock waves are usually (when the typical flow dimension is large compared with a mean-free path) so thin compared with the overall flow field that they can be treated as sheets across which flow properties change discontinuously.

In the remainder of the flow field, although the fluid properties may change by large amounts, the gradients will be small. These regions may

148

be treated by a method which is essentially a generalization to the nonlinear case of concepts developed in the description of linear-wave propagation in nondispersive media.   In the propagation of electromagnetic signals in vacuum or in acoustics a disturbance of arbitrary shape at one instant of time will have the same shape at a later time but will be displaced by a distance equal to the product of the propagation speed and the time difference (Fig. 6.1a).   Now, if this disturbance propagated through a

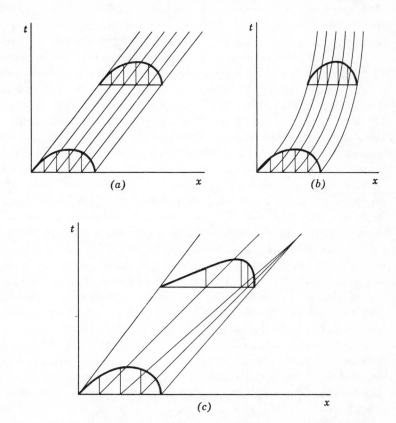

Fig. 6.1   Illustration of propagation of an arbitrary pulse in several cases.
(*a*) For linear waves in a uniform medium, the pulse shape is retained.   (*b*) For linear waves in a medium of variable propagation speed, the pulse becomes somewhat distorted; however, discontinuities do not arise.   (*c*) For a nonlinear pulse in an initially uniform medium, changes in propagation speed can lead to the formation of discontinuities or shock waves.

medium in which the propagation speed varied with position, owing to gradual changes in the index of refraction or gas temperature it would become distorted as it propagates.   For a nondispersive medium (i.e., a medium in which the propagation speed is independent of the wavelength of the disturbance) the changes in shape could be obtained by considering the disturbance to be composed of a large number of small step functions and following each step on a distance-time diagram (Fig. 6.1b). The slope of the trajectory of each step is equal to the local propagation speed. This technique would not work for a dispersive medium, since the different Fourier components of the step functions would be spread out in space. It is significant to note, however, that the applicability of the above method is not restricted to small variations in the propagation speed. Thus, arbitrarily large variations in propagation speed can be taken into account.

For the nonlinear gas dynamic problem, let us first remember that a monatomic gas is a nondispersive medium for wavelengths long compared with the mean-free path.† As an example, we imagine a long pipe containing an initially uniform gas, in which a disturbance is generated by moving a piston at one end of the pipe. This disturbance may be viewed as consisting of a large number of small-amplitude step-function waves all propagating away from the piston. Each wave will propagate through the fluid at a velocity equal to the sound speed determined by the local fluid properties. Since the medium is nonlinear, the propagation speed of a particular wave will depend upon the changes produced by the previous waves. Thus different waves will propagate at different velocities, and the disturbance will change its shape with time. As long as the waves do not cross, the number of waves preceding any particular wave is independent of time. This implies that conditions ahead of the wave and, therefore, its propagation speed are constant. The distortion of the disturbance with time is therefore determined by following each point on the disturbance along a straight line in the distance-time plane (Fig. 6.1c). Since the velocity for each point is different, some of these lines will diverge, while others converge. Note that this is fundamentally different from the linear case of wave propagation through a spatially varying medium. In that case the trajectories of all the waves composing the disturbance are parallel at a particular point in space. In the nonlinear case the converging lines can intersect. They cannot, however, cross one another, since the wave from behind is only overtaking the previous wave because of the changes in fluid properties produced by the first wave. Thus, if the second wave did go ahead of the first wave, it would then propagate more slowly. Thus, the converging waves tend to pile up and produce a large-amplitude discontinuity or shock wave.

The concepts which have been described lead in the case of converging waves to a description of the formation of shock waves. In the

† For polyatomic gases we would have to restrict ourselves to time scales that are either very long or very short compared with the relaxation times for internal degrees of freedom.

case of diverging waves they can be used to describe flow changes of arbitrarily large amplitude.   Thus far, we have discussed only the situation in which all waves are propagating in one direction.   In the more general case, for example, after the disturbance reflects from the far end of the pipe, there will be waves propagating in both directions.   In this case the general concept of following the trajectories of small step-function waves is still useful.   However, it usually requires numerical integration along a two-dimensional grid.

Mathematically, this description of gas-dynamics flows is a special case of the theory of characteristics which is applicable to certain types of hyperbolic partial differential equations.   It was first noticed by Friedrichs[1] that the mathematical theory was also applicable to the magnetohydrodynamic equations.   Much of the basic work on the description of nonlinear wave propagation in magnetohydrodynamics was covered in his initial paper.   The purpose of the present chapter is to review the present status of this theory.   In the presentation given here we shall attempt to emphasize a physical description of the subject.   In this way we hope that it may be somewhat easier for the reader who is not familiar with the formal mathematical theory of characteristics to become acquainted with its application to gas dynamics and magnetohydrodynamics.   Although we shall not assume a knowledge of the theory of characteristics in ordinary gas dynamics, the reader may wish to refer to some of the following discussions on the subject.[2]

Since the theory is an extension of the linear analysis, we shall begin by deriving the properties of the linear waves in Sec. 6.2.   We may note at this point that the magnetohydrodynamic case will be considerably more complex than the ordinary gas-dynamic case.   The presence of the magnetic field in the plasma defines a direction within a plasma.   Thus, the wave propagation speed as well as its properties will depend upon the direction in which the wave is propagating relative to the magnetic field. Furthermore, the presence of the magnetic field requires the existence of three distinct propagation modes of small-amplitude waves, as opposed to only one sound speed which exists in the absence of the field.   A simple explanation of this can be given if we imagine that the waves are produced by a piston which forms one boundary of the plasma.   The piston has 3 degrees of freedom, and therefore we would expect that a wave mode is required for each degree of freedom.   However, in the absence of a magnetic field, motions of the piston parallel to its plane are not observed in the gas (except within a diffusion layer immediately adjacent to the piston).   Thus, for an ordinary gas, only motions of the piston normal to itself produce propagating waves which are observable at appreciable distances from the piston.   However, in the case of a plasma if the magnetic field has a component normal to the plane of the piston and the piston is a conductor, the motions of the piston in its plane require a corresponding motion of the field lines. Since the plasma is frozen to the field lines, it must move with the piston. As a result such motions also produce waves, and therefore a wave exists for each degree of freedom of the piston motion.

In Sec. 6.3, we shall discuss the extension from the linear to the nonlinear case. In particular, we shall show that two of the modes of linear propagation lead to the formation of shock waves. The third mode, on the other hand, is linear even for large amplitude. As a result this third mode does not steepen to form shock waves, but even for large-amplitude waves the shape is maintained as the wave moves through the plasma.

In Sec. 6.4 we shall discuss the resulting two types of shock waves. We may note at this point that, whereas ordinary gas-dynamics shock waves represent an interchange of energy between flow kinetic energy and thermal energy, in the plasma magnetic energy is also present. Thus, there are situations in which some of the flow kinetic energy is converted into magnetic energy, reducing the amount of thermal energy which is produced. On the other hand, there are also shock waves for which the basic energy source is the magnetic field, and thus the magnetic energy is converted into thermal energy and flow kinetic energy.

In Sec. 6.5, we shall discuss briefly the internal structure of shock waves.   Finally, in Sec. 6.6 we shall discuss the application of the theory in two cases; first to the production of high-temperature plasma samples in the laboratory, and second to the rate at which magnetic energy can be converted to plasma energy. The latter case has possible interest in several astrophysical situations.

Throughout the body of the text (with the exception of a portion in Sec. 6.5) we shall assume that the fluid can be described by the magneto-hydrodynamic equations.   The delineation of the range of plasma conditions over which this is a valid assumption is at present not clearly understood.   Over an interesting but somewhat limited range of temperatures and densities the validity can be justified with reasonable certainty on the basis of rapid Maxwellization by binary collisions.   There are, however, theoretical arguments as well as some experimental evidence which strongly suggest that the actual range of validity is considerably larger, including very high-temperature plasma conditions in which binary collisions are rare provided only that the gyro radii of the ions and the Debye length of the plasma are small compared with the lengths in which the overall properties of the flow field change significantly.   Further remarks on the justification of the validity of the magnetohydrodynamic equations are given in the Appendix (Sec. 6.7).

In what follows we shall not in general give detailed references to the specific papers in which a particular conclusion was originally reached. An extensive though not complete Bibliography appears at the end of the chapter. This chapter does not contain any concepts which have not been published elsewhere; however, the presentation may have been modified.

6.2 LINEAR WAVES

The magnetohydrodynamic equations for a nondissipative medium (i.e., infinite electrical conductivity and vanishing viscosity and heat conductivity) are

Continuity: $\dfrac{\partial \rho}{\partial t} + \nabla \cdot \rho \mathbf{v} = 0$ (6.1)

Momentum: $\rho \dfrac{\partial \mathbf{v}}{\partial t} + \rho \mathbf{v} \cdot \nabla \mathbf{v} + \nabla p - \dfrac{(\nabla \times \mathbf{B}) \times \mathbf{B}}{4\pi} = 0$ (6.2)

Induction: $\dfrac{\partial \mathbf{B}}{\partial t} - \nabla \times (\mathbf{v} \times \mathbf{B}) = 0$ (6.3)

Entropy: $\dfrac{\partial (p/\rho^\gamma)}{\partial t} + \mathbf{v} \cdot \nabla \dfrac{p}{\rho^\gamma} = 0$ (6.4)

$$\nabla \cdot \mathbf{B} = 0 \qquad (6.5)$$

where ρ = density
  $\mathbf{v}$ = fluid velocity
  p = plasma pressure
  $\mathbf{B}$ = magnetic field intensity
  γ = ratio of specific heats C_p/C_v

The neglect of dissipative terms requires that the gradients be small. Thus these equations will not apply to shock waves, and for cases in which steepening occurs the resulting shock waves must be isolated and treated separately.

As indicated earlier, the theory of characteristics describes the nonlinear flow as composed of small-amplitude step-function waves. We shall therefore begin by examining the small-amplitude waves which result from Eqs. (6.1) through (6.5). We also noted that the scheme will apply only to a nondispersive medium.   It can be seen by examining the equations that it is impossible to form either a basic length or a basic time from the quantities in the equations.   Therefore, if we examined the linear waves by the more usual method of assuming sinusoidally shaped waves, we would find that the phase velocity of these waves is independent of wavelength; i.e., the medium is nondispersive.   The extension to the nonlinear case is more direct if we begin by deriving the linear wave properties in terms of step functions.

It is convenient to analyze the wave in a coordinate system in which the wave is stationary.   In this coordinate system the fluid will of course be moving on both sides of the wave.   The fluid velocity ahead of the wave must be equal and opposite to the propagation speed of the wave relative to the fluid in order for the wave to be stationary.   We shall denote this velocity by c to indicate that it is the characteristic propagation speed of the wave.   Let us further orient our coordinate system such that the wave normal is in the x direction and such that the magnetic field ahead of the wave is in the xy plane; that is, $B_z = 0$ ahead of the wave.   In our coordinate system the time derivatives are zero and the ∇ operator reduces to a derivative in the x direction.   If we now integrate Eqs. (6.1) through (6.5) across the wave, we obtain relations for the changes in the flow properties across the wave.   Keeping terms only to first order in the changes, we thus obtain for the continuity equation

$$\delta(\rho v_x) = -c\ \delta\rho + \rho\ \delta v_x = 0 \tag{6.6}$$

The three components of the momentum equation become

$$-\rho c\ \delta v_x + \delta p + \frac{B_y\ \delta B_y}{4\pi} = 0 \tag{6.7}$$

$$\rho c\ \delta v_y + \frac{B_x\ \delta B_y}{4\pi} = 0 \tag{6.8}$$

$$\rho c\ \delta v_z + \frac{B_x\ \delta B_z}{4\pi} = 0 \tag{6.9}$$

The components of Eq. (6.3) become

$$B_y\ \delta v_x - c\ \delta B_y - B_x\ \delta v_y = 0 \tag{6.10}$$
$$c\ \delta B_z + B_x\ \delta v_z = 0 \tag{6.11}$$

where we have already made use of Eq. (6.13) below.   Equation (6.4) becomes

$$c\left(\delta p - \frac{\gamma p}{\rho}\ \delta\rho\right) = 0 \tag{6.12}$$

and finally Eq. (6.5) becomes

$$\delta B_x = 0 \tag{6.13}$$

The unknowns in the above equations are the speed of propagation of the wave, c, and the various quantities which change across the wave. Counting equations and unknowns in this manner, we find one more unknown than equations.   The equations are, however, homogeneous in the unknowns corresponding to the changes across the wave.   Therefore, as one would expect for the linearized case, the differential equations do not determine the amplitude of the wave.   We do, however, have the appropriate number of equations and unknowns to determine the speed of propagation of the wave and the changes of flow properties across the wave in terms of the wave amplitude.   We shall proceed to do this in a somewhat disorderly fashion by observing that some of the equations form subsets whose solutions can be easily determined.

6.2.1. Entropy Discontinuities

We shall first discuss a solution which is not really a wave, since it corresponds to zero propagation speed.   We may observe by inspection that substituting $c = 0$ into Eqs. (6.6) through (6.13) is a solution which allows a change in density but requires that there be no change in all the other flow properties, i.e., velocity, magnetic field, and gas pressure. This is a hydrostatic equilibrium corresponding to having different density, entropy, and temperature on the two sides of the discontinuity, but maintaining the pressure constant.   Since there is no flow through the wave, this result is of course consistent with our entropy conservation law [Eq. (6.4)], which stated that, following a fluid element, the entropy was conserved.   Thus, if initially adjacent elements of fluid have different entropies, this discontinuity in entropy will be maintained.   A practical case in which entropy discontinuities are of importance is a flow situation in which shock waves exist whose strength is not constant.   The entropy change across a shock depends upon its strength.   Thus, fluid elements going through the shock at slightly different times will have slightly different entropies.

Since we have determined the only solution which can exist for zero propagation speed, we may, in looking for the other solutions, assume that the propagation speed is nonzero.   Thus, we may take the bracket in Eq. (6.12) to be equal to zero.   This corresponds simply to the statement that, if there is a flow through the wave and if entropy is conserved along a streamline, then the entropy on both sides of the wave must be the same.

6.2.2. Intermediate Waves

As noted earlier, we expect three propagating waves corresponding to the 3 degrees of freedom of motions of the boundary.   In this subsection we shall derive the properties and propagation speed of one of these modes. It so happens that the three modes can be conveniently classified as fast, intermediate, and slow according to the magnitudes of their speeds of propagation.   The wave to be discussed in this subsection corresponds, as will be shown later, to the intermediate propagation speed.

The only equations which contain δv_z and δB_z are Eqs. (6.9) and (6.11).   Since these equations also do not contain the changes in any of the other properties, we may solve them separately.   Doing this, we obtain for the propagation speed

$$c_i{}^2 = \frac{B_x{}^2}{4\pi\rho} \tag{6.14}$$

where the subscript i indicates the intermediate wave.   Substituting this result back into the full set of equations, we obtain the following relations for the changes in flow properties across the wave:

$$\delta v_z = \pm \frac{\delta B_z}{\sqrt{4\pi\rho}} \tag{6.15}$$
$$\delta v_x = \delta\rho = \delta p = \delta B_y = \delta v_y = 0$$

The sign in the top equation depends upon whether the direction of propagation is parallel or antiparallel to the normal component of magnetic field.   Since δv_x is zero, this wave is purely transverse.   Also, since the change in magnetic field is perpendicular to the original field, there is to first order only a change in the direction of the magnetic field, but no change in magnitude.   (We shall see later that a large-amplitude intermediate wave also changes only the direction and not the magnitude of the magnetic field.)   It is interesting to note that there are no changes in the thermodynamic variables across such a wave.   The only changes are in the tangential velocity and the direction of the magnetic field.

A simple physical explanation of this wave is frequently given in terms of a vibrating string.   Since the wave is purely transverse, one would expect that the wave propagation speed should be the square root of the tension divided by the density.   Since the tension in the direction of the wave normal is $B_x{}^2/4\pi$, we see that this description is in complete agreement with the result given in Eq. (6.14).   It should be emphasized, however, that, whereas the vibrating string can vibrate in any direction perpendicular to the string, the intermediate wave corresponds to changes in magnetic field only in the z direction.   Thus, the ordinary vibrating string really corresponds to two modes which have the same propagation velocity.   In the plasma, only one of these modes corresponding to magnetic field changes perpendicular to the original magnetic field gives rise to the intermediate propagation speed.   Waves in which a change in B_y also exists will give rise to some longitudinal stresses, and as a result their propagation speed is modified.   Such waves are then either the fast or the slow mode.

6.2.3. Fast and Slow Waves

The propagation speed of the remaining two modes can be obtained by eliminating the quantities which change across the wave from Eqs. (6.6),

(6.7), (6.8), (6.10), and (6.12).   The resulting relation for the propagation speed may be written in the following two completely equivalent forms,

$$(c^2 - a^2)(c^2 - b_x{}^2) = c^2 b_y{}^2 \tag{6.16}$$
$$(c^2 - a^2)(c^2 - b^2) = a^2 b_y{}^2 \tag{6.17}$$

where we have introduced shorthand notations for the ordinary sound speed and the Alfvén speed

$$a^2 = \frac{\gamma p}{\rho} \qquad b^2 = \frac{B^2}{4\pi\rho} \tag{6.18}$$

and also $b_x{}^2 = B_x{}^2/4\pi\rho$, $b_y{}^2 = B_y{}^2/4\pi\rho$.   So Eq. (6.16), as well as (6.17), is seen to be biquadratic in the propagation speeds and thus it corresponds to two modes each of which can propagate in two directions.   Let us first observe some of the properties of the propagation speeds resulting from this relation.   We shall return to physical interpretation of these results somewhat later when we discuss the changes which occur across the waves.

Since, in both forms of the dispersion relation which have been written, the right-hand sides are positive definite, it follows that the quantities in parentheses on the left-hand side must be either both positive or both negative.   It therefore follows from Eq. (6.16) that, for one of the solutions, the square of the propagation speed must be greater than both a^2 and $b_x{}^2$ and, for the other solution, it must be less than both these quantities.   Since, by definition, the slow speed is the slower of these, it follows that

$$c_s{}^2 \le b_x{}^2 \qquad c_s{}^2 \le a^2 \tag{6.19}$$

where the subscript s denotes the slow speed.   A somewhat more stringent condition on the fast propagation speed is obtained from Eq. (6.17),

$$c_f{}^2 \ge b^2 \qquad c_f{}^2 \ge a^2 \tag{6.20}$$

where the subscript f denotes the fast speed.   Since, in the present notation the intermediate speed is b_x, we note from the above relations that the wave which we have labeled fast always travels at a speed greater than or equal to the intermediate speed, while the one which has been labeled slow is always slower than or equal to the intermediate speed, thus justifying naming the wave modes in terms of their relative propagation speeds.

The solutions of this dispersion relation have been plotted in Fig. 6.2 for several ratios of a to b.   These plots are given in the form suggested by Friedrichs.   The magnetic field direction is taken to be horizontal.   For any point on the lines, the distance from the origin is proportional to the velocity of the wave, and the angle which the line connecting the origin to the point makes with the axis corresponding to the magnetic field is the direction of the wave normal relative to the magnetic field. For the fast mode, the propagation speed is relatively insensitive to the direction of propagation.   For propagation perpendicular to the magnetic

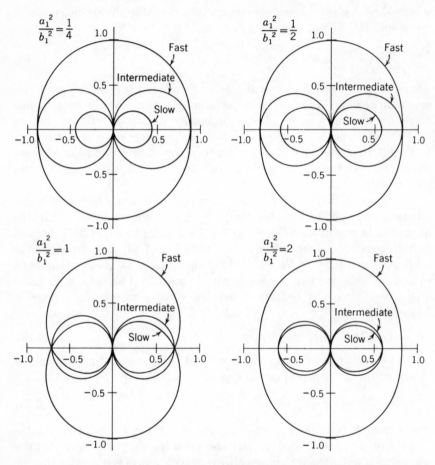

Fig. 6.2   Friedrichs diagram.   Polar plot showing the dependence of the propagation speeds of the three linear-wave modes on the angle between the wave normal and the magnetic field, for several values of the ratio of sound speed a to Alfvén speed b.   Speeds have been normalized with respect to $\sqrt{a^2 + b^2}$.

field, the propagation speed is $\sqrt{a^2 + b^2}$.   For propagation along the magnetic field, the propagation speed is either a or b depending upon which is larger.   The intermediate propagation speed is also shown and corresponds to two circles of radius $b/2$ which are tangent at the origin and whose line of centers lies in the magnetic field direction.   Since the slow propagation speed is less than the intermediate speed, the slow speed must be found inside these circles.   Thus, we see that both the slow and the intermediate speeds are zero for propagation perpendicular to the magnetic field.   For propagation along the magnetic field, the slow speed is either a or b depending upon which is smaller.   We note from the above remarks that, for propagation along the magnetic field, the intermediate speed is always equal to either the slow speed or the fast speed.   Thus, for $a < b$, the intermediate and fast speeds are equal for this direction of propagation, while, for $a > b$, the intermediate and slow speeds are equal.

Let us examine briefly the limiting values of the solutions of the dispersion relation for the cases of $a \gg b$ and $a \ll b$.   These correspond, respectively, to gas pressure much larger than and much less than the magnetic pressure.   For $a \gg b$, the fast propagation speed must also be much greater than b, and therefore, from Eq. (6.16), we conclude that the propagation speed approaches a.   This conclusion is in accordance with what one would expect, since for weak magnetic fields one would not expect the magnetic field to alter the sound speed appreciably.   In the same limit, the slow speed will be very small compared with a, and therefore, from Eq. (6.17), we conclude that the propagation speed is equal to b_x.   Thus, in this limit the slow speed approaches the intermediate speed. In other words, as the ratio of gas pressure to magnetic pressure is increased, the slow and intermediate speeds come closer together.

In the opposite limit, $a \ll b$, the fast speed approaches b, while the slow speed approaches ab_x/b.   Thus, the slow propagation speed corresponds to two circles whose line of centers lies in the direction of the magnetic field and which are tangent at the origin.

A certain degree of symmetry exists in the above dispersion relation, namely, that the fast and slow propagation speeds are unchanged if the magnitudes of a and b are interchanged.   This follows from Eq. (6.17), since the equation is unchanged when a and b are interchanged if one remembers that b_y is the product of the magnitude of b and the sine of the angle between the magnetic field and the wave normal.   This conclusion is somewhat surprising, since, as we shall see below, the changes in flow properties across the waves are quite different depending upon whether a or b is larger.   One should, of course, remember that, as illustrated in Fig. 6.2, the intermediate propagation speed depends only upon b.

Let us now examine the changes in flow properties which occur across the waves.   These are obtained by substituting the solutions of the dispersion relation [Eq. (6.16) or (6.17)] back into the equations for the changes [Eqs. (6.6) through (6.12)].   Doing this quantitatively in terms of the explicit expressions for the characteristic speeds which result from the dispersion relation is somewhat cumbersome.   However, several interesting results about the wave modes can be obtained from the properties of the dispersion relation already enumerated.

Equations (6.9) and (6.11) permit solutions with nonzero changes in v_z and B_z only if the characteristic speed is equal to the intermediate speed. Since, except for very special points, the fast and slow speeds always differ from the intermediate speed, we may conclude that across fast and slow waves

$$\delta v_z = \delta B_z = 0 \tag{6.21}$$

Thus, for these modes the magnetic field stays in the plane determined by the magnetic field ahead of the wave and the wave normal. The fast and slow waves therefore change only the magnitude of the tangential component of magnetic field but not its direction. On the other hand, we concluded earlier that the intermediate wave changes only the direction but not the magnitude of the tangential component of the magnetic field. The normal component of the magnetic field from Eq. (6.13) is, of course, constant in all cases.

For the fast and slow waves there are, in general, finite changes in both v_x and v_y. These waves are therefore partially longitudinal and partially transverse. Even though we cannot separate the wave modes into purely longitudinal and purely transverse, some insight can be gained by examining the longitudinal and transverse aspects of the waves separately. The longitudinal aspects are contained primarily in the x component of the momentum equation (6.7). Making use of the continuity and entropy equations (6.6) and (6.12), we can rewrite this equation as

$$c^2 = \frac{\delta(p + B_y{}^2/8\pi)}{\delta\rho} = a^2 + \frac{\delta B_y{}^2/8\pi}{\delta\rho} \tag{6.22}$$

The propagation speed squared is therefore the change in the longitudinal stress, resulting from both the gas pressure and the magnetic pressure divided by the change in density. This equation may also be viewed as determining the ratio of the change in B_y to the change in density. Since, by Eq. (6.20), the fast propagation speed is greater than the ordinary sound speed, it follows that the change in $B_y{}^2$ is of the same sign as the change in density for a fast wave. Conversely, for a slow wave the changes are of opposite sign. Thus the propagation speed of the fast wave is greater than the ordinary sound speed, since the magnetic and gas pressures act together, while the propagation speed of the slow wave is slower than the ordinary speed, since the magnetic and gas pressures counteract each other. The fact that the gas and magnetic pressures support each other in the fast wave and counteract each other in the slow wave is an important distinguishing feature between the two modes.

The transverse motions do not lead to as simply interpretable results. However, if we examine these by making use of the y component of momentum conservation [Eq. (6.8)] and use also Eqs. (6.10) and (6.6), we can obtain the result

$$c^2 = \frac{b_x{}^2}{1 - (B_y/\rho)(\delta\rho/\delta B_y)} \tag{6.23}$$

Thus, as one might expect, if there are no changes in the longitudinal velocity, which implies no change in density, the propagation speed approaches the intermediate speed, which is a purely transverse wave. The departures from this speed are then given by the correction term in the denominator of the above expression. This expression is, of course, consistent with the conclusions reached earlier that for a fast wave, i.e., a wave in which the gas- and magnetic-pressure changes are of the same sign, the propagation speed is greater than the intermediate speed and, conversely, for a slow wave the propagation speed is less than the intermediate speed.

It is also of some interest to examine the direction of the velocity change across a wave. Using Eqs. (6.6), (6.7), (6.8), and (6.12), we can obtain the result

$$\frac{\delta v_y}{\delta v_x} = \frac{B_x}{B_y}\left(\frac{a^2}{c^2} - 1\right) \tag{6.24}$$

As can be seen from the quadratic expression (6.16) or (6.17), the two roots of the dispersion relation are related by the following equations:

$$\begin{aligned} c_f{}^2 + c_s{}^2 &= a^2 + b^2 \\ c_f{}^2 c_s{}^2 &= a^2 b_x{}^2 \end{aligned} \tag{6.25}$$

Making use of these relations and Eq. (6.24), one can show that

$$\left(\frac{\delta v_y}{\delta v_x}\right)_f = -\left(\frac{\delta v_x}{\delta v_y}\right)_s \tag{6.26}$$

which states that the changes in velocity across the fast and the slow waves are perpendicular to each other. Remembering also that the change in velocity across the intermediate wave was in the z direction, we reach the conclusion that the changes in velocity across the three waves are in mutually perpendicular directions.

If we imagine the waves to be produced by the motion in an arbitrary direction of a piston which forms one boundary of the gas, the relative amplitude of the three waves which are produced will depend upon the components of the piston velocity along the velocity vectors produced by each of the individual waves. An attempt to illustrate this and some of the other properties of the waves which we have derived is given in Fig. 6.3.

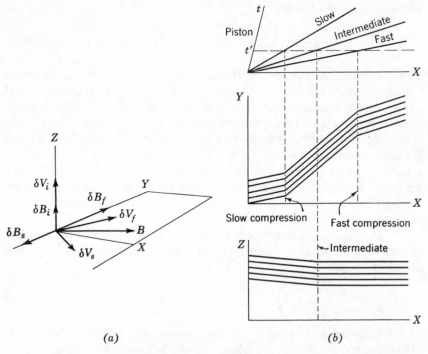

(a) (b)

Fig. 6.3   Sketch of flow resulting from the instantaneous acceleration of a piston to a small velocity.   In general, three waves will be emitted which separate with time as shown on the xt diagram.   The projections of the magnetic field lines on the xy and xz planes at a time t' are also shown for the case in which both the fast and slow waves are compressions.   The changes in velocity and magnetic field across the three waves are illustrated in the vector diagrams.   The initial magnetic field δv_s, and δv_f are in the xy plane. δv_f must lie within the acute angle between the magnetic field and the y axis.   The three velocity changes are mutually perpendicular.   The signs of δB_s and δB_f were also chosen for compression waves.

It is also instructive to look at some of the wave properties in the limits of large and small ratios of a to b.   In the limit of $a \gg b$, the fast propagation speed, as well as the changes across the fast wave, reduce to those for an ordinary sound wave.   This is to be expected, since in this case the magnetic pressures are too small to play any role.   The fast wave, therefore, approaches a purely longitudinal wave.   In the same limit, the slow wave becomes a purely transverse wave.   This can be seen most easily from Eq. (6.24), which shows that the change in the y component of velocity becomes very large as compared with the change in the

x component of velocity. It follows physically from the fact that a very small change in the longitudinal velocity produces a change in density and, therefore, a change in gas pressure which is very large compared with the magnetic pressure. Thus, only very small changes in the longitudinal velocity are required to balance the changes in the magnetic pressure. For this wave and in this limit, the fluid may therefore be considered as virtually incompressible. It then follows from Eq. (6.23), and as we have concluded earlier, that the slow propagation speed approaches the intermediate speed.

In the opposite limit, that is, $a \ll b$, the waves do not break up into purely longitudinal and purely transverse. In this limit, the slow wave is most easily understood. Since, in this limit, both the gas pressure and the dynamic pressure ρv^2 are small compared with the magnetic pressure, we cannot have appreciable changes in the magnetic field across the wave. Thus, the magnetic field lines will have virtually no change in direction across the wave. The plasma flow, on the other hand, is strongly coupled to these field lines. Thus, the plasma is constrained to flow in a direction parallel to the magnetic field lines. We have already observed that the propagation speed of the slow wave in this limit is equal to the sound speed multiplied by the cosine of the angle between the magnetic field and the wave propagation direction. This then corresponds to a sound wave traveling along the magnetic field lines, and therefore moving more slowly in the direction of the wave normal. The slow wave therefore becomes purely transverse for propagation perpendicular to the magnetic field and purely longitudinal for propagation along the magnetic field. On the other hand, since the velocity change across the fast wave is perpendicular to that across the slow wave, we conclude that in this limit the fast wave is purely longitudinal for propagation perpendicular to the magnetic field, while it is purely transverse for propagation along the magnetic field.

6.2.4. Summary

We may summarize the major conclusions which have been reached concerning these waves as follows:

1.   There are three distinct wave propagation modes which can be conveniently classified according to the magnitude of their propagation speed as fast, intermediate, and slow. The velocity changes across the three waves are mutually perpendicular.

2.   For fast and slow waves, both the velocity and the magnetic field remain in the plane defined by the magnetic field ahead of the wave and the wave normal. On the other hand, for the intermediate wave both the velocity and the magnetic field changes are purely in the direction perpendicular to this plane.

3.   For the fast mode, the magnetic pressure increases when the density increases.   For the slow mode, an increase in magnetic pressure corresponds to a decrease in density.   Across an intermediate wave, neither the magnetic pressure nor the density changes.

6.3   LARGE - AMPLITUDE ISENTROPIC WAVES AND SHOCK FORMATION

The solution to a nonlinear flow problem can be built up by considering it as a series of small-amplitude waves, each propagating through a medium which has been modified by previous waves.   In this manner, it is possible to discuss problems with arbitrarily large amplitudes.   The concept of a large number of isentropic small-amplitude waves describing the flow breaks down in the case where shock waves are formed.   However, the nonlinear isentropic solutions can be used to predict *when* shock waves occur.   The shock waves themselves will be discussed in the next section.   In this section, we shall consider the nonlinear waves related to each of the linear wave propagation modes.   We shall consider only the case in which the waves are all propagating in one direction, i.e., as though they were generated at the boundary of a semi-infinite plasma.   For the special case in which the boundary condition is changed suddenly, the fact that the three propagation speeds are different separates the resulting nonlinear waves.   Thus, for this case the nonlinear description of the individual modes can be used to obtain a general solution for an arbitrary instantaneous change in the boundary condition.   The more general case in which several wave modes exist at the same place or waves of the same mode exist in the same place propagating in opposite directions will not be considered.   Problems of this kind can also be treated by a generalization of the procedures to be described; however, in most cases, they involve considerable labor.

We shall show that compression waves for both the fast and the slow modes tend to steepen to form shock waves, whereas the expansion waves for these two modes tend to spread out with time so that the gradients become less steep.   The intermediate wave, on the other hand, has the rather surprising property that, even for large amplitudes, it remains a linear wave.   Thus, even for large amplitude, an intermediate wave of arbitrary shape will retain its shape as it propagates through the medium.

6.3.1.   Intermediate Large-amplitude Waves

Let us imagine a semifinite uniform plasma bounded by a piston.   At time zero the piston is moved so as to produce a step-function small-amplitude intermediate wave. A short time later the medium will still be undisturbed ahead of the region to which the wave has propagated, i.e., for distances greater than $b_x t$ from the piston. In the region between the piston and the instantaneous location of the wave, the medium will again be uniform, but in a slightly different condition than the condition existing ahead of the wave. If at this time the piston velocity is again changed instantaneously so as to produce a second intermediate wave, we may examine the propagation speed of this second wave. To do this, we must determine the conditions behind the first wave. Since, as we concluded in the previous section, there is no change in density, normal component of velocity, or normal component of magnetic field across an intermediate wave, the propagation speed for intermediate waves remains unchanged. Thus, the second wave will move at precisely the same speed as the first wave. We may now consider a third and fourth wave generated by the piston, and it follows from the same argument that the propagation speeds of all these waves will be precisely equal. Since we can consider an arbitrary pulse of intermediate waves to be composed of a series of step functions, it follows that, provided that the piston motion is constrained to produce only intermediate waves, the wave shape will be retained as the entire large-amplitude disturbance propagates through the fluid. Thus we obtain neither steepening to form a shock wave nor spreading out as in the case of expansion fans.

The restriction on the piston motion which is required to produce a pure intermediate wave is easily seen from the condition that the change in velocity across a small-amplitude intermediate wave must be perpendicular to the plane defined by the magnetic field and the wave or piston normals. Thus, the instantaneous changes in velocity or acceleration of the piston must always be perpendicular to the magnetic field at the surface of the piston.

The changes in flow properties across a large-amplitude intermediate wave are obtained by summing the changes across each of the component small step-function waves, which in turn are considered as differential elements. It follows immediately from Eq. (6.15) that across the large-amplitude wave the changes in normal velocity, density, and pressure will be zero. In evaluating the change in magnetic field we must remember that our coordinate system was chosen such that B_z was zero ahead of each small-amplitude wave. Equation (6.15) therefore states that the differential change in magnetic field is in the plane of the wave front and perpendicular to the local field. Integrating a number of such changes gives the result that the magnitude of the magnetic field is unchanged across a large-amplitude intermediate wave; however, the magnetic field vector can be rotated through an arbitrary large angle about an axis perpendicular to the wave front. The change in tangential velocity across the wave is from Eq. (6.15) in the direction of the change in magnetic field and is equal to $\Delta \mathbf{B}/\sqrt{4\pi\rho}$. Although such a wave produces no change in the thermodynamic quantities, the normal velocity, or the magnitude of the magnetic field, it is still a large-amplitude wave in the sense that the angle of rotation of the magnetic field and the change in tangential velocity can be large, i.e., of the order of radians and the propagation speed, respectively.

We may anticipate that, since for small-amplitude fast and slow waves the magnetic field remains in the plane defined by the wave normal and the magnetic field ahead of the wave, it will also remain in this plane for large-amplitude fast and slow waves. The intermediate wave will therefore be required in flow fields in which the boundary conditions require a rotation of the plane of the magnetic field. The particular case of rotation through 180° is frequently overlooked. In this case the magnetic field appears to stay in the same plane, but its tangential component changes sign. As we shall see, neither fast nor slow expansion waves or shock waves can change the sign of the tangential component; thus the intermediate wave will also appear in cases where such a sign change is required by the boundary conditions.

6.3.2. Fast and Slow Waves

Let us now consider the case in which the piston motions are such as to generate two successive small-amplitude waves which are either both fast or both slow waves. Differences in the propagation speeds of the two waves result from two causes. First, there is a change in the propagation speed of the wave relative to the fluid, due to the changes in magnetic field and density across the wave. Second, there is a change in fluid velocity across the first wave, which means that the second wave is riding on a fluid which is already moving. If the first wave is a compression wave, the fluid behind it is moving at a slow velocity in the direction of propagation of the wave. Thus, the second wave, even if it had the same propagation speed relative to the fluid, would tend to catch up with the first wave. In most cases, we shall find that the change in fluid velocity is the predominant effect.

A quantitative determination of the conditions under which the two waves will catch up with each other can be obtained by examining the expression $\dfrac{\rho}{c}\dfrac{\delta(v_x + c)}{\delta\rho}$. $\delta(v_x + c)$ is the propagation speed of the second wave relative to the first. If this is of the same sign as the propagation speed of the first wave, the second wave will overtake the first. Thus $\delta(v_x + c)/c$ positive implies steepening. Therefore, if the above expression is positive, steepening occurs for positive $\delta\rho$, that is, compression waves. If it is negative, steepening occurs for rarefaction waves. By making use of the jump conditions across the wave and the dispersion relation, this quantity with some algebraic manipulation, can be written as

$$\frac{\rho}{c} \frac{\delta(v_x + c)}{\delta\rho} = 1 + \frac{1}{2} \frac{(\gamma - 1)a^2b_y{}^2 + (c^2 - a^2)^2}{a^2b_y{}^2 + (c^2 - a^2)^2} \tag{6.27}$$

It can easily be seen that this expression is always positive. It is interesting to note that this quantity, which describes the rate at which waves catch up with one another, is insensitive to the gas conditions. It is always between $(\gamma + 1)/2$ and $\frac{3}{2}$.

If we now extend this argument to a smooth pressure pulse which we consider to be made up of a large number of small-amplitude waves, we see that, for both the fast and the slow mode, the compression parts of the pulse tend to steepen to form shock waves, while the rarefaction portions tend to separate. Thus in terms of the illustration given in a general sense in Fig. 6.1c, we may now interpret the pulse shape which is shown as the density or pressure profile. The maximum-density point overtakes the lower-density regions ahead of it while it moves away from those behind. As mentioned earlier, across the discontinuities which are formed we can no longer assume that the gradients are too small for viscous dissipation and Joule heating to be important. Thus, there will be an entropy change, and we must consider such shock waves separately. For rarefaction waves, on the other hand, the lines of constant conditions spread apart, and thus the gradients become less steep as time progresses. The isentropic theory is therefore applicable to all rarefaction waves and to compression waves provided that they have not yet steepened to form shocks.

Nonlinear waves consisting only of one wave mode propagating in only one direction are generally referred to as *simple waves*. In a simple wave, plasma conditions are constant along lines moving at the propagation speed of the individual small-amplitude waves, $v_x + c$. (Such lines do not cross any waves.) Thus, if conditions are known along one boundary, for example, a piston which we imagine to be producing the flow, then conditions at later time can be determined by projecting constant properties along these lines. Since the relative changes in the flow properties across the component small-amplitude waves are given by Eqs. (6.6) through (6.13), integration of these equations allows one to express all the flow properties in terms of one of them. Thus, the changes which can occur across simple waves can be determined independently of the piston motion. The rate of change of velocity of the piston will determine at what point in the flow field these changes in flow properties actually occur.

Although some of these relations can be integrated formally, a somewhat more surveyable graphical representation of the integrals was suggested by Shercliff.[3] For the case of one-dimensional time-dependent flows (B_x constant), the differential equation relating B_y and ρ contains only these two variables.   This can easily be seen by eliminating c^2 from Eqs. (6.22) and (6.23) and remembering that sound speed is related to the density by the isentropic law.   Upon introducing nondimensional variables

$$B_y' = \frac{B_y}{B_x} \qquad \rho' = \left(\frac{a}{b_x}\right)^{2/\gamma} = \left(\frac{4\pi\gamma}{B_x{}^2}\frac{p}{p^\gamma}\right)^{1/\gamma} \qquad p = \frac{\rho}{\rho_0} \qquad (6.28)$$

the resulting differential equation can be written in the form

$$\left(\frac{dB_y'}{d\rho'}\right)^2 - \left(\frac{1 + B_y'^2 - \rho'^\gamma}{\rho'B_y'}\right)\frac{dB_y'}{d\rho'} - \rho'^{(\gamma-2)} = 0 \qquad (6.29)$$

The variable ρ' is indeed proportional to the first power of the density since from the isentropic law ρ_0 is a constant. ρ_0 may be regarded as the density to which the gas must be expanded or compressed isentropically in order to reach the condition $a = b_x$.   The above equation is quadratic in $dB_y'/d\rho'$ and therefore may be factored into two first-order differential equations.   These two equations correspond to the trajectories in the $B_y' - \rho'$ plane across fast and slow waves, respectively.   The numerical solutions of these equations for the case $\gamma = \frac{5}{3}$ are shown in Fig. 6.4a. As can be seen, these solutions correspond to a one-parameter family of curves.   The parameter could be expressed as the value of ρ' which occurs at $B_y' = 0$.   An arbitrary initial gas condition determines a point in this plane.   Through each point there are two lines corresponding to the trajectories along fast and slow simple waves.   The changes in the remaining flow properties along these trajectories can be obtained by first solving the dispersion relation [Eq. (6.16)] for the speed of propagation, c, in terms of B_y' and ρ'.   Then, upon integrating Eqs. (6.6) and (6.8), the changes in the x and y components of velocity can be determined.   These results are shown in Fig. 6.4b to d, with the velocities nondimensionalized with respect to the velocity $B_x/\sqrt{4\pi\rho_0}$.   The equations determining the change in velocity do not depend upon the magnitude of the velocity. Thus, the integration gives only the change in velocity between two points along a trajectory and not the absolute magnitude.   In Fig. 6.4c and d, the velocity coordinates are to be regarded as the change in that component of velocity which would occur between a given point on the trajectory and the point at which B_y' is equal to zero.

Figure 6.4 contains all the information required to determine conditions across fast and slow simple waves.   If, for example, the variation of density with time is known at a fixed position, then the variation in B_y, v_x, v_y, and c can be determined from Fig. 6.4.   Conditions in the

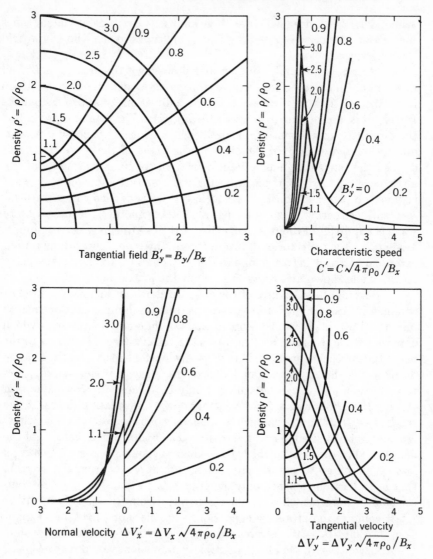

Fig. 6.4 Changes in fluid properties across one-dimensional time-dependent fast and slow simple waves for $\gamma = \frac{5}{3}$. Curves are labeled according to the value of ρ' at $B_y' = 0$; for fast waves this is less than 1, and for slow waves it is greater than 1. Both components of flow velocity have been taken to be zero at $B_y' = 0$; thus, these curves should be considered as the change in velocity from this point on the trajectory. Only the fast characteristic speed is given along a fast wave and only the slow characteristic along a slow wave.

remainder of the flow field are then known to be constant along lines moving at a velocity of $c + v_x$. Thus, conditions in the whole flow field are determined.

Across large-amplitude fast and slow waves the change in B_z and v_z is zero, since it is zero across the small-amplitude waves. Upon remembering that our choice of the y and z directions was such as to make B_z zero, it is appropriate to consider the B_y' axis in Fig. 6.4a as representing the tangential component of magnetic field. Thus, if we chose our coordinates relative to a fixed direction in the plane of the wave front and defined the tangential magnetic field in terms of its magnitude and the angle $\phi = \tan^{-1}(B_z/B_y)$, the trajectories across fast and slow waves are confined to planes of constant ϕ and have the shapes shown in Fig. 6.4a. As mentioned earlier, to go from one value of ϕ to another, the intermediate wave is required. From the properties of the intermediate wave enumerated above, the trajectories of intermediate waves on a three-dimensional extension of Fig. 6.4a correspond to the lines obtained by taking any point on the plane shown and rotating it about the ρ' axis.

Several other features of the diagram shown in Fig. 6.4a are worth commenting on. First, as was apparent from the jump conditions across individual waves or, what is equivalent, the differential equations which determined these curves, for fast waves the density always increases as the magnitude of the magnetic field increases, while for slow waves the density decreases as the magnitude of the magnetic field increases. Expansion to a vacuum, $\rho = 0$, can be achieved only by means of a slow expansion wave. If we followed a fast trajectory toward decreasing densities, we reach $B_y = 0$ at a finite value of ρ. There is one exception to this, which is somewhat disguised in our diagram by the nondimensionalization of the density. For the case in which B_x is equal to zero, it is possible to have a fast expansion go to zero density. It is also apparent from the diagram that $B_y = 0$ can be reached only by either a fast expansion or a slow compression. Conversely it is possible to go away from $B_y = 0$ only by either a fast compression or a slow expansion. Furthermore, for a fast expansion, the point $B_y = 0$ is always reached at a value of ρ' which is less than 1. And, conversely, the slow isentropic compressions would reach $B_y' = 0$ only at ρ' greater than 1. This latter relation turns out also to be valid across the corresponding slow shock waves.

In our discussion we have restricted ourselves to a considerable extent to the case of one-dimensional time-dependent flows. Simple waves also exist in steady two-dimensional flows in which the flow velocity is larger than the wave propagation speed. Such flows will not be discussed in detail here except to point out that the major conceptual difference between the two cases lies in determining the trajectories of the elementary small-amplitude waves. To achieve a steady flow situation, the waves must propagate relative to the fluid at the same speed at which they are blown back by the fluid velocity.   Thus, the angle which the wave normal makes with respect to the flow velocity is determined by the conditions that the normal component of velocity be equal to the propagation speed c of the wave.   Several references on two-dimensional flows are included in the Bibliography.

6.3.3. Instantaneous Piston Acceleration

Two possible flow patterns which can occur when the boundary conditions are changed instantaneously to a different constant value are illustrated in Fig. 6.5.   In this case all the waves are generated at the boundary at the time at which conditions there are changed.   Thus, flow properties are constant along lines in the xt plane emanating from this point.   Since

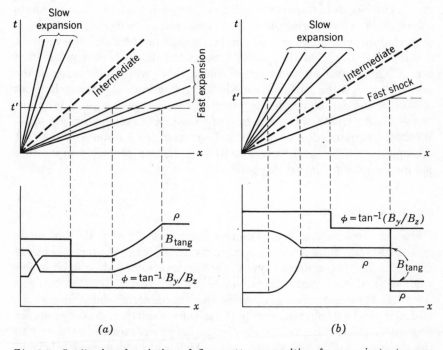

(a)   (b)

Fig. 6.5 Qualitative description of flow patterns resulting from an instantaneous change in boundary conditions to a new constant condition. The lower curve shows fluid properties as a function of position for a particular time t'. In (a) both the fast and slow waves are expansions, and the flow properties follow the trajectories given in Fig. 6.4.   Note that there are four separate regions within which the flow is completely uniform.   In (b) the fast wave is a compression; thus, the changes across it occur suddenly and are obtained from the shock relations discussed in Sec. 6.5.

the propagation speeds of the three wave modes are different, they will separate as time progresses.   Thus, the entire flow can be described as consisting of three simple waves corresponding to the three modes.   Since, by definition, the fast mode has the fastest propagation speed, a fluid element some distance from the boundary will first experience a change corresponding to a fast, simple expansion fan, then an intermediate, and finally a slow one.   The regions occupied by the fast and slow expansion fans widen as time progresses, since the propagation speed at the back edge of a rarefaction wave is slower than that at the front edge.   Since, in general, there is a finite difference between the propagation speeds of the different wave modes, there will be regions between the different waves which also spread with time in which conditions are completely uniform. Since the propagation speed of the intermediate wave does not change as the fluid moves through the wave, the intermediate wave will remain a discontinuity.

In case the boundary conditions are such that one of the waves corresponds to a compression wave, a shock will be formed.   For the case of instantaneous change of the boundary condition the shock will move at constant velocity, and thus conditions behind it will be independent of time.   The jump conditions across shock waves will be discussed in the next section.   If the other wave mode is a rarefaction wave, the techniques of this section can be applied to calculate the changes and flow property across the expansion wave.   A case of this type is sketched in Fig. 6.5b. We shall return to a discussion of this diagram in Sec. 6.6 when we consider the application of shock waves to the production of high-temperature plasma samples for laboratory study.

6.4 SHOCK WAVES

From our previous discussion compression waves of either the fast mode or the slow mode steepen to form discontinuities.   We may therefore expect to find two different types of shock waves depending upon which mode of small-amplitude waves formed the shock.

The changes in flow properties which occur across shock waves may be determined without reference to the specific dissipation processes occurring within the shock wave.   The dissipation processes are related only to the structure within the shock wave, which will be discussed briefly in Sec. 6.5.   For the present discussion, we shall consider the thickness of the shock wave to be infinitesimally small compared with the scale of the flow field.

Let us go to a coordinate system moving with the instantaneous shock velocity and draw two planes parallel to the plane of the shock, one on either side of the shock wave.   Since the shock wave itself is very thin, these two planes can be very close together in terms of the scale of the overall flow field.   For a steady flow there can be no net rate of accumulation of mass, momentum, energy, or magnetic field in the region between the two planes.   Thus, if a shock wave is to exist between these two planes, the conditions at the planes must satisfy the restrictions that the fluxes of the above quantities are the same across both planes.   The relations equating these two fluxes then determine the jump conditions which are allowed across shock waves.   If the planes are very close together, the time that it takes the fluid to go from one plane to the other is very small compared with the time scale on which the overall flow field changes.   Thus, even for an unsteady flow, we may consider this small portion of the flow as being steady.   The resulting shock relations can therefore be used in unsteady flows.

Let us denote the two planes by the subscripts 1 and 2.   Further, let us choose a coordinate system such that the x direction is the normal to the wave front, and let us assume that the coordinate system is moving at a velocity such that $v_{y1} = v_{z1} = 0$ and again orient our coordinate system so that $B_{z1} = 0$.   The equations which we shall obtain are in essence the nonlinear versions of the jump conditions across small-amplitude waves which we discussed earlier.   One exception to this is that entropy is not conserved.   This equation must be replaced by the conservation of energy flux.

Let us consider first the conservation of flux of the magnetic field. The statement that no magnetic field is accumulated between the two planes is equivalent to the statement that for a steady state $\nabla \times \mathbf{E} = 0$. Thus, the tangential components of electric field, E_y and E_z, must be the same on both sides of the shock front.   Since our two control planes are outside the shock, the current density is small at these points and we may use the relation

$$\mathbf{E} + \frac{\mathbf{v} \times \mathbf{B}}{c_e} = 0 \tag{6.30}$$

where c_e = velocity of light
Making use of this equation and the fact that E_y and E_z do not change across the shock, we obtain the equations

$$v_{x2}B_{y2} - v_{y2}B_x = v_{x1}B_{y1} \tag{6.31}$$
$$v_{x2}B_{z2} - v_{z2}B_x = 0 \tag{6.32}$$

Note that no subscript is required on B_x, since it must be the same on both sides of the shock.

The requirement of the conservation of mass leads to the relation

$$\rho_2 v_{x2} = \rho_1 v_{x1} \tag{6.33}$$

The momentum flux across a plane is composed of the flow of fluid momentum across the plane (i.e., the product of mass flow and velocity), the gas pressure, and the magnetic stresses acting on the plane.   Thus, we obtain for the three components of the momentum equation

$$\rho_2 v_{x2}^2 + p_2 + \frac{B_{y2}^2}{8\pi} = \rho_1 v_{x1}^2 + p_1 + \frac{B_{y1}^2}{8\pi} \tag{6.34}$$

$$\rho_2 v_{x2} v_{y2} - \frac{B_x B_{y2}}{4\pi} = -\frac{B_x B_{y1}}{4\pi} \tag{6.35}$$

$$\rho_2 v_{x2} v_{z2} - \frac{B_x B_{z2}}{4\pi} = 0 \tag{6.36}$$

Finally, the energy equation consists of the flow of thermal and kinetic energy of the fluid as well as the flow of electromagnetic energy, which is given by the Poynting vector.   Making use of Eq. (6.30), we obtain the relation

$$\rho_2 v_{x2} \left(\frac{\gamma}{\gamma - 1} \frac{p_2}{\rho_2} + \tfrac{1}{2} v_{x2}^2 + \tfrac{1}{2} v_{y2}^2 \right) + \frac{B_{y2}}{4\pi} (B_{y2} v_{x2} - B_x v_{y2})$$
$$= \rho_1 v_{x1} \left(\frac{\gamma}{\gamma - 1} \frac{p_1}{\rho_1} + \tfrac{1}{2} v_{x1}^2 \right) + \frac{B_{y1}^2 v_{x1}}{4\pi} \tag{6.37}$$

The above equations are sufficient so that, if all the conditions on one side of the shock are known, the conditions on the other side are determined. Note that conditions ahead of the shock include the flow velocity relative to the shock.   Thus the shock speed as well as the thermodynamic and magnetic field conditions must be specified.

The solution of these equations results largely from straightforward but somewhat tedious algebraic manipulation.   In the discussion which follows we shall first describe, but not necessarily prove, some of the general results which apply to the jump conditions across shock waves.   Following this, we shall discuss graphs of the jump conditions for various special cases.   These graphs have been chosen to give the reader some feeling for the changes which can occur across shock waves.   They are not intended to provide a complete summary of the shock relations for computational purposes.

Shock waves are always compression waves.   This conclusion is to be expected from the shock-formation arguments given in the previous section, which showed that compression waves tended to steepen to form discontinuities, while rarefaction waves spread apart.   It does not follow immediately from the equations written above, since the sign of the velocity may be changed throughout without changing the equations. Thus, solutions of the above equations could correspond to the flow going either from low density to high density or from high density to low density. If, however, one examines the entropy on the two sides of the shock, it can be shown that the entropy is always higher on the higher-density side.[4] Since the entropy must increase with time, it follows that the flow must go from the low-density side toward the high-density side.

The physically realizable solutions of these equations can be divided into two categories, which have been named *fast* and *slow shocks*. The limits of weak fast and slow shock waves are the fast and slow small-amplitude disturbances discussed in Sec. 6.2. The magnetic field changes across these shock waves are qualitatively the same as they would be for fast and slow small-amplitude compression waves. The tangential component of the magnetic field increases across fast shocks and decreases across slow shocks. Furthermore, as will be shown below, the magnetic field behind the shock is in the plane defined by the wave normal and the magnetic field ahead of the shock.

The fast disturbance speed ahead of a fast shock is always less than the normal component of flow velocity, while the fast disturbance speed corresponding to conditions behind the shock is always greater than the normal component of the flow velocity behind the shock. The corresponding statement can also be made for slow shocks relative to the slow disturbance speed. The proof of this for weak shock waves follows directly from the arguments used to show that compression waves steepen. Consider one of the waves used in the steepening argument to be a weak shock. It was then shown that a small-amplitude wave behind this wave would overtake it, thus showing that the small-amplitude disturbance speed behind the waves was greater than the flow velocity. Correspondingly the wave under consideration would overtake a small-amplitude wave ahead of it, thus showing that its velocity relative to the fluid was larger than the propagation speed ahead of it. A plausibility argument for this statement for stronger shock waves can also be given on the basis of the shock-steepening analysis. If we imagine a shock to be formed from the steepening of a gradual pressure pulse, then as the first few sound waves cross, the shock will gradually become stronger. Each increase in shock strength is then directly related to a small-amplitude wave overtaking the shock. Thus we would not expect to be able to produce shock waves in this manner for which small-amplitude disturbances from behind cannot catch up. If the flow velocity ahead of the shock were less than the disturbance speed, then a wave coming from behind would not pile up at the shock, but would go out ahead. Thus, we also require that the flow velocity ahead be greater than the disturbance speed. Direct verification of this restriction on the velocities ahead and behind shock waves can be obtained by examining the algebraic solutions of the shock relations. A further restriction on the flow speeds which will be discussed in more detail below is that for fast shocks the flow velocity must be greater than the intermediate speed on both sides of the shock, while for slow shock waves the flow velocity must be less than the intermediate disturbance speed on both sides of the shock.

The proof of the fact that the magnetic field behind the shock must lie in the plane defined by the magnetic field ahead and the wave normal proceeds as follows: Eliminating v_{y2} between Eqs. (6.31) and (6.35) and eliminating v_{z2} between Eqs. (6.32) and (6.36), we obtain the relations

$$\left(\frac{v_{x2}^2}{b_{x2}^2} - 1\right) B_{y2} = \left(\frac{v_{x1}^2}{b_{x1}^2} - 1\right) B_{y1} \tag{6.38}$$

$$\left(\frac{v_{x2}^2}{b_{x2}^2} - 1\right) B_{z2} = 0 \tag{6.39}$$

It follows from Eq. (6.39) that either the z component of magnetic field behind the wave is zero or the flow velocity must be equal to the intermediate propagation speed. The latter case requires from Eq. (6.38) that either the flow velocity ahead be also equal to the intermediate speed ahead of the wave or that the y component of magnetic field ahead of the wave be zero. If B_{y1} is zero, then the plane of the magnetic field ahead of the wave is not defined and it does not make sense to distinguish between the y and z components of the magnetic field behind the wave. If the flow velocity is equal to the intermediate speed both ahead and behind the wave, this simply corresponds to the large-amplitude intermediate wave discussed in the previous section. Since intermediate waves do not steepen to form discontinuities and since there is no entropy change across them, it does not seem appropriate to refer to them as shock waves. It should be remembered, however, that an almost discontinuous rotation of the plane of the magnetic field can occur across such a wave if it is initiated sufficiently rapidly. Thus, excluding the discontinuous intermediate wave, a finite value of B_{z2} cannot occur behind a shock wave. Hence, the magnetic field and the velocity are entirely in the xy plane on both sides of the shock front, and Eqs. (6.32) and (6.36) need not be considered in further discussion of the conservation equations.

We shall now show that shock waves with velocities greater than the intermediate speed ahead and less than the intermediate speed behind cannot occur. As can be seen from Eq. (6.38), this statement is completely equivalent to the statement that the sign of the tangential component of the magnetic field cannot change across a shock wave. There has been considerable confusion on this point in the literature, since solutions of the conservation equations (6.31) through (6.37) which violate the above condition do exist. These solutions, however, cannot occur in nature and should therefore be regarded as extraneous.† A portion of the

---

† In the literature these solutions are usually referred to as *unstable* or *nonevolutionary shock waves*. We prefer the term *extraneous*, since it implies more directly that these solutions cannot occur in nature even for very short times.

confusion is related to the fact that a number of workers who have discussed solutions of the conservation equations did not recognize that some of these solutions were extraneous and discussed them at considerable length.   Since the extraneous solutions are considerably more complicated than the real ones, this leaves the overall impression that the solution of the shock equations is much more complex than it actually is.   Further confusion resulted from the fact that these extraneous solutions were originally referred to as unstable shock waves.   It was recognized that this was not an ordinary instability growing exponentially with time but rather that it corresponded to a sudden disintegration of the shock wave. Numerous papers have been written on the waves produced when such a shock wave disintegrated.   In the present discussion we shall use essentially the same arguments which were initially proposed as leading to the instability of these solutions.   The logical conclusion of the argument as presented here is, however, that these solutions cannot occur in nature even for very short times and thus that they are extraneous solutions of the mathematics which do not correspond to physical reality.

Let us imagine that a flow containing an extraneous shock does exist and focus our attention on a small region of this flow in the neighborhood of the shock wave.   This small region may be regarded as having been produced by the instantaneous acceleration of a piston to a velocity corresponding to conditions behind the shock wave.   Since for all solutions of the conservation equations B_{z2} and v_{z2} are zero, the required piston motion is purely in the xy plane.   If we now consider a second case in which the boundary conditions are changed slightly so as to require an arbitrarily small z component of velocity at the piston, we shall find that no neighboring solutions of the flow problem exist.   The only wave mode which can produce the required z component of velocity or magnetic field is the intermediate wave.   We would therefore expect that a small-amplitude intermediate wave would also be emitted from the piston at the time at which it is accelerated.   There is, however, no place in the flow field where the intermediate wave can exist.   The velocity of the extraneous shock relative to the fluid ahead of it is greater than the intermediate propagation speed.   Thus, the intermediate wave cannot propagate ahead of the shock wave.   On the other hand, the velocity of the extraneous shock wave relative to the fluid behind it is less than the intermediate propagation speed.   Thus, the intermediate wave cannot remain behind the shock wave.   Furthermore, since the plane of the magnetic field cannot rotate across the shock wave, the intermediate wave can also not exist in the middle of the shock wave.   Thus a flow containing an extraneous shock does not have any neighboring solutions corresponding to a small change in the boundary conditions which requires an arbitrarily small angle of rotation of the plane of the magnetic field.

This difficulty does not occur for the allowed fast or slow shocks. For the fast shock the flow velocity is greater than the intermediate propagation speed on both sides of the shock, and thus the required intermediate wave could exist behind the shock wave. Correspondingly for the allowed slow shock solutions the flow velocity is less than the intermediate speed on both sides of the shock, and thus the intermediate wave could exist ahead of the shock wave. As has already been observed across the extraneous shock solutions the tangential component of magnetic field changes sign, i.e., rotates through 180°. We shall now show that boundary conditions which require such a 180° rotation of the tangential component of the magnetic field can always be satisfied by flows which contain allowed fast and slow waves and a 180° intermediate wave. Thus there is no consistent set of boundary conditions which would require the existence of an extraneous shock in the solution. One might argue that, for the case in which the rotation in the intermediate wave is precisely 180°, the flow solution is not unique. However, since the lack of uniqueness exists only over an immeasurably small range of angles, one would not expect to be able to produce an extraneous shock in any physical situations.

To show that an alternate solution always exists, let us consider again a piston on which boundary conditions are changed instantaneously. For the present discussion, it will be more convenient to consider the piston as an insulator. In this case, the tangential component of flow velocity at the surface is not necessarily the velocity of the piston. Thus, the y and z components of velocity cannot be controlled directly by the piston motion. However, since the piston is an insulator, it is possible to control the magnetic field at the surface. The appropriate boundary conditions are therefore the x component of velocity and the y and z components of the magnetic field at the surface. A one-to-one correspondence exists between flows specified by the above boundary conditions and the flows resulting from conducting-piston boundary conditions which are the specification of the three components of velocity. Let us consider, first, the set of cases in which B_z is zero at the piston and B_y is of the same sign as it is before the boundary conditions are changed. Solutions must be possible for arbitrary values of v_x and arbitrary positive values of B_y. Furthermore, since the sign of B_y is unchanged, these solutions do not contain either an intermediate wave or an extraneous one. The case of two extraneous shocks which would change the sign of B_y twice is also not possible, since they would overtake one another. The solution in this case must therefore consist only of the allowed waves. We now change the boundary conditions so that the magnitude of the tangential component of B at the piston remains the same but its direction is changed. Then the flow solution remains identical except that an intermediate wave which rotates through the appropriate angle must be inserted between the fast and slow waves. This follows from the fact that across the intermediate wave the magnitude of the magnetic field, the normal component of the flow velocity, and the density and pressure are all unchanged. Thus, the quantities which are relevant for determining the changes across the fast and slow waves are unchanged by the presence of the intermediate wave.   In terms of these quantities the flow solution for an arbitrary angle of rotation is known once it is known for a particular angle of rotation. The magnitude of the tangential component of velocity is not given quite as directly, since it can change across an intermediate wave.   We can, however, conclude that, given a flow solution corresponding to no rotation of the magnetic field, solutions also exist for all other angles of rotation of the magnetic field.   In particular the case of 180° rotation is included. Thus boundary conditions which might suggest the existence of an extraneous shock can also be satisfied by a solution of the type just described.

In the process of eliminating the extraneous solutions from consideration we have derived two restrictions on the allowed shock solutions. These may be conveniently summarized as the statement that for fast shocks the flow velocity on both sides of the shock must be greater than the intermediate propagation speed, while for slow shocks the flow velocity on both sides must be less than the intermediate propagation speed. Using Eq. (6.38) this statement also shows that the sign of the tangential component of magnetic field is unchanged across either a fast shock wave or a slow shock wave.   It also follows immediately from the velocity restriction that a slow shock cannot overtake a fast shock, while a fast shock necessarily overtakes a slow shock.

6.4.1. Fast Shocks

Let us now turn to a more quantitative description of the solution of the shock equations.   The conditions behind a shock are specified if conditions ahead, including the flow velocity relative to the shock, are known. Even in nondimensional terms there are four parameters which are required to define a particular shock, a_1/b_1, M_1, θ_1, and γ.   The subscript 1 will be used to define conditions in the low-density stream ahead of the shock. a_1/b_1 may be regarded as defining the ratio of gas to magnetic pressure ahead of the wave.   The Mach number M_1 specifies the shock strength and is defined as the flow velocity in shock coordinates divided by the appropriate (fast for fast shocks and slow for slow shocks) small-amplitude disturbance speed. $\theta_1 = \tan^{-1}(B_{y1}/B_x)$ defines the angle of the magnetic field relative to the shock normal.   The thermal properties of the gas can be defined in terms of the ratio of specific heats γ.   A complete description of the shock relations for the entire range of these

parameters clearly requires numerous graphs.   We shall restrict ourselves
to considering only $\gamma = \frac{5}{3}$ and shall present graphs only for selected values
of the other parameters.   The intent is to give the reader some impression
of the significance of these parameters in determining conditions behind
the shock rather than to present a complete survey suitable for use in the
computation of flow fields.

In Fig. 6.6 the conditions behind fast shocks propagating perpen-

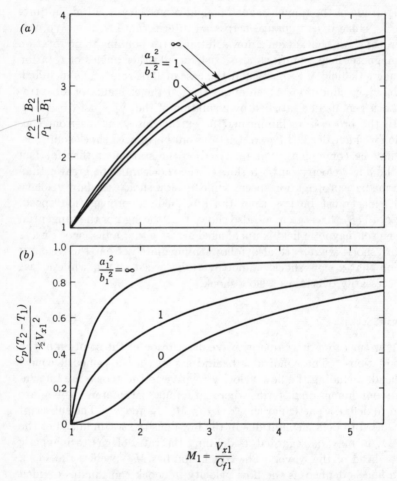

Fig. 6.6   Fast shocks propagating perpendicular to the magnetic field.
(a) Density and magnetic field ratio; (b) change in enthalpy
normalized with respect to the flow kinetic energy ahead
of the shock. Both are plotted against the shock Mach
number defined as the ratio of shock velocity to the fast disturb-
ance speed ahead. $c_{f1} = \sqrt{a_1{}^2 + b_1{}^2}$ for this case. $(\gamma = \frac{5}{3}.)$

dicular to the magnetic field ($B_x = 0$) are plotted with a_1/b_1 as a parameter. The density ratio across the shock is relatively insensitive to a_1/b_1. For weak shocks, the density ratio must, of course, approach unity. For strong shocks, all the curves approach the limiting density ratio of 4 for the case of $\gamma = \frac{5}{3}$. In the strong shock limit, the curves must approach one another, since the flow kinetic energy is large compared with either the thermal energy or the magnetic energy, and thus the ratio becomes insignificant. The temperature change across the shock has been normalized with respect to the flow kinetic energy ahead of the shock wave. In the limit of strong shock waves, virtually all ($\frac{15}{16}$) of the flow energy ahead of the shock becomes converted to thermal energy behind. In the limit of weak shocks the temperature change must of course approach zero. It is interesting to notice that, for intermediate-strength shocks, the temperature change is significantly smaller for the case of a strong magnetic field ahead of the shock ($a_1/b_1 < 1$) than it is for the case of weak magnetic fields ahead of the shock. This difference is accounted for by the fact that the magnetic energy density behind the shock is higher than it is ahead of the shock; thus, in the case of strong magnetic fields, some of the initial kinetic energy is converted into magnetic energy, and a smaller amount remains for thermal energy. It is a fairly general property of fast shock waves that the presence of the magnetic field tends to reduce the temperature change because of the energy which must go into the magnetic field.

Figure 6.7 shows conditions behind fast shocks propagating along the magnetic field ($B_{y1} = 0$). Since ordinary hydrodynamic shock waves produce no transverse velocity, one might expect that in this situation one would obtain only the ordinary hydrodynamic solutions. This, however, is not necessarily the case. If the magnetic field is sufficiently large, the ordinary shock solution may give a flow velocity behind the shock which is less than the intermediate propagation speed. In terms of our previous discussion, this would correspond to an extraneous solution and is thus not allowed. Under these conditions, the allowed fast-shock solution has a tangential component of the magnetic field behind the shock although it is zero ahead of the shock. Such shocks are referred to as *switch-on shocks*, since the tangential field is switched on by the shock. In this case, it follows from Eq. (6.38) that the flow speed behind the shock is precisely equal to the intermediate speed. This conclusion leads to an apparent paradox, since it would seem to be perfectly reasonable to accelerate a piston along magnetic field lines to a velocity which corresponds to such an extraneous ordinary shock solution. In this case one might expect to find the ordinary shock rather than the switch-on shock. This paradox can be resolved by the fact that there is also a slow *switch-off shock wave*. For a switch-off shock wave, the tangential component of the

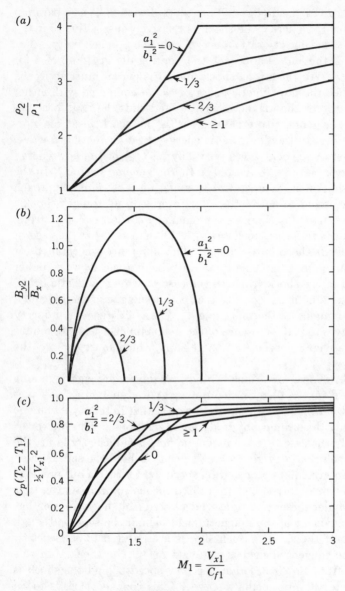

Fig. 6.7 Fast shocks propagating along the magnetic field. (*a*)
Density ratio; (*b*) ratio of tangential component of
magnetic field behind the shock to normal component;
(*c*) change in enthalpy, all plotted against shock Mach
number.   In this case the fast disturbance speed ahead
is given by $c_{f1} = a_1$ for $a_1/b_1 > 1$ and $c_{f1} = b_1$ for
$a_1/b_1 < 1$.   ($\gamma = \frac{5}{3}$.)

magnetic field is finite ahead of the shock and zero behind it.   Again from Eq. (6.38), the velocity ahead of a switch-off shock wave is equal to the intermediate speed ahead of the shock.   Thus, for the particular case just given, the switch-on and switch-off shock waves would travel at precisely the same speed.   If both are created at the same instant of time, the net result of a switch-on shock followed immediately by a switch-off shock would be indistinguishable from the ordinary hydrodynamic shock solution.   Although the distinction between a composite shock, made up of a switch-on and a switch-off shock, and an ordinary hydrodynamic shock seems somewhat artificial for the case in which the shock waves are propagating precisely along the magnetic field lines, the distinction does have some significance if the wave propagation is at a slight angle to the magnetic field.   In the latter case, the fast shock will be not quite a switch-on shock, and thus the flow velocity behind the shock will be slightly greater than the intermediate speed behind the shock.   Similarly, the propagation speed of the almost switch-off shock will be slightly less than the intermediate propagation speed.   Thus, the slow shock will move slightly more slowly than the fast shock, and as time progresses, the two shock waves will separate.

For the case $a_1/b_1 = 0$, the ordinary shock solutions do apply for Mach numbers greater than 2 in the case $\gamma = \frac{5}{3}$.   However, in the range of Mach numbers between 1 and 2, the appropriate fast shock solutions are switch-on shock waves.   For $a_1/b_1 = 1$, the ordinary shock solutions are possible down to Mach number 1 without violating the condition of not crossing the intermediate speed.   Thus, no switch-on shock waves exist for $a_1 > b_1$.   For all of the curves in Fig. 6.7 with $a_1/b_1 < 1$, the ordinary shock solution applies above some critical Mach number which lies between 1 and 2.   Note that the Mach number which is used as abscissa in these curves is the ratio of the flow velocity ahead of the shock to the small-amplitude disturbance speed and not the ordinary sound speed.   This difference in the definition of the Mach number accounts for the fact that the curves are not identical in the range in which they satisfy the ordinary shock equations.   In the range of the switch-on shock waves both the temperature and the density behind the shock are lower than they would be for the corresponding ordinary shock.   As in the previous case, this is due to the magnetic-energy density and magnetic pressure behind the shock.

The magnitude of the tangential component of the magnetic field which is switched on in the shock wave must be zero both for weak shock waves $M \rightarrow 1$ and for the critical Mach number at which the transition from the ordinary shock to the switch-on solution occurs.   Thus, the magnitude of the tangential component of the magnetic field behind the shock has a maximum at some intermediate-shock Mach number.   It

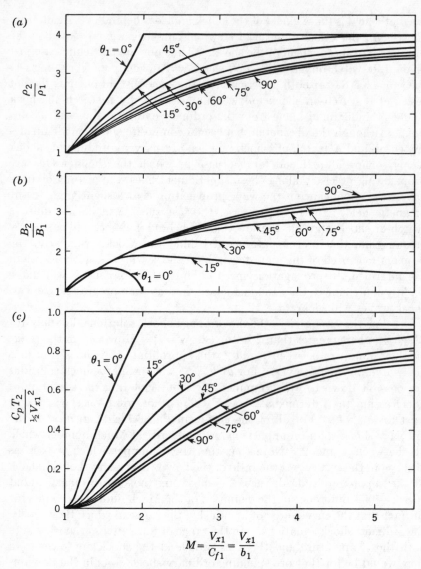

Fig. 6.8   Fast shocks propagating into a cold gas $a_1/b_1 = 0$ for arbitrary propagation direction relative to the magnetic field.   (*a*) Density ratio; (*b*) ratio
of magnitude of magnetic field; (*c*) enthalpy behind shock, all plotted
against shock Mach number. $\theta_1 = \tan^{-1}(B_{y1}/B_x)$.   ($\gamma = \frac{5}{3}$.)

may be seen that the magnitude of the tangential component of the magnetic field can in some cases be slightly larger than the normal component of the magnetic field.

In order to give some feeling for the variation with initial angle of the magnetic field, the properties behind fast shock waves have been plotted in Fig. 6.8, with the initial angle as a parameter for the special case $a_1/b_1 = 0$. As can be seen, there are no dramatic new effects occurring. The various curves between 0 and 90° fall smoothly between these limiting cases, which we have already discussed. It is of some interest to note that the curves corresponding to large angles tend to bunch together. Thus, angles over a fairly wide range in the neighborhood of 90° can be approximated fairly well by the simpler 90° calculations. The ratio of the magnitude of the magnetic field behind the shock to that ahead is also shown. It can be seen by inspection that this is always less than the density ratio. However, as is obvious for switch-on shock waves, the ratio of the tangential component may be larger than the density ratio. In the limit of very strong shock waves the tangential component increases by the same ratio as the density. However, as the shock strength decreases, the density ratio decreases, while for small angles the ratio of the magnitude of the magnetic field increases. Thus the ratio of the tangential component must become larger than the density ratio.

6.4.2. Slow Shocks

Some of the properties of slow shock waves are shown in Figs. 6.9 and 6.10. In Fig. 6.9 the properties of switch-off shock waves are shown as a function of the initial angle of the magnetic field relative to the wave normal for various values of a_1/b_1. Since switch-off shock waves propagate at the maximum allowable speed for slow shocks, namely, the intermediate speed, the switch-off shock may be regarded as the strongest possible slow shock for a particular plasma condition. Since the limit of weak shocks was covered in the consideration of small amplitude disturbances, it would seem instructive to examine the opposite limit of strong shocks. For propagation along the magnetic field, there is no tangential component of the magnetic field to switch-off. Thus, for a_1/b_1 less than unity, the shock reduces to an ordinary hydrodynamic shock with a Mach number equal to b_1/a_1. For a_1/b_1 greater than unity, the intermediate and slow speeds are equal for propagation along the magnetic field. Thus, the propagation speed of the switch-off shock must be equal to the slow small-amplitude disturbance speed. The shock wave is therefore weak, and the temperature and density ratios must be equal to unity. For large angles between the wave normal and the magnetic field, the propagation velocity becomes small, since the intermediate speed depends on the normal component of the magnetic field.   In this limit, the pressure balance across the shock wave approaches a hydrostatic balance.   That is, the dynamic pressure becomes small, and thus the gas-pressure change must balance the change in magnetic pressure associated with the change in the tangential component of the magnetic field.   As a_1/b_1 becomes large, the gas pressure becomes large compared with the magnetic pressure.   Thus, only small density changes are required to change the gas pressure by an amount equal to the magnetic pressure, and

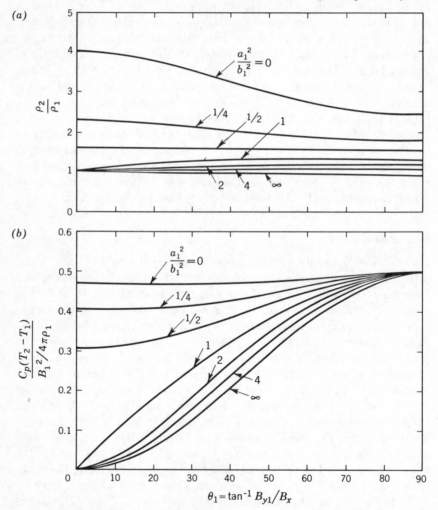

Fig. 6.9   Maximum-strength slow shocks (switch-off, $B_{y2} = 0$).   (a) Density ratio; (b) change in enthalpy normalized with respect to available magnetic energy per unit mass plotted against direction of magnetic field relative to wave normal.   ($\gamma = \frac{5}{3}$.)

the density ratio across the shock waves becomes small.   That is exhibited in Fig. 6.9 by the fact that for a_1/b_1 greater than about 1 the density ratio across the switch-off shock at all initial angles of the magnetic field is fairly small.   We may conclude that the switch-off shock and thus all slow shocks have a fairly small amplitude for a_1/b_1 greater than about 1.

The change in thermal energy across the shock wave has been non-dimensionalized with respect to the available magnetic energy per particle. In the limit of propagation normal to the magnetic field for arbitrary values of a_1/b_1, and in the limit of $a_1/b_1 \to \infty$ for all angles of propagation, the change in thermal energy is precisely equal to one half of the available magnetic energy.   For other conditions, the change in thermal energy is always greater.   This additional energy may be regarded as coming from the kinetic energy of the flow associated with the normal velocity ahead of the shock.   In the two limiting cases just mentioned, the other half of the available magnetic energy goes into producing tangential velocity behind the shock wave.

In Fig. 6.10, the changes in flow properties across slow shocks for a range of shock velocities, but all at an arbitrarily chosen angle of 45° between the direction of propagation and the initial magnetic field, are plotted.   All these curves stop abruptly when the switch-off shock is reached, i.e., when $B_{y2}/B_{y1} = 0$.   In the discussion of small-amplitude disturbances, we mentioned that, in the limit of a_1/b_1 much less than 1, the magnetic field is so stiff that the field lines remain straight, and thus the disturbance may be considered as an ordinary sound wave which is constrained to move along the field lines.   The same consideration applies to moderate-strength shock waves.   As long as the shock velocity is small compared with the intermediate propagation speed, the conditions behind the shock are the same as they would be for an ordinary shock propagating along the magnetic field lines at a velocity $v_{x1}/\cos \theta_1$.   As the shock speed approaches the intermediate speed, the changes in the magnetic field do become significant in changing the flow properties behind the shock. Thus, for example, the density ratio appears to reach a maximum somewhat before the limiting shock strength, namely, the switch-off shock, is reached.

6.5 SHOCK STRUCTURE

The structure of shock waves depends not only upon the various parameters used to describe the macroscopic properties of the shock waves in the previous section but also upon the plasma conditions, such as degree of ionization and the ratio of the mean-free path to the gyroradius.   The entire subject, therefore, covers a vast range of phenomena, many of which are not fully understood at the present time.   A complete discus-

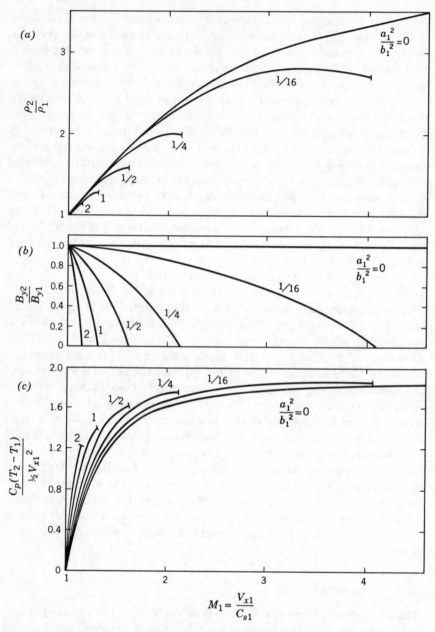

Fig. 6.10   Slow shocks propagating at 45° relative to the magnetic field direction. (a) Density ratio; (b) ratio of tangential component of magnetic field; (c) change in enthalpy, all plotted against Mach number, ratio of shock velocity to slow disturbance speed. ($\gamma = \frac{5}{3}$.)

sion of the present state of knowledge on this subject is beyond the scope of this chapter. We shall restrict ourselves to a very brief and highly simplified discussion of some of the salient features.

There are two predominant reasons for interest in the structure of shock waves. On the one hand all the theory described in the earlier part of this chapter depends upon the assumption that the shock waves which are formed in the flow can be considered as extremely thin. We must therefore have some estimate of the thickness of the shock wave in order to determine under what conditions this assumption is justified. On the other hand, there are many conditions in plasmas, particularly when the mean-free path becomes large compared with the gyroradius, where the basic dissipation mechanisms which occur in the plasma are relatively poorly understood. Since the conservation equations for shock waves require the existence of dissipation, the study of the shock structure can help to elucidate the basic dissipation processes which occur.

6.5.1. Collision-dominated Shock Waves

Let us consider first cases in which the transport coefficients are assumed to be known on the basis of the usual kinetic-theory arguments. This assumption can be reasonably well justified in the case of high densities and low temperatures where the mean-free path is small compared with the gyroradius. However, as we shall see later, when the mean-free path becomes large the usual kinetic-theory approximations are no longer justified for shock waves.

The formal procedure for solving the shock structure when the transport coefficients are known involves writing down the steady-flow equations, including the transport terms. Given a uniform flow of known conditions at minus infinity, these equations allow a transition to only one other uniform flow condition, corresponding to the other side of the shock wave. The solution of the equations then determines the detailed shock structure. For the present discussion we shall not elaborate on this procedure. We shall restrict ourselves to rough arguments which give an estimate of the overall thickness of the shock wave.

Since the total rate of dissipation per unit area of the shock front is specified by the conservation equations, the shock structure must adjust to a thickness such that the required dissipation is produced. The rate of dissipation per unit volume is usually proportional to the square of the gradient. For example, viscous dissipation depends upon the square of the velocity gradient, and Joule heating depends upon the square of the current, which is equivalent to the square of the magnetic field gradient. Since the changes in velocity and magnetic field across the shock are known, the volume rate of dissipation varies inversely as the square of the shock thickness. This dissipation exists over a region of the order of the shock thickness, and hence the total rate of dissipation within the shock front varies inversely as the first power of the shock thickness. Thus, as a pressure pulse steepens to form a shock wave, the total dissipation rate is initially low and the steepening process continues until a thickness is reached such that sufficient dissipation is produced. On the other hand, if we imagined a very sharp discontinuity to be formed initially which is thinner than the required shock structure, then the dissipation would be too high and the discontinuity would spread out until the steady-state shock structure was reached.

Let us first obtain a very crude estimate of the shock thickness in the case in which viscosity is the predominant dissipation mechanism. If Φ is the dissipation rate per unit area of the shock front divided by the rate of flow of kinetic energy ahead of the shock wave, then the above remarks lead to the conclusion

$$\Phi \approx \frac{\mu \left(\dfrac{\mathbf{v}_2 - \mathbf{v}_1}{\delta}\right)^2 \delta}{\frac{1}{2}\rho_1 v_{x1}^3} \approx \frac{\mu}{\rho v_{x1}\delta} \cdot \frac{2(\mathbf{v}_2 - \mathbf{v}_1)^2}{v_{x1}^2} \tag{6.40}$$

where μ = viscosity

δ = shock thickness

Since both Φ and the parentheses on the right-hand side are known in terms of the overall shock conditions, the above relation gives the Reynolds number $\rho v_{x1}\delta/\mu$ based on the shock thickness in terms of the overall shock properties. This equation therefore gives an estimate of the shock thickness.

For a strong fast shock, the energy dissipated is essentially all the thermal energy in the gas behind the shock. This in turn is roughly equal to the kinetic energy of the flow ahead of the shock. Since the change in velocity is of the order of the velocity ahead of the shock, Eq. (6.40) reduces to the statement that the Reynolds number based on the shock thickness is of the order of unity. If we use the kinetic-theory formula for the viscosity, the statement is equivalent to the statement that the shock thickness is of the order of the mean-free path. For weaker shock waves the energy dissipated decreases more rapidly than the change in velocity across the shock wave. Thus, the Reynolds number, or the shock thickness measured in mean-free paths, becomes larger as the shock strength decreases.

The same argument applied to the case in which Joule heating is the predominant dissipation process leads to the relation

$$\Phi \approx \frac{[c_e^2\delta/(4\pi)^2\sigma][(B_{2y} - B_{1y})/\delta]^2}{\frac{1}{2}\rho_1 v_{x1}^3}$$

$$\approx \frac{c_e^2}{4\pi\sigma v_{x1}\delta}\frac{B_{y1}^2}{2\pi\rho v_{x1}^2}\left(\frac{B_{y2} - B_{y1}}{B_{y1}}\right)^2 \tag{6.41}$$

where c_e = speed of light

σ = electrical conductivity

This relation specifies the order of magnitude of the magnetic Reynolds number $4\pi\sigma v_{x1}\delta/c_e^2$ which would be required to provide the appropriate dissipation in terms of the macroscopic shock parameters. The shock thickness can therefore be estimated from Eq. (6.41) if the predominant dissipation process is Joule heating.

If both dissipation coefficients are finite, one might expect that the appropriate shock thickness is simply given by choosing the larger of the two thicknesses given by Eqs. (6.40) and (6.41). While this view is correct under some conditions, in general it is somewhat oversimplified. For example, in a strong fast shock propagating in a plasma with a low electrical conductivity ($4\pi\sigma\mu/\rho c_e^2 \ll 1$), there are actually two characteristic thicknesses associated with the shock. Near the front of the shock, the magnetic field rises with very little change in flow velocity in a distance such that the magnetic Reynolds number is of the order of unity. Following this there is a more abrupt change in the flow velocity with very little change in the magnetic field. The change in flow velocity occurs in a region whose thickness is such as to make the ordinary Reynolds number of the order of unity. A rough criterion for the conditions under which both dissipation mechanisms are important in determining the shock structure may be obtained as follows: For a fast shock propagating perpendicular to the magnetic field the current at any point within the shock is proportional to the electric field in a coordinate system moving with the gas and may be written as

$$j = \frac{\sigma}{c_e}(v_{x1}B_{y1} - v_x B_y) \tag{6.42}$$

Since the flow velocity decreases monotonically, through the shock, the current density will always be less than what is obtained by replacing v_x by v_{x1}. The current density is the curl of the magnetic field, and thus we may rewrite Eq. (6.42) as

$$\frac{1}{\delta} = \frac{1}{B_y - B_{y1}}\frac{d(B_y - B_{y1})}{dx} \le \frac{4\pi\sigma v_{x1}}{c_e^2} \tag{6.43}$$

This equation then states that the minimum distance in which the magnetic field can rise is such that the magnetic Reynolds number based on that distance be of the order of unity. Returning to Eq. (6.41), we see that for strong shocks the thickness required to produce all the dissipation in the shock wave is less than this minimum thickness. Thus, for strong shock waves, Joule heating alone is incapable of producing the required amount of dissipation. When this occurs, there must also be a portion of the shock wave in which the velocity gradients are sufficiently steep so that viscous dissipation can account for the remainder of the required dissipation.

A somewhat more precise criterion for when both dissipation mechanisms are required in the shock structure can be obtained as follows: Let us look at the shock structure from the viewpoint of the shock formation process. We know that, for a fast shock, a small-amplitude disturbance behind this shock propagating at the fast propagation speed will overtake the shock. However, as the gradients in the wave that is catching up increase, the magnetic field will no longer follow the density changes and the propagation speed of the wave will decrease until it becomes equal to the ordinary hydrodynamic sound speed. A wave with such a steep gradient may or may not be able to overtake the shock, depending upon whether the flow velocity behind the shock is less than or greater than the ordinary sound speed. For the case in which the flow velocity behind the shock is greater than the ordinary sound speed the steepening process would stop when gradients are reached such that the propagation speed is reduced to the flow speed. In this case one would not obtain gradients so steep that viscous dissipation is required within the shock structure. On the other hand, if the flow velocity is less than the ordinary sound speed, the steepening process can continue even when the gradients have a characteristic length so short that there is no change in magnetic field. The steepening would then continue until another dissipation mechanism, such as viscosity, inhibits further steepening. Thus, in this case there would be two characteristic lengths associated with the shock structure as discussed above. Applying this criterion to a shock wave propagating perpendicular to the magnetic field into a plasma with $a_1/b_1 = 0$, we find that the shock structure can be based purely on Joule dissipation for Mach numbers less than about 2, while for larger Mach numbers viscous dissipation is also required. As a_1/b_1 increases, this critical Mach number is reduced.

The above discussion has been extremely restricted. We have considered the competition between two dissipation processes only for the case of fast shocks. Furthermore, we have considered only two possible dissipation mechanisms, electrical conductivity and viscosity. In general, other dissipation processes such as heat conduction, temperature relaxation between electrons and ions, collisions with neutrals, and ionization of neutrals may also be important. Conditions such as those of Eqs. (6.40) and (6.41) with the appropriate dissipation mechanism can be useful in giving an order-of-magnitude estimate of the shock thickness. Caution, however, should be observed to ensure that the dissipation process which is being considered can indeed provide all the required dissipation.

6.5.2. Collision-free Shock Waves

Let us now turn to a case in which ordinary dissipation mechanisms associated with interparticle collisions cannot produce the required dissipation in the shock wave. Consider a plasma in which $a/b \ll 1$ but the temperature is high enough so that the mean-free path is very large compared with the ion gyroradius. For a fast wave propagating through such a medium the changes in gas pressure will be small as compared with the changes in dynamic pressure and magnetic pressure. To a first approximation we may therefore neglect the plasma pressure entirely. Within this approximation the relations determining the small-amplitude wave propagation are independent of whether or not the length scale is larger or smaller than the mean-free path. For length scales in the flow field larger than the mean-free path the distribution function of particle velocities will actually be isotropic and Eqs. (6.1) through (6.5) will be valid in detail. For length scales small compared with the mean-free path the pressure in Eq. (6.2) should become a tensor and Eq. (6.4) becomes invalid. However, if the pressure is sufficiently small, it can be neglected in Eq. (6.2) whether it is a tensor or a scalar. Furthermore, in this case Eq. (6.4) is no longer required. Thus the relations determining the propagation of fast waves are independent of whether the length scale is larger than or smaller than the mean-free path. It follows immediately that the arguments which were used to show that a compression pulse steepens toward a shock wave are valid even when the length scale of the compression pulse is small compared with the mean-free path. We may therefore expect that the shock wave which forms will have a structure containing length scales much smaller than the mean-free path. As a result binary collisions will probably not provide the required dissipation in the shock.

The length scale at which the above argument for steepening ceases to apply is the ion gyroradius based on a velocity equal to the Alfvén speed. We may therefore anticipate that the actual shock structure will contain length scales of this order of magnitude. This limitation arises from the fact that Eq. (6.3) becomes invalid when such steep gradients are reached [see (A.9) and (A.10)]. This equation is based on the assumption that $\mathbf{E} + (\mathbf{v} \times \mathbf{B})/c_e = 0$. In the absence of collisions this is merely the statement that individual particles drift at the E/B velocity. This, however, is only valid as long as the gyroradius is small compared with the length scale of the changes in the magnetic field, i.e., when the ions move adiabatically. Thus, when length scales comparable with the ion gyroradius are reached, the basic equations become invalid and a different dispersion relation is obtained. The steepening process is therefore modified at this point.

Several attempts at theoretical predictions of the structure of such shock waves have been made.   A clear-cut resolution of the differences between these various theories and the range of plasma conditions over which they apply must await more detailed experimental evidence.   In all these theories the shock wave contains a fine structure with scales of the order of the ion gyroradius or less.   For weak shock waves it is possible to obtain a fairly regular solution consisting of a long train of large-amplitude waves which are steady in a coordinate system moving with the shock wave.   In the presence of any finite rate of damping due to either collisions or Landau damping these oscillations will eventually damp to the uniform conditions behind the shock wave.   The scale length of the individual waves is of the order of the ion gyroradius for general directions of propagation of the shock wave relative to the magnetic field.   However, for the special case, which has received the most attention, of propagation perpendicular to the magnetic field the scale length is smaller than this by the square root of the mass ratio between electrons and ions.

For strong shock waves the microstructure is probably better characterized as a random turbulent structure.   In this case, the flow energy goes initially into turbulent motions and later damps into actual thermal velocities of the ions and electrons.   For such a turbulent shock wave the changes in the average density, average flow velocity, and average magnetic field strength can occur in a distance in which the energy which must be dissipated in the shock wave is put into some form of random energy.   Thus, the thickness of the shock wave as defined by the average quantities does not have to be as long as the thickness required for the turbulent energy to damp into particle motions.   It is therefore possible to have a shock wave in which the major changes in density, flow velocity, and magnetic field occur in a region in which the random energy goes into turbulent magnetic energy.   The region in which this turbulent energy is damped into particle motions may be significantly larger but may have only small changes in density, flow velocity, and average magnetic field associated with it.

The best experimental evidence for collision-free shock waves at the present time has been obtained from recent measurements of the IMP satellite.[5]   The interplanetary plasma has a flow velocity which is of the order of 5 to 10 times the fast propagation speed.   The mean-free path in this plasma is of the order of 10^8 km.   The interaction of this wind with the earth's magnetic field is found to produce a bow shock wave with a thickness of the order of 1,000 km or less.   This thickness is therefore many orders of magnitude less than the mean-free path and is not too different from the ion gyroradius, which is of the order of 100 km.   These results are fairly recent, and it is not clear that the experiment has sufficient resolution to have observed a shock wave much thinner.   The shock wave is characterized by a sudden jump in the average magnetic field strength, as well as an increase in the turbulent fluctuations in the magnetic field strength.

6.6 APPLICATIONS

In this section we shall cite briefly two examples which illustrate applications of the theory discussed in previous sections.   We shall also discuss briefly some experimental evidence which supports the theory.   Unfortunately, at the present time there are very few experimental data available.

6.6.1. Magnetic Shock Tubes

The distance-time diagram shown in Fig. 6.5*b* corresponds to a fast shock followed by a uniform region and a slow expansion fan.   If such a flow configuration can be produced in the laboratory, the uniform gas sample between the fast and slow waves provides a possibility of achieving a high-temperature plasma of known conditions for study in the laboratory. Plasma conditions behind the shock are known in terms of the shock velocity and the conditions ahead of the shock.   Thus, the relatively simple measurement of a shock velocity determines the average magnetic field, enthalpy, and density in the hot-plasma sample.

A schematic diagram of the magnetic annular shock tube in which such a flow has been achieved is shown in Fig. 6.11.   Initially a gas and a

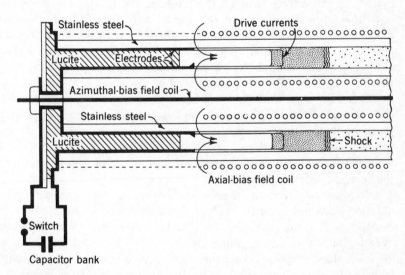

Fig. 6.11 Schematic cross-sectional view of magnetic annular shock tube.

quasisteady magnetic field are present in the thin annular region between the conducting cylinders.   When the condenser bank is discharged, the magnetic field at the insulator is changed and waves propagate along the device.   If the condenser bank is arranged so that the current is essentially a step function, then the diagram shown in Fig. 6.5b (with no intermediate wave) with plasma conditions a function of x/t only is appropriate.   The boundary conditions which must be applied are that for an ideal insulator there is no mass flow through the insulating surface and that the tangential component of magnetic field at the insulating surface is specified by the known current from the condenser bank.   The current from the condenser bank flows partially in the shock wave and partially in the slow-expansion fan.

It has been demonstrated that, under appropriate conditions, a shock wave is produced which travels at the calculated velocity, that there is a region of uniform plasma flow behind the shock wave, and that the density and magnetic field strength in this region correspond to the calculated conditions behind the shock.[6]   Some uncertainties still exist as to the temperature behind the shock, and the expansion fan has not been clearly identified.

It should be pointed out that a number of criteria must be satisfied in order to achieve such operation.   Most experiments which attempt to produce magnetically driven shock waves do not produce a clean shock with a uniform test sample behind it.   First the shock velocity must be sufficiently high so that the gas behind does, in fact, have a high electrical conductivity.   For low conductivity behind the shock the driving magnetic field will diffuse through the uniform test sample and thus disturb both the uniform region and the shock characteristics.   In the experiments mentioned above the gas ahead of the shock wave was at room temperature.   It was found that for sufficiently strong shock waves the electric field ahead of the shock was small enough to justify the assumption that $\mathbf{E} + \mathbf{v} \times \mathbf{B}/c_e$ was zero ahead of the shock.   It has been found by several authors[7] that when one attempts to produce a·slower shock the electric field ahead of the shock is no longer negligible compared with the cross product of the flow velocity and magnetic field, ahead of the shock.   In this case, other modes of operation are found.   For example, it is no longer necessary that the initial disturbance travel at a speed greater than the fast propagation speed.

A second criterion for reasonable operation of such a device is that ablation does not occur.[8]   Ablation, which seems to be predominantly associated with the insulating wall, adds mass to the region between the cylinders and thus can significantly slow down the resulting shock wave.

A third criterion is that the annular region between the cylinders be sufficiently small compared with the radius of the cylinders.[9]   The azimuthal component of the magnetic field which is to be regarded as the y component in our previous discussions will for a force-free field fall off inversely as the radius.   Thus, there is a difference between this field at the inner and the outer walls.   Unless the radius ratio is kept sufficiently close to unity, this nonuniformity will destroy the one-dimensional nature of the flow.

A series of experiments has also been conducted which have verified the linear propagation of the fast mode propagating along the magnetic field in a plasma with a/b small compared with 1.[10]   These experiments have recently been extended to a larger-amplitude disturbance and have shown that such large-amplitude pulses do indeed steepen toward shock waves.[11]

6.6.2. Conversion of Magnetic to Plasma Energy

It was mentioned in the discussion of slow shock waves that these shock waves tend to convert magnetic energy to plasma kinetic or thermal energy.   This may be a significant mechanism by which energy is converted in a number of situations which may occur in nature.   Let us imagine an interface between two regions of plasma in which the magnetic field has significantly different orientation.   The overall pressure balance across the interface will be adjusted rapidly by the fast-wave mode.   If there is no component of the magnetic field normal to the surface, then the intermediate and slow propagation speeds are zero and one would expect the magnetic fields on the two sides to diffuse into one another at a speed determined by the conductivity of the medium.   If, on the other hand, a small component of the magnetic field normal to the interface exists, then the propagation of intermediate and slow waves is possible. It is always possible to find a combination of intermediate and slow waves of appropriate strengths propagating in both directions from the boundary to satisfy the boundary conditions of appropriate change in direction and magnitude of the magnetic field.   Furthermore, since the components of the magnetic field which are oppositely directed on the two sides of the boundary can cancel one another, the resulting configuration will have a lower magnetic energy.   For a high-conductivity medium the wave propagation process will obviously convert the magnetic energy at a much higher rate than the diffusion process, since the wave propagation speeds do not decrease with increasing conductivity.   In the example we are at present discussing, the waves would propagate at a constant speed, and the region in which the magnetic energy has decreased would increase linearly with time.

It has also been suggested[12] that a steady two-dimensional configuration is possible in which the predominant mechanism for the conversion of magnetic energy to plasma energy is the existence of standing intermediate and slow waves.   In this case an x-type neutral point is formed in the flow.   The flow on both sides of the boundary is toward the boundary and leaves the region by flowing along the boundary, in opposite directions on either side of the neutral point.   In such a configuration standing waves are possible along the boundary, with the exception of a small region in the immediate vicinity of the neutral point.   In this region, since the magnetic field is small, the propagation speed vanishes and therefore diffusion must still be important.   However, the region over which diffusion is important is significantly smaller than it would be if one had assumed no normal component of the magnetic field anywhere along the boundary.   As a result, the net rate at which flow goes toward the boundary and the rate at which magnetic energy is converted to plasma energy decrease only logarithmically with increasing conductivity or magnetic Reynolds number.   Since pure diffusion would lead to an inverse-square-root dependence on conductivity, the wave mechanism leads to significantly higher rates for high-magnetic-Reynolds-number situations.

This result is probably of particular significance in a number of astrophysical situations where the length scale is so large that the magnetic Reynolds number is usually very large.   Several examples where such a flow configuration may be of interest can be given.   The origin of solar flares is believed to be a storage of magnetic energy above the photosphere of the sun which is then released suddenly.   A rapid mechanism for conversion of this magnetic energy such as the one just described seems to be required in order to account for the observed suddenness of solar flares.   At the boundary of the magnetosphere the magnetic field direction on the solar-wind side and on the earth's side are significantly different.   As a result one may expect that the boundary would resolve itself into a combination of intermediate and slow waves.   This in turn would result in a significant rate of reconnection of the field lines in the solar wind to the dipole field lines on the earth, which, in turn, can cause motions within the magnetosphere.   Thus far, there have been no direct measurements from satellites at boundary crossings to indicate this resolution of the boundary.   However, the flow rate observed indirectly inside the magnetosphere from ionspheric currents and auroral motions is in rough agreement with the predicted rate of reconnection at the boundary.   Finally, a dilemma is raised by the fact that the solar wind continually drags field lines away from the sun; however, the net field strength at the surface of the sun cannot increase indefinitely.   Thus, these field lines must eventually be broken so that they may return to the sun.   On the basis of simple diffusion arguments this rate of breaking would be much too slow.   It seems likely that a mechanism involving wave propagation such as the one just discussed could give a sufficiently rapid rate of breaking to avoid this increase in field strength at the solar surface.

6.7 APPENDIX: RANGE OF APPLICABILITY OF BASIC EQUATIONS

The limitations on the validity of the basic equations (6.1) to (6.5) arise principally from two causes. First, since heat conduction and viscosity were neglected, the process by which the particle-distribution function is made isotropic must be rapid compared with the typical time scale in the flow. Second, the equations assume infinite conductivity, thus implying that currents can flow freely. If the predominant process for the randomization of the particle distributions is scattering by binary collisions, these two requirements at first sight seem somewhat contradictory. The requirement of rapid achievement of isotropy implies a mean-free time for collisions short compared with the flow time, while the requirement of high conductivity implies a long mean-free time. As we shall see, these two requirements are not mutually exclusive, however. Thus, for a fully ionized plasma of a given density and with a given length scale to the flow field, there exists a minimum temperature below which the conductivity becomes too low and a maximum temperature above which the mean-free time becomes too large for isotropization to be achieved sufficiently rapidly by binary collisions.

These limits will be evaluated quantitatively below. The actual range of validity of the equations, however, is probably considerably larger than implied by these limits. Several types of nonisotropic particle distributions in collision-free plasmas are known to be unstable (cf. Chap. 5). As such an instability grows, it must lead to the production of a more isotropic particle distribution. Thus, if the growth times of these instabilities are sufficiently short, isotropy of the particle distribution may be achieved by this mechanism rather than by binary collisions. This would imply that the above equations may be valid at temperatures or mean-free times longer than the limits derived from assuming randomization by binary collisions.

At the present time, no concise and quantitative estimate of the limits to the validity of the equations based on randomization by the growth of instabilities has been given. It should be borne in mind, however, that the limits which we shall now derive based on particle collisions probably underestimate the region of validity.

Let us now turn to discussion of the limitations on the individual equations. Equations (6.1) and (6.5) are generally valid. In the momentum equation (6.2), the relation

$$\nabla \times \mathbf{B} = \frac{4\pi \mathbf{j}}{c_e} \tag{A.1}$$

has been used to eliminate the current density **j**.   As before, c_e is the velocity of light.   Thus, the assumption has been made that the displacement current is negligible.   This implies that both the fluid velocities and the wave velocities must be small compared with the velocity of light. Furthermore, the body force due to the interaction of the electric field with any net charge density in the plasma has been neglected.   This assumption is valid if the Debye length is small compared with the length scale of the flow field.   Neither of the above assumptions is very restrictive for the range of plasma conditions which are usually of interest.   In writing the pressure as a scalar we have assumed that the gradients are sufficiently gentle so that viscous terms are not important.   Quantitatively, this condition may be given as

$$\mu \frac{v}{L} \ll p \quad \text{or} \quad \lambda \ll L \tag{A.2}$$

where μ = viscosity of plasma

L = characteristic length of flow field

λ = mean-free path for particle collisions

The second form of this restriction can be obtained from the first by making use of the ordinary kinetic-theory formula for the viscosity in the absence of a magnetic field and making the assumption that the flow velocity is of the order of the thermal velocity of the ions.   It could also be obtained more directly from the condition that, in order to maintain the particle distribution isotropic, there must be frequent randomizing collisions and therefore the collision distance must be small compared with the distance in which the flow properties change appreciably.

We may note in passing that there are the two limits in which the above condition is overrestrictive.   One of these occurs when the ion gyroradius is small compared with the mean-free path and the gradients are primarily in a direction perpendicular to the magnetic field.   Under these conditions, the particle orbits will be turned in a distance appreciably less than the mean-free path.   The distribution function for motions perpendicular to the magnetic field then becomes isotropic in a smaller distance than the mean-free path.   Alternatively, we could say that, under these conditions, the viscosity becomes a tensor whose components perpendicular to the magnetic field become appreciably reduced.   The second limit occurs when both the dynamic pressure ρv^2 and the magnetic pressure $B^2/8\pi$ are large compared with the plasma pressure.   In this case, the entire pressure tensor can be neglected in the momentum equation, and the restriction (A.2) is irrelevant.   For the sake of simplicity we shall assume throughout the remainder of the discussion in this section that the three pressures above are comparable,

$$\rho v^2 \approx p \approx \frac{B^2}{8\pi} \tag{A.3}$$

and that the tangent of the angle between the magnetic field and the direction of the gradients is of order unity.

Equation (6.3) is obtained from Maxwell's equation

$$\nabla \times \mathbf{E} = -\frac{1}{c_e}\frac{\partial \mathbf{B}}{\partial t} \tag{A.4}$$

where the electric field $\mathbf{E}$ has been eliminated by making use of Ohm's law,

$$\mathbf{E} + \frac{\mathbf{v} \times \mathbf{B}}{c_e} = \sigma^{-1} \cdot \mathbf{j} \approx 0 \tag{A.5}$$

under the assumption that the conductivity tensor σ is so large that the right-hand side may be taken to be negligibly small.   If the electron gyrofrequency is small compared with the electron mean-free time, then the electrical conductivity is a scalar.   In this case the above assumption is justified if

$$\frac{\mathbf{j}}{\sigma} \ll \frac{\mathbf{v} \times \mathbf{B}}{c_e} \tag{A.6}$$

Upon using Eq. (A.1) to eliminate the current density, this condition may be written roughly as

$$\frac{4\pi\sigma v L}{c_e^2} \gg 1 \tag{A.7}$$

On the other hand, if the electrons make many gyroorbits between collisions, then their motions are equivalent to motions of a free electron in an electric and magnetic field.   Thus, Ohm's law becomes equivalent to the statement

$$\mathbf{E} + \frac{\mathbf{v}_e \times \mathbf{B}}{c_e} = 0 \tag{A.8}$$

where $\mathbf{v}_e$ is the electron drift velocity.   This equation is equivalent to Eq. (A.5) with the right-hand side set equal to zero, if the difference between the electron velocity and the fluid velocity is small compared with the fluid velocity.   Making use of the fact that the current density is the product of the particle density n, the electronic charge e, and the difference between the electron and ion velocities and that for a fully ionized gas the ion velocity is approximately the flow velocity, we can write the above condition in the following form,

$$\frac{v_e - v_i}{v} \approx \frac{J}{Nev} \approx \frac{c_e B}{4\pi N e v L} \approx \frac{m_i c_e v}{eBL}\frac{B^2}{4\pi\rho v^2} \ll 1 \tag{A.9}$$

where m_i is the ion mass.   By using Eq. (A.3), this reduces to

$$\frac{m_i c_e v}{eBL} \ll 1 \tag{A.10}$$

which states that the ion gyroradius must be small compared with the scale length of the flow field.

Equation (6.4) is the statement that a fluid element changes its state isentropically. This implies that dissipation must be negligible. If we examine the dissipation associated with viscous stresses and Joule heating, we find that it is small if the conditions (A.2) and (A.7) are satisfied. There is, however, another and somewhat more restrictive condition associated with thermal conductivity. The divergence of the heat-flow vector corresponds to a heat addition to the fluid and therefore to an entropy change. We must therefore require that the divergence of the heat flux multiplied by the characteristic flow time L/v be small compared with the thermal energy of the plasma.

$$k \frac{T}{L^2} \frac{L}{v} \ll \rho C_p T \tag{A.11}$$

where k = heat-conduction coefficient
$\quad C_p$ = specific heat at constant pressure
$\quad T$ = temperature
Upon making use of the kinetic-theory formula for heat conduction in the absence of a magnetic field but remembering that the heat conduction is primarily due to the electrons because of their high thermal velocity, this equation can be written in the form

$$L \gg \sqrt{\frac{m_i}{m_e}} \lambda \tag{A.12}$$

This equation differs from Eq. (A.2) only by a numerical factor. However, Eq. (A.12) is the more restrictive one.

The limitations on the range of validity of the basic equations is thus given by Eqs. (A.7), (A.10), and (A.12). By use of Eq. (A.12) and the kinetic-theory formula for the electrical conductivity of the plasma, Eq. (A.7) may be written as

$$L \gg \frac{r_i^2}{\sqrt{m_i/m_e} \, \lambda} \tag{A.13}$$

where r_i = ion gyroradius
It is easily seen that if conditions (A.12) and (A.13) are satisfied Eq. (A.10) is automatically justified. The two remaining conditions are therefore Eqs. (A.12) and (A.13). By using the relation for the mean-free path in a fully ionized plasma,[13] these equations may be rewritten as

$$L \gg \frac{3 \times 10^{14} T^2}{N} \tag{A.14}$$

$$L \gg \frac{2}{T^2} \tag{A.15}$$

where the length L is to be measured in centimeters, the temperature is in electron volts, and the particle density is in particles per cubic centimeter.

Equation (A.15), which results from the electrical conductivity, is satisfied if we are dealing with length scales of several centimeters and temperatures greater than a few electron volts. In using this condition for very large lengths, it should, of course, be borne in mind that we have discussed only fully ionized plasmas and that complete ionization is not expected at temperatures below about 1 eV. The condition (A.14), on the other hand, is violated at very high temperatures or low densities. Thus, for a particular length scale and a sufficiently high density a temperature range exists in which the equations are valid between a temperature determined by Eq. (A.15) and that determined by Eq. (A.14).

As we have tried to indicate throughout this section, the limitations given by Eqs. (A.14) and (A.15) should be regarded as defining the minimum range of validity of the basic equations. Several limiting cases corresponding to violation of the condition (A.3) and the assumption that the gradients are in a direction which makes an angle of the order of 1 rad with the magnetic field exist in which the range of validity would be wider. Furthermore the possibility that isotropic particle distributions are achieved by the growth of instabilities rather than by binary collisions suggests that the restriction given by Eq. (6.9) may in many cases be removed entirely.

In particular recent satellite data give clear evidence that the flow of the solar wind over the magnetic field of the earth exhibits several magnetohydrodynamic phenomena in spite of the fact that the mean-free path is several orders of magnitude larger than the scale of the region in which the interaction takes place.[5] The data indicate clearly that a shock wave is formed some distance ahead of the actual boundary between the solar-wind plasma and the earth's magnetic field. The distance between the shock wave and the interface agrees with calculations based on a magnetohydrodynamic model. This distance also expands as one would expect as one moves to points away from the stagnation point. Within such a flow there are certainly phenomena occurring which cannot be explained directly by the magnetohydrodynamic model, for example, the existence of considerable turbulence which is presumably related directly to the dissipation in the shock and also the production of a non-Maxwellian tail of high-energy particles. The magnetohydrodynamic theory does, however, seem capable of describing the gross properties of the flow field such as changes in average magnetic field, density, and flow velocity.

REFERENCES

1. Friedrichs, K. O.: Nonlinear Wave Motion in Magnetohydrodynamics, *Los Alamos Rept.* LAMS-2105 (Physics), written September, 1954, dis-

tributed Mar. 8. 1957; K. O. Friedrichs and H. Kranzer: Notes on Magneto-hydrodynamics.   VIII. Nonlinear Wave Motion, *N.Y. Univ. Rept.* NYO 6486-VIII, 1958.

2. Courant, R., and K. O. Friedrichs: "Supersonic Flow and Shock Waves," Interscience, New York, 1948; L. D. Landau and E. M. Lifshitz: "Fluid Mechanics," Addison-Wesley, Reading, Mass., 1959; A. R. Kantrowitz: One-dimensional Treatment of Nonsteady Gas Dynamics, in H. W. Emmons (ed.), "Fundamentals of Gas Dynamics," vol. 3, High Speed Aerodynamics and Jet Propulsion Series, Princeton, Princeton, N.J., 1958.

3. Shercliff, J. A.: One-dimensional Magnetogasdynamics in Oblique Fields, *J. Fluid Mech.*, **9,** 481 (1960).

4. Ericson, W. B., and J. Bazer: Hydromagnetic Shocks, *Astrophys. J.*, **129,** 758 (1959).

5. Ness, N. F., C. S. Scearce, and J. B. Seek: Initial Results of the IMP-1 Magnetic Field Experiment, *J. Geophys. Res.*, **69,** 3531 (1964).

6. Patrick, R. M.: High Speed Shock Waves in a Magnetic Annular Shock Tube, *Phys. Fluids*, **2,** 589 (1959); R. M. Patrick and M. Camac: Shock Waves and Collision Free Plasmas, in D. Bershader (ed.), "Plasma Hydromagnetics," Stanford, Stanford, Calif., 1962.

7. Wilcox, J. M., F. I. Boley, and A. W. DeSilva: Experimental Study of Alfvén Wave Properties, *Phys. Fluids*, **3,** 15 (1960); M. H. Brennan, I. G. Brown, D. D. Millar, and C. N. Watson-Munro: Some Experimental Observations of the Characteristics of Hydrodynamic Ionizing Fronts, *J. Nuclear Energy*, **C5,** 229 (1963); J. M. Wilcox, E. Pugh, A. Dattner, and J. Eninger: Experimental Study of the Propagation of an Ionizing Wave in a Coaxial Plasma Gun, to be published in *Phys. Fluids*.

8. Keck, J. C.: Ablation of an Insulator by a Current Parallel to Its Surface, *Bull. Am. Phys. Soc.*, **8** (5), 436 (1963); *Avco-Everett Research Lab. Research Rept.* 152, 1963.

9. Keck, J. C.: Current Distribution in a Magnetic Annular Shock Tube, *Phys. Fluids*, **5,** 630 (1962); F. J. Fishman and H. E. Petschek: Flow Model for Large Radius-Ratio Magnetic Annular Shock Tube Operation, *Phys. Fluids*, **5,** 632 (1962); F. J. Fishman and H. E. Petschek: Deviation of Magnetic Annular Shock Tube Operation from One-dimensional Model, *Phys. Fluids*, **5,** 1188 (1962).

10. Wilcox, J. M., A. W. DeSilva, and W. S. Copper, III: Experiments on Alfvén Wave Propagation, *Phys. Fluids*, **4,** 1506 (1961).

11. Boley, F. I., and P. R. Forman: Steepening of Large Amplitude Alfvén Waves, *Phys. Rev. Letters*, **12,** 385 (1964).

12. Petschek, H. E.: Magnetic Field Annihilation, in W. N. Hess (ed.), *AAS-NASA Symposium on Phys. Solar Flares*, NASA SP-50, 1964, *Avco-Everett Research Lab.* AMP 123; R. H. Levy, H. E. Petschek, and G. L. Siscoe: Aerodynamic Aspects of the Magnetospheric Flow, presented at the AIAA Meeting, New York, Jan. 20, 1964, *Avco-Everett Research Lab. Research Rept.* 170.

13. Spitzer, Jr., L.: "Physics of Fully Ionized Gases," Interscience, New York, 1962.

BIBLIOGRAPHY

Characteristics and Shock Waves

Akhiezer, A. I., G. I. Liubarskii, and R. V. Polovin: The Stability of Shock Waves in Magnetohydrodynamics, *Soviet Phys. JETP*, **8**, 507 (1959).

Anderson, J. E.: "Magnetohydrodynamic Shock Waves," M.I.T., Cambridge, Mass., 1963.

Bazer, J.: Resolution of an Initial Shear Flow Discontinuity in One-dimensional Hydromagnetic Flow, *Astrophys. J.*, **128**, 686 (1958).

Chu, C. K., and Y. M. Lynn: Steady Magnetohydrodynamic Flow Past a Non-conducting Wedge, *AIAA J.*, **1**, 1062 (1963).

De Hoffman, F., and E. Teller: Magnetohydrodynamic Shocks, *Phys. Rev.*, **80**, 692 (1950).

Grad, H.: Propagation of Magnetohydrodynamic Waves without Radial Attenuation, in D. Bershader (ed.), "The Magnetodynamics of Conducting Fluids," Stanford, Calif., 1959; Third Symposium on Magnetohydrodynamics, Palo Alto, Calif., 1958.

Helfer, H. L.: Magnetohydrodynamic Shock Waves, *Astrophys. J.*, **117**, 177 (1953).

Kemp, N. H., and H. E. Petschek: Theory of the Flow in the Magnetic Annular Shock Tube, *Phys. Fluids*, **2**, 599 (1959).

Kontorovich, V. M.: On the Interaction between Small Disturbances and Discontinuities in Magnetohydrodynamics and the Stability of Shock Waves, *Soviet Phys. JETP*, **8**, 851 (1959).

Liubarskii, G. I., and R. V. Polovin: The Disintegration of Unstable Shock Waves in Magnetohydrodynamics, *Soviet Phys. JETP*, **9**, 902 (1959).

Lust, R.: Magnetohydrodynamische Stosswellen in einem Plasma unendlicher Leitfahigkeit, *Z. Naturforsch.*, **8a**, 277 (1953).

Lust, R.: Stationare Magnetohydrodynamische Stosswellen beliebiger Starke, *Z. Naturforsch.*, **10a**, 125 (1955).

Mimura, Y.: Magnetohydrodynamic Flow Past a Wedge with a Perpendicular Magnetic Field, *AIAA J.*, **1**, 2272 (1963).

Polovin, R. V.: Contribution to the Theory of Simple Magnetohydrodynamic Waves, *Soviet Phys. JETP*, **12**, 326 (1961).

Polovin, R. V.: Shock Waves Magnetohydrodynamics, *Soviet Phys. Uspekhi*, **3**, 677 (1961).

Syrovatskii, S. I.: Magnitnoigidrodinamike, *Uspekhi Fiz. Nauk.*, **62**, 247 (1957).

Syrovatskii, S. I.: The Stability of Shock Waves in Magnetohydrodynamics, *Soviet Phys. JETP*, **8**, 1024 (1959).

Shock Structure

Auer, P. L., H. Hurwitz, and R. W. Kilb: Low Mach Number Magnetic Compression Waves in a Collision-free Plasma, *Phys. Fluids*, **4**, 1105 (1961).

Auer, P. L., H. Hurwitz, Jr., and R. W. Kilb: Large-amplitude Magnetic Compression of a Collision-free Plasma. II. Development of a Thermalized Plasma, *Phys. Fluids*, **5**, 298 (1962).

Camac, M., A. R. Kantrowitz, M. M. Litvak, R. M. Patrick, and H. E. Petschek: Shock Waves in Collision Free Plasmas, *Nuclear Fusion Suppl.*, pt. 2, p. 423, 1962.

Colgate, S. A.: Collisionless Plasma Shock, *Phys. Fluids*, **2**, 485 (1959).

David, L., R. Lust, and A. Schluter: The Structure of Hydromagnetic Shock Waves, *Z. Naturforsch*, **13a**, 916 (1958).

Fishman, F. J., A. R. Kantrowitz, and H. E. Petschek: Magnetohydrodynamic Shock Wave in a Collision Free Plasma, *Revs. Modern Phys.*, **32**, 959 (1960).

Gardner, C. S., H. Goertzel, H. Grad, C. S. Morawetz, M. J. Rose, and H. Rubin: Hydromagnetic Shock Waves in High Temperature Plasmas, *Proc. Second UN Intern. Conf. on the Peaceful Uses of Atomic Energy*, Geneva, September, 1958, vol. 31, pp. 230–237.

Germain, P.: Shock Waves in Shock Structure in Magneto-fluid Dynamics, *Revs. Modern Phys.*, **32**, 951 (1960).

Gerry, E. T., R. M. Patrick, and H. E. Petschek: Magnetohydrodynamic Shock Structure in a Partially Ionized Gas, presented at the Sixth International Symposium on Ionization Phenomena in Gases, Paris, July 8–13, 1963.

Karpman, V. I.: Structure of the Shock Front Propagation Angle to a Magnetic Field in a Low-density Plasma, *Soviet Phys.–Tech. Phys.*, **8**, 715 (1964).

Kulikovskii, A. G., and G. A. Liubimov: Structure of Oblique Magnetohydrodynamic Shock Waves, *Appl. Math. Mech. (PMM)*, **25**, 125 (1961).

Ludford, G. S. S.: The Structure of a Hydromagnetic Shock in Steady Plane Motion, *J. Fluid Mech.*, **5**, 67 (1959).

Marshall, W.: The Structure of Magnetohydrodynamic Shock Waves, *Proc. Roy. Soc. (London)*, **A233**, 367 (1955).

Montgomery, D.: Development of Hydromagnetic Shocks from Large Amplitude Alfvén Waves, *Phys. Rev. Letters*, **2**, 36 (1959).

Morawetz, C. S.: Magnetohydrodynamic Shock Structure without Collisions, *Phys. Fluids*, **4**, 988 (1961).

Morawetz, C. S.: Modification for Magnetohydrodynamic Shock Structure without Collisions, *Phys. Fluids*, **5**, 1447 (1962).

Morton, K. W.: Finite Amplitude Compression Waves in a Collision-free Plasma, *Courant Inst. Math. Sci. N.Y. Univ. Rept.* NYO-10434, 1964.

Parker, E. N.: Plasma Dynamical Determination of Shock Thickness in an Ionized Gas, *Astrophys. J.*, **129**, 217 (1959).

Petschek, H. E.: Aerodynamic Dissipation, *Revs. Modern Phys.*, **30**, 966 (1958).

Wilson, T. A.: Structure of Collision-free Magnetohydrodynamic Waves, *Phys. Fluids*, **5**, 1451 (1962).

7

HYDROMAGNETIC FLOW

COLE

JULIAN D. COLE, *Professor of Aeronautics, California Institute of Technology, Pasadena, California*

7

7.1 INTRODUCTION

Hydromagnetic flow is that realm of continuum mechanics which is described by the combination of classical fluid mechanics and (usually nonrelativistic) electrodynamics. This combination leads typically to new phenomena not appearing in fluid mechanics or electrodynamics alone.

The basic interaction occurs because the fluid conducts electricity, according to Ohm's law, because fields are induced, and because body forces are exerted on the fluid. This gives the possibility of control of the flow or generation of power.

In various technical applications this description of a conducting fluid interacting with electric and magnetic fields suffices. An obvious case is that of a liquid metal, but even partially ionized gases, so long as the mean-free path is small in comparison with the Larmor radius and with some significant geometric dimension, can be described as conducting fluids.

In this section of the book a rather brief treatment of the subject is presented, only a few topics being chosen which can be treated easily and which are fairly close to applications. First, however, the general equations are presented, not so much with the aim of solving them as with the aim of showing which physical phenomena are accounted for and how they appear. Various dimensionless parameters are introduced and

208

significant limiting cases discussed. The effect of finite electric conductivity on some basic flows, due to compression (piston problems) and due to shear (Couette flow), are discussed briefly. An example of magnetically driven shock waves is analyzed, and the basic ideas of channel flow are discussed.

There are, of course, many more problems which can be worked out under the hypotheses stated here. Flow past bodies in the presence of a magnetic field is a typical omission. Some idea of the scope of these problems is provided in a brief Bibliography at the end of this chapter. All quantities are expressed in the mks system of units.

7.2 EQUATIONS OF MOTION

The model considered is a single conducting fluid having local average properties such as density ρ, velocity $\mathbf{u}$, pressure p, etc. The basic equations are conservation laws for mass, momentum, and energy and an induction equation for the magnetic field. The fluid is neutral, and relativistically small effects such as displacement currents and space charge are neglected. An equation of state is specified, usually that for a perfect gas; occasionally the effect of ionization on the equation of state is accounted for in equilibrium (Saha equation, cf. Chap. 3) or by adding another unknown, degree of ionization, and a rate equation for it. The dissipative processes in the fluid are described by a set of transport parameters (cf. Chap. 4),

$$\sigma = \text{electric conductivity} = \frac{\text{mho}}{\text{m}}$$

$\eta, \lambda =$ coefficients of viscosity
$\mathrm{k} =$ thermal conductivity

These parameters are taken to be scalars dependent on the local state of the gas, in particular on pressure and temperature. They are defined through a set of linear relations between generalized fluxes and driving forces. These relations are Ohm's law, stating that the local current density is proportional to the electric field as observed in the moving fluid, the constant of proportionality being σ; a linear isotropic relation between stress and rate of strain, and Fourier's heat-conduction law. From our point of view these are taken as phenomenological laws, although a certain basis for them can also be given by kinetic theory, as indicated in Chap. 4.

The magnetic permeability μ is taken to be constant.

Thus the general equations of motion are

$$\text{Continuity:} \qquad \frac{\partial \rho}{\partial t} + \nabla \cdot (\rho \mathbf{u}) = 0 \qquad (7.1)$$

Momentum: $\quad \dfrac{\partial(\rho\mathbf{u})}{\partial t} + \nabla \cdot (\rho\mathbf{u}\mathbf{u} + p\mathbf{I} - \mathbf{T}_M) = \nabla \cdot \mathbf{\tau}$ (7.2)

Entropy: $\quad \dfrac{\partial(\rho s)}{\partial t} + \nabla \cdot (\rho\mathbf{u}s) + \nabla \cdot \left(\dfrac{\mathbf{Q}}{T}\right)$

$$= \dfrac{\Phi}{T} + \dfrac{(\nabla \times \mathbf{B})^2}{\mu^2\sigma T} + \dfrac{Q^2}{KT^2} \quad (7.3a)$$

Energy: $\quad \dfrac{\partial p}{\partial t} + \dfrac{\partial}{\partial t}(\rho h_0 + W) + \nabla \cdot (\rho\mathbf{u}h_0 + \mathbf{S})$

$$= \nabla \cdot (\mathbf{\tau} \cdot \mathbf{u} - \mathbf{Q}) \quad (7.3b)$$

Induction: $\quad \dfrac{\partial\mathbf{B}}{\partial t} + \nabla \times (\mathbf{B} \times \mathbf{u}) = -\nabla \times \left(\dfrac{\nabla \times \mathbf{B}}{\mu\sigma}\right)$ (7.4)

In the equations above $\mathbf{T}_M$ is the relevant part of the Maxwell stress tensor, that depending on the magnetic field $\mathbf{B}$, the part due to electric field being relativistically small.

$$T_{M_{ik}} = -\dfrac{B^2}{2\mu}\delta_{ik} + \dfrac{B_iB_k}{\mu} \quad (7.5)$$

$\mathbf{\tau}$ is the part of the stress tensor due to viscosity.

$$\tau_{ik} = \lambda\delta_{ik}(\nabla \cdot \mathbf{u}) + \eta\left(\dfrac{\partial u_i}{\partial x_k} + \dfrac{\partial u_k}{\partial x_i}\right) \quad (7.6)$$

$\mathbf{Q}$ is the heat-flux vector, e.g., watts/m^2,

$$\mathbf{Q} = -K\nabla T \quad (7.7)$$

Φ is the dissipation function, expressing dissipation to heat due to viscosity,

$$\Phi = \sum_{i,k} \tau_{ik}\dfrac{\partial u_i}{\partial x_k}$$

The electric current density is related to the magnetic field by one of the Maxwell equations, displacement current $\mathbf{j}_d = \epsilon(\partial\mathbf{E}/\partial t)$ being neglected, and to the flow by Ohm's law,

$$\mathbf{j} = \dfrac{1}{\mu}(\nabla \times \mathbf{B}) = \sigma(\mathbf{E} + \mathbf{u} \times \mathbf{B}) \quad (7.8)$$

The continuity equation is merely the differential expression for mass conservation.   The momentum equation balances the increase of momentum with the divergence of the various stresses, pressure, viscous, and Maxwell stress.   The divergence of the Maxwell stress is equivalent to volume force $\mathbf{j} \times \mathbf{B}$ exerted on currents since

$$\mathbf{j} \times \mathbf{B} = \dfrac{1}{\mu}(\nabla \times \mathbf{B}) \times \mathbf{B} = \dfrac{1}{\mu}\left(\mathbf{B} \cdot \nabla\mathbf{B} - \nabla\dfrac{B^2}{2}\right)$$

$$= \dfrac{1}{\mu}\nabla \cdot \left(\mathbf{B}\mathbf{B} - \mathbf{I}\dfrac{B^2}{2}\right) \quad (7.9)$$

The entropy equation shows the production of entropy by dissipation terms due to viscosity, ohmic heating, and heat flow on the right-hand side. The other form of the energy equation involves the total specific enthalpy

$$h_0 = h + \frac{u^2}{2}$$

the magnetic energy per volume

$$W = \frac{B^2}{2\mu}$$

and the Poynting vector

$$\mathbf{S} = \frac{1}{\mu} \mathbf{E} \times \mathbf{B}$$

An equation of state, of the form

$$p = f(\rho, s) \tag{7.10a}$$

or

$$\text{Perfect gas:} \quad p = R\rho T \tag{7.10b}$$

makes the system (7.1) to (7.4) a complete system for ρ, $\mathbf{u}$, p, and s = specific entropy or h = enthalpy = $h_0 - u^2/2$. With suitable boundary conditions, for example, continuity of T, $\mathbf{u}$, $\mathbf{B}$ at boundaries, various problems can be solved for sufficiently simple geometries. In general, however, the equations must be simplified to bring out the dominant effects. This is expressed best in terms of certain dimensionless parameters. First consider some terms which have already been neglected, such as displacement current. A typical argument for neglecting $\mathbf{j}_d$ assumes that:

A characteristic magnetic field B_0 is given.
A characteristic speed U is given.
A characteristic length L is given.

It also assumes that the electric field induced $E \sim UB_0$. Then

$$\mathbf{j} = \frac{\nabla \times \mathbf{B}}{\mu} \sim \frac{B_0}{\mu L}$$

$$\mathbf{j}_d = \epsilon \frac{\partial \mathbf{E}}{\partial t} \sim \epsilon_0 \frac{UB_0}{L/U}$$

$$\therefore \quad \left| \frac{j_d}{j} \right| \sim \frac{U^2}{c^2} \quad c^2 = \mu\epsilon_0 = (\text{light speed})^2$$

7.3 DIMENSIONLESS PARAMETERS[1]

Various dimensionless characteristic parameters appear in the basic equations if the physical quantities are expressed in dimensionless units;

different ways of choosing dimensionless variables are suitable for different problems. In all cases, however, the parameters appearing express ratios of physical quantities, for example, energies or fluxes; further, there is some analogy between the old parameters of ordinary gas dynamics and the new ones introduced because of electromagnetic effects. Thus

Mach number M: $M^2 = \dfrac{U^2}{a^2} \sim \dfrac{\rho U^2}{p} \sim \dfrac{\text{kinetic energy/vol}}{\text{thermal energy/vol}}$

Alfvén number M_A: $M_A{}^2 = \dfrac{U^2}{b^2} \sim \dfrac{\rho U^2}{B^2/\mu}$

$$\sim \frac{\text{kinetic energy/vol}}{\text{magnetic energy/vol}}$$

where a = sound speed = $\sqrt{\dfrac{\gamma p}{\rho}}$ b = Alfvén speed = $\dfrac{|B|}{\sqrt{\mu\rho}}$

a is typical for sound wave propagation, and b is typical for wave propagation with magnetic effects. Qualitatively different flows can be expected to occur if $M_A < 1$ or $M_A > 1$ just as subsonic and supersonic flows with $M < 1$ or $M > 1$. This is actually the case, although the details are rather complicated (cf. Refs. 7, 8, 11, 13). Analogously, viscous and electric dissipative effects lead to

Reynolds number $R = \dfrac{(U/L)(\rho U^2)}{\eta(U/L)^2}$

$$= \frac{\text{convected kinetic energy}}{\text{viscous dissipation}} \sim \frac{\rho UL}{\eta}$$

Magnetic Reynolds number $R_M = \dfrac{(U/L)(B^2/2\mu)}{(1/\sigma)(B/\mu L)^2}$

$$\sim \frac{\text{convected magnetic energy}}{\text{ohmic dissipation}} \sim \mu\sigma UL$$

Combinations of these parameters also appear. If the entropy equation is written in dimensionless form, the viscous entropy production $\sim M^2/R$, and the magnetic entropy production $\sim M^2/M_A{}^2 R_M = M_b{}^2/R_M$, where

$$M_b{}^2 = \frac{M^2}{M_A{}^2} \sim \frac{B^2/\mu}{p} \sim \frac{\text{magnetic energy/vol}}{\text{thermal energy/vol}}$$

Generally nonlinear effects produce the steepening of wave front and vortex cores, while viscosity, thermal, and electric conductivity tend to diffuse any discontinuities, leading, in certain cases, as shock waves, to a balance. In ordinary fluid mechanics there are many problems, such as wave propagation and flow past bodies, in which it is very fruitful to study the limiting case of dissipationless fluid $R \to \infty$; it is understood that the effects of dissipation are then confined to narrow zones such as the interior of shock waves and boundary layers.   The analogous procedure is useful conceptually for hydromagnetic problems.

7.4 DISSIPATIONLESS LIMIT

The dissipationless limit occurs when ($\sigma \to \infty$, λ, η, $K \to 0$) or in dimensionless terms where R, $R_M \to \infty$.   The limit is a good approximation for highly conducting fluids outside local regions where dissipation may be important.   In this part of the flow the entropy and induction equations become

$$0 = \frac{\partial(\rho s)}{\partial t} + \nabla \cdot (\rho \mathbf{u} s) = \rho \left(\frac{\partial s}{\partial t} + \mathbf{u} \cdot \nabla s \right) \equiv \rho \frac{Ds}{Dt} \tag{7.11}$$

$$\frac{\partial \mathbf{B}}{\partial t} + \nabla \times (\mathbf{B} \times \mathbf{u}) = 0 \tag{7.12}$$

Equation (7.11) states that the entropy following the fluid particles is conserved; this is strictly true so long as the particle does not enter or cross a dissipation zone where (7.11) does not apply.   Equation (7.12) can be cast into an integral form

$$\frac{D}{Dt} \int_c \mathbf{B} \cdot \mathbf{n} \, dA = 0 \qquad \nabla \cdot \mathbf{B} = 0 \tag{7.13}$$

where c is an integral attached to the particles of fluid.   The result (7.13) states that in a perfect conductor the magnetic flux through a contour attached to the particles of fluid is conserved.   There is an analogy here with vorticity in a dissipationless fluid (no entropy gradients or magnetic effects) since for the vorticity $\boldsymbol{\zeta} = \nabla \times \mathbf{u}$ momentum conservation shows that

$$\frac{\partial \boldsymbol{\zeta}}{\partial t} + \nabla \times (\boldsymbol{\zeta} \times \mathbf{u}) = 0 \qquad \nabla \cdot \boldsymbol{\zeta} = 0 \tag{7.14}$$

$$\frac{D}{Dt} \int_c \boldsymbol{\zeta} \cdot \mathbf{n} \, dA = 0 \tag{7.15}$$

There is analogy between vortex lines and field lines; both are frozen into the flow in the dissipationless limit.   Inequalities of vorticity are diffused by viscosity and those of field by resistivity.

The concept of *frozen-in-field* appears in the hydromagnetic shock relations; shocked fluid particles come closer together so that the field is compressed if $\sigma = \infty$.   In fact an estimate of hydromagnetic shock thickness for finite σ is obtained by balancing the magnetic diffusion and compression.   The magnetic Reynolds number R_M based on the jump in velocity Δu and shock thickness δ is $0(1)$.

$$R_M \sim \mu \sigma \, \Delta u \, \delta \sim 1 \qquad \therefore \; \delta \sim \frac{1}{\mu \sigma \, \Delta u} \tag{7.16}$$

where σ is some average conductivity.

In the next section a linearized problem is sketched to show how finite conductivity affects the idealized dissipationless waves.

7.5 PISTON PROBLEM WITH FINITE CONDUCTIVITY[2]

Consider a semi-infinite plane piston of conducting material (conductivity $\bar{\sigma}$) initially occupying the half space $x < 0$ and conducting gas (conductivity σ) occupying the half space $x > 0$. Consider a uniform magnetic field B_0 parallel to the piston face to exist initially at $t = 0$, and let the piston move into the gas with constant velocity u_p. For small piston speeds a hydromagnetic sound wave is sent out in the gas. Some idea of the effect of finite conductivity on this problem can be obtained by neglecting viscous effects and treating σ as constant. Linearizing the basic equations, just as is done in acoustics, leads to the following system (perfect gas, isentropic flow) in the gas,

$$\frac{\partial \rho'}{\partial t} + \frac{\partial u}{\partial x} = 0 \tag{7.17}$$

$$\frac{\partial u}{\partial t} + a_0^2 \frac{\partial \rho'}{\partial x} + b_0^2 \frac{\partial b}{\partial x} = 0 \tag{7.18}$$

$$\frac{\partial b}{\partial t} + \frac{\partial u}{\partial x} = \frac{1}{\mu\sigma} \frac{\partial^2 b}{\partial x^2} \tag{7.19}$$

where $u \ll a_0, b_0$

$$a_0 = \sqrt{\frac{\gamma p_0}{\rho_0}} \qquad b_0 = \frac{B_0}{\sqrt{\mu\rho_0}} \qquad \rho = \rho_0(1 + \rho') \qquad B = B_0(1 + b)$$

The perturbation field is parallel to the initial field, and the perturbation current

$$j = \frac{B_0}{\mu} \frac{\partial b}{\partial x} \tag{7.20}$$

flows at right angles to B and x. In the solid piston

$$\frac{1}{\mu\bar{\sigma}} \frac{\partial^2 b}{\partial x^2} - \frac{\partial b}{\partial t} = 0 \tag{7.21}$$

The equations of motion in the gas can be combined into a single fourth-order equation,

$$\frac{1}{\mu\sigma} \frac{\partial^2}{\partial x^2} \left(\frac{\partial_2}{\partial x^2} - \frac{1}{a_0^2} \frac{\partial^2}{\partial t^2} \right) u - \frac{c_0^2}{a_0^2} \frac{\partial}{\partial t} \left(\frac{\partial^2}{\partial x^2} - \frac{1}{c_0^2} \frac{\partial^2}{\partial t^2} \right) u = 0 \tag{7.22}$$

Since the motion is across the field lines, the combined speed obtained by addition of magnetic and gas pressures (cf. Chap. 6),

$$c_0 = \sqrt{a_0^2 + b_0^2}$$

enters the equations.   The boundary conditions, linearized to $x = 0$, are

$$u(0,t) = u_p$$
$$b(0+,t) = b(0-,t)$$
$$\frac{1}{\sigma} b_x(0+,t) = \frac{1}{\bar{\sigma}} b_x(0-,t)$$

(7.23)

The last boundary condition follows from the continuity of the tangential electric field at the piston face and the use of Ohm's law.

The wave operator with speed a_0 appearing in (7.22) shows that a discontinuity appears on $x = a_0t$, but because of the lower-order terms of (7.22) this discontinuity dies out with time.   The combined speed c_0 appearing in the lower-order terms is more significant because the main wave front appears near $x = c_0t$; the higher-order terms act to diffuse this wave front.   A rigorous solution of (7.21) and (7.22) subject to boundary conditions (7.23) can be constructed by Laplace transformation, and the features just mentioned appear from the asymptotic behavior of the solutions.   As far as the solid is concerned we have only a diffusion process for the magnetic field.   To obtain an idea of the behavior near the wave front, introduce subcharacteristic coordinates

$$\xi = t - \frac{x}{c_0} \qquad \eta = t + \frac{x}{c_0}$$

and assume that the rates of change across the front are large, or

$$\frac{\partial}{\partial \eta} \gg \frac{\partial}{\partial \xi}$$

Then (7.22) is approximately satisfied if

$$\frac{1}{\mu\sigma} \frac{b_0^2}{c_0^4} \frac{\partial^2 u}{\partial \xi^2} - 4 \frac{\partial u}{\partial \eta} = 0$$

(7.24)

This shows how a discontinuity in u or b at the wave front diffuses across the front just like heat.   The asymptotic form of the solution actually is

$$\frac{u(x,t)}{u_p} \sim \frac{1}{2} \operatorname{erfc}\left(\frac{x - c_0 t}{b_0} c_0^{3/2} \sqrt{\frac{2\mu\sigma}{x}} \right)$$

(7.25)

$$\frac{b(x,t)c_0}{u_p} \sim \begin{cases} \frac{1}{2} \operatorname{erfc}\left(\dfrac{x - c_0 t}{b_0} c_0^{3/2} \sqrt{\dfrac{2\mu\sigma}{x}} \right) & \\ \qquad - \dfrac{G - 1}{G} \operatorname{erfc}\left(\dfrac{c_0^2 x}{2a_0^2} \sqrt{\dfrac{\mu\sigma}{t}} \right) & x > 0 \\ \dfrac{1}{G} \operatorname{erfc}\left(-\dfrac{x}{2} \sqrt{\dfrac{\mu\bar{\sigma}}{t}} \right) & x < 0 \end{cases}$$

(7.26)

where G depends on the ratio of piston conductivity to gas conductivity,

$$G = 1 + \sqrt{\frac{\sigma}{\tilde{\sigma}}} \frac{b_0}{c_0}$$

Notice that all the solutions have the behavior of diffusion solutions, some of the diffusion accounting for a wave structure and some for a current and field boundary layer at the piston face.

A qualitative picture of the behavior of the solution is sketched in Fig. 7.1. There is a gradient of field and current flows in the neighborhood of the wave front and the piston face. The total current flowing is

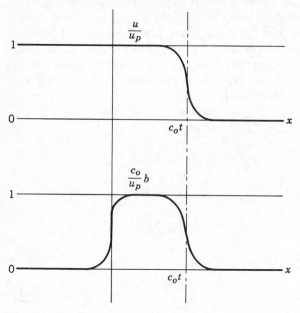

Fig. 7.1 Fluid flow and magnetic field of piston problem
in MHD with finite conductivity.

zero. As the conductivities σ, $\tilde{\sigma} \rightarrow \infty$, these current zones approach the current sheets of the dissipationless limit. The asymptotic force on the piston is also of interest, since it is composed partly of Maxwell stress and partly of gas pressure. The force per area P due to the piston motion is asymptotically

$$P \sim \rho_0 \frac{u_p}{c_0} \left(a_0^2 + \frac{b_0^2}{G} \right) \tag{7.27}$$

For a highly conducting piston $\tilde{\sigma} \rightarrow \infty$, $G \rightarrow 1$,

$$P \sim \rho_0 u_p c_0 \tag{7.28}$$

the usual formula of acoustic theory with $a_0 \to c_0$, since no currents flow in the piston in this case.   All the asymptotic results discussed here should be valid for $\mu\sigma c_0 t$ sufficiently large.   The thickness of the diffusion zone at the wave front is of the order of $\sqrt{t/\mu\sigma}$ owing to the linearization; non-linear effects would produce a shock whose thickness was indicated previously.

7.6   COUETTE FLOW[3]

The previous section discussed some effects of finite conductivity on compressive motion across the field lines.   In this section shear motion is discussed for an example in which the geometry is sufficiently simple to allow many real fluid effects such as variable viscosity to be taken into account.

The physical setup is illustrated in Fig. 7.2.

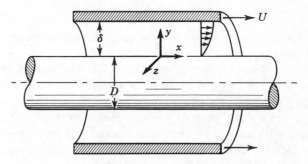

Fig. 7.2   Couette flow between sliding coaxial cylinders.

Consider the motion of a steady, viscous, and compressible fluid contained in the annular region between two coaxial circular cylinders of infinite length, sliding relative to each other along their common axis. The thickness δ of the annular region is small compared with the cylinder radii.   This configuration corresponds to a boundary layer on the slender afterbody of a body of revolution.   The stationary wall is referred to as the *wall*, and all quantities there are denoted by the subscript w.   The upper wall ($y = \delta$) corresponds to the *free stream* and is denoted by the subscript ∞.   (The *free-stream velocity* is denoted by U.)

We assume a uniform, externally applied magnetic field in the radial direction.   This is an approximation valid for $\delta \ll D$.

Couette flow produced between the coaxial cylinders rather than that produced between two infinite flat plates is considered.   In this case the current flows in closed circuits around the cylinder.   This fact will help simplify the necessary computations.   The electric field **E** may be set equal to zero.   From the symmetry of the problem it is evident that the velocities in the y and z directions will be zero (that is, $v = w = 0$). In other words all physical quantities can be at most functions of y only. The outer wall of the annular region is assumed to be insulating.

Considering the problem as a limiting case of a two-dimensional problem (that is, $\delta \ll D$), the following two-dimensional equations derived from (7.1) to (7.4) may be used.   Continuity is identically satisfied. There is no momentum flow, so that only the stresses balance,

$$\nabla \cdot (p\mathbf{I}^\circ - \mathbf{T}_M) = \nabla \cdot \boldsymbol{\tau} \tag{7.29}$$

or

x component: $\quad \dfrac{d}{dy}\,(T_{M_{xy}} + \tau_{xy}) = 0 \tag{7.30}$

y component: $\quad \dfrac{d}{dy}\,(p - T_{M_{yy}}) = 0 \tag{7.31}$

The third component vanishes by symmetry.   Integration of (7.30) yields

$$\tau + T_{M_{xy}} = \tau_\delta = \text{const} \tag{7.32}$$

When $T_{xy} = 0$, then $\tau = \tau_\delta = $ const, a well-known result in plane Couette flow.

On the other hand, Eq. (7.31) shows that the pressure gradient normal to the direction of flow does not vanish by itself, as usually happens in Couette or boundary-layer flow, but is always balanced by the corresponding part due to magnetic pressure.   This normal pressure gradient, as will be seen later, introduces complications into the energy relation.   Since T_{yy} is in general small, it will be assumed in the following analysis that the pressure is constant.

Next the basic induction law and Ohm's law (7.9) become

$$\mu\mathbf{j} = \nabla \times \mathbf{B} = \mu\sigma(\mathbf{u} \times \mathbf{B}) \tag{7.33}$$

while $\nabla \cdot \mathbf{B} = 0$ implies

$$\frac{dB_y}{dy} = 0 \qquad B_y = B_0 = \text{const} \tag{7.34}$$

In components, (7.33) is

x component: $\qquad \dfrac{dB_z}{dy} = 0$

y component $\qquad\quad\; 0 = \mu\sigma u B_z$

z component: $\qquad -\dfrac{dB_x}{dy} = \mu\sigma u B_y$

Thus $\quad \dfrac{dB_x}{dy} = -B_0\mu\sigma u(y)$

$\qquad B_y = B_0 = \text{const}$

$\qquad B_z = 0$

$$\tag{7.35}$$

By integration

$$B_x(y) - B_x(0) = -\mu B_0 \int_0^y \sigma u \, dy \tag{7.36}$$

The boundary condition to be imposed on B_x requires some discussion. In Fig. 7.2, it is clear that the axial velocity u and the radial magnetic field B_0 will lead to a circumferential current density given by

$$j = -\frac{1}{\mu}\frac{dB_x}{dy} = \sigma B_0 u \tag{7.37}$$

Since δ is constant around the circumference, the total current flowing in the annulus is the same at all positions around the circumference and no electric field is generated. If no electric field is imposed, it is evident that this configuration satisfies the condition $\mathbf{E} = 0$. Now, note that this configuration is essentially a multilayer coil of infinite length, the magnitude and distribution of the current being known, of course, only after the problem is solved. It is well known that such a coil induces a magnetic field only in the axial direction and that the induced magnetic field is zero at the outer edge of the windings and is constant everywhere inside the coil. Thus, the boundary condition on B_x is

$$B_x = 0 \qquad y = \delta$$

Since the current has the same sign throughout the annulus, B_x increases monotonically from zero at the moving cylinder to a maximum value at the stationary cylinder.

Equation (7.37) then gives

$$B_x(0) = \mu B_0 \int_0^\delta \sigma u \, dy \tag{7.38a}$$

$$B_x(y) = \mu B_0 \int_y^\delta \sigma u \, dy \tag{7.38b}$$

Also, the magnetic stresses are

$$T_{M_{xy}} = \frac{B_x B_y}{\mu} = B_0{}^2 \int_y^\delta \sigma u \, dy \tag{7.39a}$$

$$T_{M_{xy}}(\delta) = 0 \qquad T_{M_{xy}}(0) = B_0{}^2 \int_0^\delta \sigma u \, dy > 0 \tag{7.39b}$$

Since $T_{M_{xy}}$ is the only magnetic stress that remains, we shall denote it simply by T_M and (7.32) becomes

$$\tau + T_M = \tau_\delta$$

and at the wall ($y = 0$)

$$\tau_w + T_{M_w} = \tau_\delta$$

Thus $\quad \tau_w = \tau_\delta - T_{M_w} < \tau_\delta$

which indicates that the skin friction is less than the drag exerted by the free stream on the stationary body maintaining the imposed magnetic field. Heat transfer depends on a consideration of the energy equation (7.3b), which now reads

$$\nabla \cdot (\rho \mathbf{u} h_0 + \mathbf{S} + \mathbf{Q} - \mathbf{\tau} \cdot \mathbf{u}) = 0 \tag{7.40}$$

Again there is no convection of heat, and $\mathbf{S} = 0$ so that (7.40) reduces to

$$\frac{d}{dy} (\tau u - Q) = 0$$

or $\tau u - Q = -Q_w$ \hfill (7.41)

where Q_w = heat flux at the wall $(u_w = 0)$. Equation (7.41) is identical in form with the equivalent gas-dynamic case. However, it differs in one important aspect: instead of the relation

$$Q = -\frac{K}{c_p} \frac{dh}{dy} \tag{7.42}$$

the complete relation

$$Q = -K \frac{dT}{dy} \tag{7.43}$$

and $\dfrac{dh}{dy} = \left(\dfrac{\partial h}{\partial T}\right)_p \dfrac{dT}{dy} + \left(\dfrac{\partial h}{\partial p}\right)_T \dfrac{dp}{dy}$ $h = h(p,T)$

should be used. Evidently (7.43) reduces to (7.42) in either of the two cases: (1) $(\partial h/\partial p)_T = 0$, that is, perfect gas; (2) $dp/dy = 0$, that is, zero normal pressure gradient. By (7.31), Eq. (7.42) is applicable to the present problem only when the term corresponding to $T_{M_{yy}}$ is neglected.

Under such assumptions, Eq. (7.41) takes the form

$$\eta \left[\frac{d}{dy} \left(\frac{u^2}{2}\right) + \frac{1}{\text{Pr}} \frac{dh}{dy} \right] = -Q_w \tag{7.44}$$

where $\eta(T)$ = viscosity coefficient and Pr = Prandtl number = $\eta c_p / K$. In the temperature range discussed, Pr may be taken as constant. Thus

$$\frac{u^2}{2} + \frac{h - h_w}{\text{Pr}} = -Q_w \int_0^y \frac{dy}{\eta} = -Q_w \int_0^u \frac{du}{\tau} \tag{7.45}$$

and $\dfrac{U^2}{2} + \dfrac{h_\delta - h_w}{\text{Pr}} = -\dfrac{Q_w U}{\tau_\delta} \displaystyle\int_0^1 \dfrac{d(u/U)}{\tau/\tau_\delta}$ \hfill (7.46)

Consider the results now applied to various special cases:

1. Insulated plate ($Q_w = 0$).   One has from (7.46)

$$\frac{U^2}{2} + \frac{h_\delta - h_r}{\text{Pr}} = 0$$

where h_r is the recovery enthalpy and is seen to be the same as in the corresponding pure-gas dynamic case.

2. Isothermal plate ($h_w = \text{const}$)

$$\frac{h_w - h_r}{\text{Pr}} = \frac{Q_w U}{\tau_\delta} \int_0^1 \frac{d(u/U)}{\tau/\tau_\delta}$$

or
$$\frac{C_D}{C_h} = 2\text{Pr} \int_0^1 \frac{d(u/U)}{\tau/\tau_\delta} = 2\text{Pr} \cdot f_1$$

where $C_D = \dfrac{\tau_\delta}{\frac{1}{2}\rho_\infty U^2} = \text{drag coefficient}$

$$C_h = \frac{Q_w}{(h_w - h_r)\rho_\infty U^2} = \text{Stanton no.}$$

The corresponding expression in terms of τ is

$$\frac{h_w - h_r}{\text{Pr}} = \frac{Q_w U}{\tau_w} \int_0^1 \frac{d(u/U)}{\tau/\tau_w}$$

and
$$\frac{C_f}{C_h} = 2\text{Pr} \int_0^1 \frac{d(u/U)}{\tau/\tau_w} = 2\text{Pr} \cdot f_2$$

where $\dfrac{\tau_w}{\frac{1}{2}\rho_\infty U^2} = C_f = \text{skin-friction coefficient.}$

It is evident in view of (7.32) that

$$f_1 \begin{cases} = 1 & \text{without magnetic field} \\ > 1 & \text{with magnetic field} \end{cases}$$
$$f_2 \begin{cases} = 1 & \text{without magnetic field} \\ < 1 & \text{with magnetic field} \end{cases}$$

Hence, for the same heat-transfer coefficient C_h, the drag coefficient goes up and the skin-friction coefficient goes down as a result of the magnetic effect.

These results are shown in Fig. 7.3, based on the following assumed conditions,

$M = 20$

$T = 300°\text{K}$

$T_w = 1200°\text{K}$

$p = 10^{-2}\,\text{atm}$ (corresponds roughly to 100,000 ft altitude) and a realistic law for T.

Numerical calculation (cf. Fig. 7.3) shows that relatively weak magnetic fields produce large increases in the total drag and large reduction in the skin friction and at the same time have relatively little effect on the heat transfer.   Whereas the total drag without the magnetic field is skin-friction drag, the total drag with the magnetic field is primarily magnetic

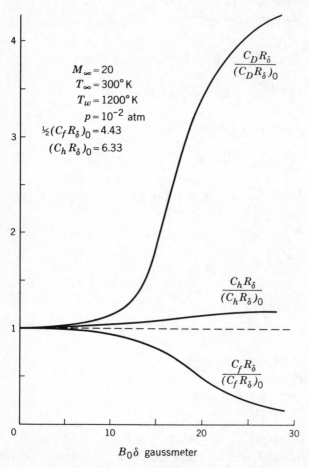

$$M_\infty = 20$$
$$T_\infty = 300°\,K$$
$$T_w = 1200°\,K$$
$$p = 10^{-2}\ \text{atm}$$
$$\tfrac{1}{2}(C_f R_\delta)_0 = 4.43$$
$$(C_h R_\delta)_0 = 6.33$$

$$\frac{C_D R_\delta}{(C_D R_\delta)_0}$$

$$\frac{C_h R_\delta}{(C_h R_\delta)_0}$$

$$\frac{C_f R_\delta}{(C_f R_\delta)_0}$$

$B_0 \delta$ gaussmeter

Fig. 7.3 Coefficients of drag, heat transfer, and skin friction vs. magnetic field.

drag.   It can be inferred from the above that, if the magnetic and non-magnetic cases are compared on the basis of the same total drag, the heat transfer for the magnetic case will be much less than for the nonmagnetic case.

It is clear, based on the consideration of Maxwell stress, that the induced magnetic component B_x, no matter how small, plays a direct role in the magnetic force on the body. The magnetic effect gives rise to an adverse pressure gradient which would tend to destabilize the boundary layer on a body.

7.7 MAGNETICALLY DRIVEN SHOCK WAVES[4]

One of the most interesting technical applications of hydromagnetic flow is the production of very strong shock waves. Such magnetic drives work successfully when a gas of high electric conductivity can be produced and can be driven by the magnetic pressure of a field which cannot effectively penetrate the gas. The situation is favorable since a strong shock wave produces a highly conducting gas behind it owing to ionization. In this section one such simple device, the inverse-pinch machine (Fig. 7.4), is analyzed. The flow produced is studied here, not on the basis of the full hydromagnetic equations, but rather on the basis of a simplified model of the flow. Essential to the model are the idea that the field cannot penetrate highly conducting fluid elements and the idea of magnetic pressure. Actually, however, because of the finite conductivity of the plasma, the field diffuses into it. For a transient problem the diffusion distance δ at time t is roughly

$$\delta \sim \left(\frac{t}{\mu\sigma}\right)^{1/2}$$

where σ is the conductivity of the gas and μ the permeability of free space. For a shock of roughly constant speed c_s, the separation Δ of the shock and the current-carrying contact front is approximately

$$\Delta \sim (c_s - u)t$$

where u is the speed of the contact front. The density ratio across the shock is

$$\frac{\rho_0}{\rho_s} = \frac{\gamma - 1}{\gamma + 1}$$

where the last equality holds for a very strong shock in a perfect gas (γ is the usual ratio of specific heats). For a real gas (with partial dissociation and ionization), $(\gamma - 1)/(\gamma + 1) > \frac{1}{15}$, so that

$$\Delta > \tfrac{1}{15}c_s t$$

The approximation should be valid so long as (upon taking an average for Δ)

$$\frac{\delta}{\Delta} \sim \frac{10}{c_s}\left(\frac{1}{\mu\sigma t}\right)^{1/2} \ll 1$$

If this is to apply to a shock traveling a distance D, then $t \sim D/c_s$ and

$$\frac{\delta}{\Delta} \sim \frac{10}{(\mu\sigma D e_s)^{1/2}}$$

Thus, a magnetic Reynolds number R_M based on the dimensions of the device and the shock speed must be fairly large,

$$R_M = \mu\sigma D c_s > 100$$

say. As a typical case, for a Mach 20 shock progressing through cold deuterium, $c_s \sim 4 \times 10^4$ m/sec, and $\sigma \sim 6 \times 10^4$ mhos/m, so that the inequality required is well satisfied for dimensions D greater than about 2 in.

A particularly useful simplification is the so-called snow-plow model. It is assumed, in this model, that all the mass swept up by the shock is compressed into a very thin layer immediately behind the shock, so that the contact front and shock are almost together. The motion of the interface is determined from the principle that the time rate of change of momentum of the accumulated mass is equal to the force on the interface.

It is interesting to note that the snow-plow theory can be derived from the full hydromagnetic equations in the limit shock strength $\rightarrow \infty$, specific heat ratio $\rightarrow 1$. This limit is called *Newtonian theory*. The model should therefore be a valid approximation for strong shocks in gases in which ionization and dissociation are taking place, since the additional degrees of freedom provided by these processes result in a specific heat ratio which approaches unity. It is clear that, under these conditions, the compression $(\gamma + 1)/(\gamma - 1)$ becomes very large, as required by the theory.

In the device to be analyzed below, the flow is approximately one-dimensional; i.e., the location of the shock at a time t is describable by a single coordinate $X(t)$. If the mass of gas which has been swept up by the shock is denoted by $M(t)$, the basic equation of snow-plow theory can be written

$$\frac{d}{dt}\left(M\frac{dX}{dt}\right) = F \tag{7.47}$$

where F is the force on the interface.

In the inverse pinch a current is passed through the gas and returns along the central conducting rod. This current produces an azimuthal magnetic field which pushes the gas away from the rod, leaving behind a cylindrical vacuum region. In such a device, the gas can be preionized, and an axial magnetic field, produced by an external solenoid, can be trapped in the plasma region. The resulting shock wave will then be a transverse magnetohydrodynamic one. This more general case will in fact be considered.

The additional geometrical idealization will be made of replacing the central rod by a line. If the plasma is initially at a uniform pressure

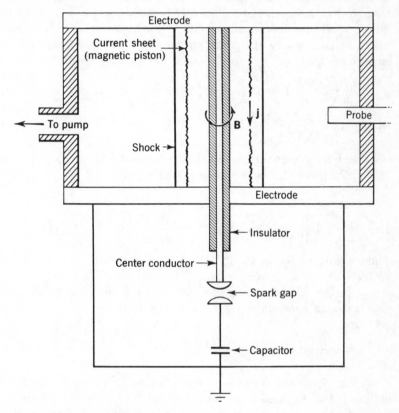

Fig. 7.4 The inverse-pinch experiment.

p_0 and density ρ_0 and in a uniform axial magnetic field of strength B_0, the accumulated mass per unit length of the sheath is

$$M = \pi \rho_0 X^2 \tag{7.48}$$

where X is the distance of the sheath from the axis, while the net outward pressure (mks units) on the sheath is

$$p = \frac{B_\theta^2}{2\mu} - \frac{B_0^2}{2\mu} - p_0 = \frac{\mu I^2}{8\pi^2} - \frac{B_0^2}{2\mu} - p_0 \tag{7.49}$$

The first term on the right-hand side of Eq. (7.49) is the magnetic pressure of the azimuthal field B_θ produced by the pinch current I, while the second term is the magnetic pressure of the external field.

It is now necessary to make some assumption concerning the form of $I(t)$. Although under typical experimental conditions the current is usually sinusoidal, the ringing time of the circuit can be made sufficiently long so that the current rises linearly over a large portion of the time of interest. If one then assumes a linear pinch current $I = I_0\omega t$ and introduces as parameters the sound speed $a_0 = (\gamma p_0/\rho_0)^{1/2}$ and the Alfvén speed $b_0 = (B_0^2/\mu\rho_0)^{1/2}$, Eq. (7.47) with $F = 2\pi X p$ becomes

$$\frac{d}{dt}\left(X^2 \frac{dX}{dt}\right) = c_0^4 \frac{t^2}{X} - X\left(b_0^2 + \frac{2}{\gamma}a_0^2\right) \tag{7.50}$$

where

$$c_0 = \left(\frac{\mu I_0^2\omega^2}{4\pi^2\rho_0}\right)^{1/4} \tag{7.51}$$

is a characteristic quantity with the dimensions of a speed. The solution of Eq. (7.51) which passes through $X(0) = 0$ is

$$X = kt$$

$$k^2 = \tfrac{1}{4}\left\{\left[\left(b_0^2 + \frac{2}{\gamma}a_0^2\right)^2 + 8c_0^4\right]^{1/2} - \left(b_0^2 + \frac{2}{\gamma}a_0^2\right)\right\} \tag{7.52}$$

Thus, according to the theory, the front moves with the constant speed $dX/dt = k$.

In the limit of strong shocks, where the snow-plow theory should give reasonably accurate results, $a_0^2/c_0^2 \ll 1$, and $b_0^2/c_0^2 \ll 1$; Eq. (7.52) becomes

$$k \sim \frac{c_0}{2^{1/4}}\left(1 - \frac{b_0^2 + a_0^2/\gamma}{2^{5/2}c_0^2}\right) \tag{7.53}$$

From Eq. (7.53) and the definition of c_0 the scaling laws for the device are obtained; the shock speed scales as the square root of the rate of current rise and inversely as the fourth root of the density. A comparison of the snow-plow result (for $a_0 = b_0 = 0$) with experiments of Liepmann and Vlases[5] is shown in Fig. 7.5. The agreement is very good, verifying the scaling law with respect to both current rise and initial density.

In addition to providing the scaling laws for the device, the theory also provides a fair idea of the shock speeds to be expected under typical experimental conditions. For example, if the working gas is deuterium at 100 μ initial pressure and if the rate of current rise is 250 kA/μ-sec, the predicted shock speed is about 9 cm/μsec, again in good agreement with experiment.

Finally, although the theory provides no information about the detail of the flow, the internal energy in the gas, within the framework of the theory, may still be calculated. This is the energy which is available

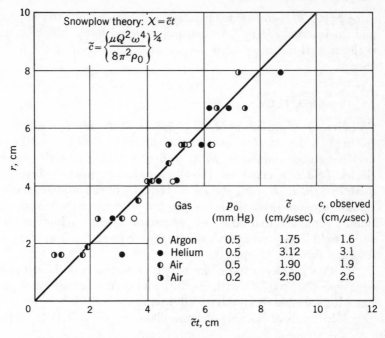

Fig. 7.5 Inverse-pinch results of Liepmann and Vlases.

to heat, dissociate, and ionize the gas. The total energy per unit length, $E(t)$, delivered to the gas in the time t is just the work done on it by the magnetic piston, i.e.,

$$E(t) = \int_0^X F \, dX = \int_0^{X(t)} \frac{d}{dt}\left(M \frac{dX}{dt}\right) dX$$
$$= M\left(\frac{dX}{dt}\right)^2 - \int_0^X M \frac{d^2X}{dt^2} \, dX \tag{7.54}$$

With the aid of Eq. (7.47) the last integral can be transformed,

$$\int_0^X M \frac{d^2X}{dt^2} \, dX = E(t) - \int_0^M \left(\frac{dX}{dt}\right)^2 dM$$
$$E(t) = \tfrac{1}{2}M\left(\frac{dX}{dt}\right)^2 + \tfrac{1}{2}\int_0^M \left(\frac{dX}{dt}\right)^2 dM \tag{7.55}$$

The first term on the right-hand side of Eq. (7.55) is the kinetic energy of the gas, while the second term is the internal energy. Internal energy is thus acquired at the rate of $\tfrac{1}{2}(dX/dt)^2$ per unit mass. (The dissipative mechanism responsible for the production of internal energy is the shock wave.) The internal energy at any time depends on the history of the motion to that time, whereas the kinetic energy depends only on the velocity at that instant. When, as in the case under consideration, the velocity dX/dt is constant, the integral in Eq. (7.55) is trivially evaluated and the internal energy is evidently just equal to the kinetic energy. In such cases, then, there is exact equipartition between kinetic and internal energy.

7.8 CHANNEL FLOW[6]

Consider one-dimensional ($\partial/\partial y = \partial/\partial z = 0$) steady flow of a conducting gas in the x direction along a channel of uniform area. Let the magnetic field have one component $B_y(x) = B(x)$ in the y direction, and let the electric field $E = $ const and the currents $j(x)$ be in the z direction. The quantities u, E, B, etc., should be thought of as suitable average values across a cross section. This analysis provides a gross picture of the operation of a magnetohydrodynamic generator where conducting gas flowing across field lines produces an electromotive force (emf) at right angles to field and flow (cf. Chap. 11).

This problem can be formulated on the basis of the hydromagnetic equations (7.1) to (7.4), with viscosity and thermal conductivity neglected but with variable conductivity allowed for.

In terms of suitable dimensionless variables (7.1) to (7.4) become

Continuity:
$$\rho u = 1 \tag{7.56}$$

Momentum:
$$\frac{du}{dx} + \frac{dp}{dx} + \frac{1}{2M_{A_0}^2} \frac{d(B^2)}{dx} = 0 \tag{7.57}$$

Energy:
$$\frac{1}{(\gamma - 1)M_0^2} \frac{dT}{dx} + \frac{1}{2} \frac{d(u^2)}{dx}$$
$$+ \frac{E}{M_{A_0}^2} \frac{dB}{dx} = 0 \tag{7.58}$$

State:
$$p = \frac{\rho T}{\gamma M_0^2} \tag{7.59}$$

Induction, Ohm's law:
$$\frac{dB}{dx} = -R_M j = -R_M(E - uB) \tag{7.60}$$

Here B and u have been made dimensionless with respect to their values at an arbitrary station $x = 0$; x has been made dimensionless with respect to a characteristic channel length L, p with respect to $\rho_0 u_0^2$, j with respect to $\sigma_0 B_0 u_0$, E with respect to $B_0 u_0$, T with respect to T_0.

$R_M(x)$ is the magnetic Reynolds number $R_M = \mu L u_0 \sigma$.

All possibilities for the flow can be studied in a (B,u) plane. Combination of (7.57) and (7.60) shows that

$$\frac{dB}{du} = \frac{\dfrac{\gamma + 1}{\gamma - 1} \dfrac{u^2}{2} + \dfrac{EB}{M_{A_0}^2} - \left[\dfrac{1}{(\gamma - 1)M_0^2} + \dfrac{1}{2} + \dfrac{E}{M_{A_0}^2} \right]}{(u/M_{A_0}^2)\{E - [\gamma/(\gamma - 1)]uB\}} \tag{7.61}$$

Equation (7.61) determines trajectories of motion in the (B,u) plane.

Each point on a trajectory corresponds to a fixed x value from the last of Eq. (7.60). Clearly there are three particular curves of interest. They are (1) the parabola along which the numerator of Eq. (7.61) vanishes; (2) the hyperbola along which the denominator vanishes; (3) the hyperbola along which $E = Bu$. These three curves depend on the three parameters E, M_{A_0}, and M_0.

The parabola corresponds to a local Mach number of unity. The two hyperbolas represent lines of zero acceleration ($du/dx = 0$) and zero current. The first of these two descriptions follows from the fact that, along the first hyperbola, $dB/du = \infty$ while $j = 0$. Consequently dB/dx is finite, and hence it follows that $du/dx = 0$. The three curves may intersect in a number of ways, a typical example being illustrated in Fig. 7.6. For this particular method of introducing dimensionless variables,

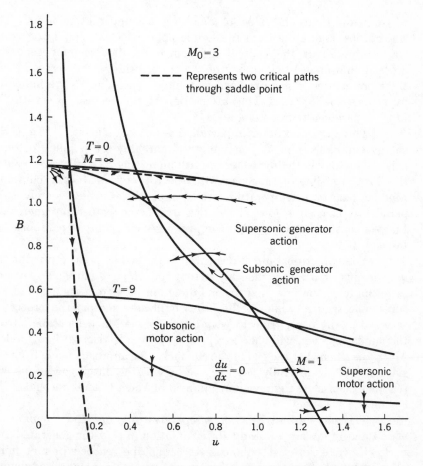

Fig. 7.6 Example of critical lines in (B,u) plane for MHD channel flow.

flow will always begin at the origin ($x = 0$, $B = u = 1$). It is in fact possible to choose different reference quantities such that the parabola and hyperbolas depend on just one parameter, but this is not done here. The conductivity enters only into the determination of the physical length to be associated with a movement along an integral curve in the (B,u) plane.

The intersections of the zero-acceleration hyperbola with the parabola are singular points, in that both the numerator and denominator of Eq. (7.61) vanish. By linearizing Eq. (7.61) in the vicinity of these intersections, it can be shown that the upper one is a saddle point, while the lower one is a center. The slopes of the two paths through the saddle point are given by

$$\frac{dB}{du} = \frac{B_1}{u_1}\left(-1 \pm \sqrt{1 - \frac{\gamma + 1}{\gamma}\frac{u_1}{B_1^2}M_{A_0}^2}\right)$$

where u_1 and B_1 are the values of u and B at the upper intersection. The trajectories originating from the saddle point enclose a family of closed trajectories about the center. Of course the flow cannot generally proceed continuously along one of the trajectories as they cross all the three boundaries; directions corresponding to increasing x are indicated by arrows on Fig. 7.6. The intersection of the vertical axis with the Mach 1 parabola represents a node.

There are also other important lines on the diagram. From the energy equations, lines of constant total enthalpy are parallel to the B axis. The total enthalpy decreases with increasing B, going to zero along the horizontal line of tangency with the top of the Mach 1 parabola. Lines of constant temperature are also parabolas which merely shift parallel to themselves down the vertical axis as the temperature increases. Lines of constant Mach number comprise a second family of parabolas, all tangent at their apex.

The two critical directions through the saddle point represent the only means of passing smoothly across the Mach 1 parabola. Acting as a "motor," the flow may in one direction pass from the sub- to the supersonic regime, while in the other (still as a motor) the process is reversed. Any other trajectory approaching the Mach 1 parabola from the supersonic side will shock (represented by a horizontal line), unless Mach 1 is reached at the end of the channel. An approach made from the subsonic side is generally not realistic in that the initial conditions will readjust themselves to prevent attainment of Mach 1 before the end of the channel.

The zero-current hyperbola separates "motor" and "generator" action and represents a steady-state condition in that the generated and back electric fields just balance. It cannot be reached in any finite physical length, the approach being asymptotic. Magnetohydrodynamic shocks must begin and end on this hyperbola; however, realization of such trajectories may generally involve ordinary hydrodynamic shocks.

This analysis is expected to give the overall behavior of generators. However, more careful analysis is needed of the validity of the one-dimensional approximation and of the size of three-dimensional effects.

REFERENCES

1. Liepmann, H. W., and J. D. Cole: Plasma Dynamics, in F. H. Clauser (ed.), "Continuum Plasma Dynamics," chap. 6, Addison-Wesley, Reading, Mass., 1960.
2. Cole, J. D.: The Magnetodynamics of Conducting Fluids, in D. Bershader (ed.), "Magnetohydrodynamic Waves," Stanford, Stanford, Calif., 1959.
3. Bleviss, Z. O.: Magnetogasdynamics of Hypersonic Couette Flow, *J. Aero/ Space Sci.*, **25** (10), 601–615 (October, 1958).
4. Cole, J. D., and C. Greifinger: Analytic Methods and Approximations of MHD Problems, *Proc. AFSWC Second Hydrodynamic Conf.*, *AFSWC* TN-61-29.
5. Vlases, G. C.: *J. Fluid Mech.*, **16**(1), 82–96 (May, 1963).
6. Cole, J. D., and J. Huth: On One Dimensional Magnetohydrodynamic Flow, *Rand Corp.* P-1827, October, 1959.

BIBLIOGRAPHY

Chu, C. K.: Magnetohydrodynamic Nozzle Flow with Three Transitions, *N.Y. Univ. Inst. Math. Sci.*, January, 1962,

Dahlberg, Erling: On the One-dimensional Flow of a Conducting Gas in Crossed Fields, *Quart. Appl. Math.*, **19**(3), 177–193 (October, 1961).

Germain, P.: Introduction a l'Étude de l'Aéromagneto Dynamique, *Cahiers Phys.*, no. 103, pp. 98–128, March, 1959.

Gourdine, Meredith C.: On the Role of Viscosity and Conductivity in Magnetohydrodynamics, *Jet Propulsion Lab. Calif. Inst. Technol. Rept.* 32-3.

Grad, Harold: Reducible Problems in Magneto-fluid Dynamic Steady Flows, *Revs. Modern Phys.*, **32** (4), 830–847 (October, 1960).

Hains, F. D.: Subsonic Hydromagnetic Flow in a Two-dimensional Channel, *Boeing Sci. Research Labs.* D1-82-0057, March, 1960.

Imai, Isao: Some Remarks on Flows of Conducting Fluids Past Bodies, *Revs. Modern Phys.*, **32** (4) (October, 1960).

Ludford, G. S. S., and J. D. Murray: On The Flow of a Conducting Fluid Past a Magnetized Sphere, *J. Fluid Mech.*, **7**(4), 516–528 (April, 1960).

McCune, J. E., and E. L. Resler, Jr: Compressibility Effects in Magneto-aerodynamic Flows Past Thin Bodies, *J. Aero/Space Sci.*, **27** (7) (July, 1960).

Marshall, W.: Structure of Magnetohydrodynamic Shock Waves, *Proc. Roy. Soc. (London)*, **233**, 367–376 (1955).

Resler, E. L., and W. R. Sears: The Prospects for Magnetoaerodynamics, *J. Aeronaut. Sci.*, **25**, 235–295 (1958), **26**, 318 (1959).

Riley, N.: A Magneto-hydrodynamic Stokes Flow, *Proc. Roy. Soc. (London)*, **A260** (1961).

Rossow, Vernon J.: On Rayleigh's Problem in Magnetohydrodynamics, *Phys. Fluids*, **3** (3) (May–June, 1960).

Stewartson, K.: Motion of Bodies through Conducting Fluids, *Revs. Modern Phys.*, **32** (4), 855–859 (October, 1960).

————: On The Motion of a Non-conducting Body through a Perfectly Conducting Fluid, *J. Fluid Mech.*, **8**, 82–96, pt. 1 (1960).

Whitham, G. B.: Some Comments on Wave Propagation and Shock Wave Structure, *Communs. Pure Appl. Math.*, **12**, 113–158 (1959).

8

WAVES IN LABORATORY PLASMAS

WHARTON (INTRODUCTION BY TRIVELPIECE)

CHARLES B. WHARTON, *Staff Member General Atomic Division, General Dynamics Corporation, John Jay Hopkins Laboratory for Pure and Applied Science, San Diego, California*

ALVIN W. TRIVELPIECE, *Associate Professor of Electrical Engineering, University of California, Berkeley, California*

8

8.1 INTRODUCTION

This chapter is intended to furnish a brief physical picture of the various modes of plasma waves and oscillations that can be expected in laboratory plasmas, to provide some of the details of the high-frequency response of a plasma to electromagnetic waves, and to describe the related experimental methods of microwave plasma diagnostics.

The introduction to this chapter will be a survey of the wave and oscillatory behavior of plasmas but will not contain details of the analysis. Only the results will be given and appropriate reference made to one of the many texts or tracts where the details are available.

A laboratory plasma is in general very complex, and a complete description of the various modes of oscillation and waves that propagate under a given set of conditions is virtually impossible. However, various idealized models have been devised to allow the study of waves and oscillations in plasmas. With these idealized models it is possible to build up a picture of what can be expected when an ionized gas is subjected to certain perturbations or excited with electromagnetic waves. The great variety of experimental conditions that are used to produce a plasma in the laboratory further complicate the situation to the extent that it is usually necessary to examine the situation for that specific configuration and set of conditions (cf. Chap. 10).

To build up a picture of the structure of oscillations and waves in plasmas, one usually begins with some assumed equilibrium situation where the plasma has known predetermined average properties such as charge density, temperature, magnetic field, percent ionization, etc., and then investigates the manner in which small disturbances from this equilibrium behave as a function of time and space.   In general, this requires the self-consistent solution of the Maxwell equations together with the Boltzmann equation.   While this is possible in principle, the resulting complications usually lead to intractable mathematics which does not lend itself to simple physical interpretation.

The most elementary model of an ionized gas or a plasma that can illustrate interesting oscillatory properties is a sea of immobile positive ions which serve as a neutralizing background for mobile electrons.   In the absence of any disturbances, the electrons are distributed uniformly and have no thermal motion.   Further, these electrons are assumed to move without collisions under the action of electric forces that may result from any nonequilibrium charge distribution.   If some of the electrons are displaced from their equilibrium position, they experience the attractive force of the electric field from the immobile ions left behind.   When these displaced electrons are released, they accelerate toward their equilibrium position.   Upon reaching this equilibrium position, they have the velocity gained during their acceleration from the position of their initial displacement.   Then, as with any simple harmonic oscillator, the electrons overshoot their equilibrium position until they achieve a displacement which converts the kinetic energy back into potential energy.   If otherwise undisturbed, this oscillatory motion persists indefinitely at the frequency $\omega_p = (ne^2/\epsilon_0 m)^{1/2}$, which is the plasma frequency.   (Practical mksa units are used in this chapter.)

This type of oscillation was probably first described by Lord Rayleigh in 1906 in a paper in which he discussed the modes of oscillation of a cloud of electrons in the Thomson model of the atom.   The term *plasma frequency* was first used by Langmuir and Tonks in 1929 in their studies of oscillations in electric discharges.[1]   A brief but excellent account of the history of plasma oscillations is given by Drummond.[2]

For the elementary one-dimensional model of a plasma, the small disturbances, which always oscillate at the plasma frequency independent of the wavelength of the disturbance, do not propagate away from their point of origin.   The reason is that, as the electrons move, their convection current is exactly canceled by the displacement current.   This means that there is no magnetic field associated with this oscillatory motion and hence no radiation; so the disturbances remain localized.

However, these disturbances will propagate if the plasma electrons have a finite temperature.   To understand this, imagine that the zero-temperature plasma is moving with some uniform velocity; then these disturbances will be transported bodily from one place to another at this speed. The propagation characteristics of disturbances on drifting electron beams, which are a good realization of the present idealized situation, were first studied by Hahn and by Ramo in 1939.[2] They showed that for steady-state harmonic time-dependent disturbances there were two waves (usually called the *fast* and *slow space-charge waves*) given by the dispersion relation

$$\beta = \frac{\omega \pm \omega_p}{v_0} \tag{8.1}$$

The wave number is $\beta = 2\pi/\lambda$, ω the frequency of the disturbance, and v_0 the average drift velocity of the electrons. One of these waves (slow space-charge wave—minus sign) has the property that it carries negative stored energy, which means that disturbances grow as the wave gives up energy. This effect is what causes one class of plasma instabilities and makes possible a wide variety of microwave amplifiers and oscillators such as klystrons.

The dispersion properties of a one-dimensional plasma with any desired velocity-distribution function can therefore be synthesized by adding together a large number of such plasma or electron streams, with an appropriate weighting factor for the number of electrons in each velocity class.

This simple picture illustrates that disturbances in plasma with a distribution of velocities do propagate away from their point of origin with a velocity comparable with the mean thermal speed. Another effect which this model can illustrate is that if the disturbance was assumed to have a fixed wavelength initially then the energy associated with this disturbance will diffuse into other wavelengths by a process of collisionless damping generally known as *Landau damping* (see Chap. 3). This collisionless damping can be explained in simple terms by observing that the distribution function of velocities is such that there is a decreasing number of particles in higher velocity classes and that as a consequence a given velocity disturbance will have more particles moving slightly slower than the wave (taking energy from it) than there are moving slightly faster than the wave (giving energy to it). The net effect is a temporal decrease of the amplitude of the wave at the original velocity of the disturbance.

A detailed study of this process is given in Chap. 3, where the appropriate statistical properties of the plasma have been correctly accounted for. The dispersion relation for a plasma in kinetic equilibrium at a temperature T is

$$\omega_r{}^2 = \omega_p{}^2 + \beta^2 \frac{3kT}{m} \tag{8.2}$$

$$\omega_i = \sqrt{\frac{\pi}{8}} \frac{\omega_p}{(\beta\lambda_D)^3} e^{-(\beta\lambda_D)^2/2} \tag{8.3}$$

where ω_i is the imaginary part of the frequency giving the collisionless damping.

The simple model of a zero-temperature plasma can also be used to explain some of the features of electromagnetic waves in an ionized gas. It is customary, although neither unique nor necessary, to consider first the response of the electrons to a harmonic time-dependent electric field and then to cast the induced current in such a form that the effect of the plasma is accounted for in terms of an equivalent charge-free frequency-dependent permittivity $[\epsilon = \epsilon_0(1 - \omega_p^2/\omega^2)$ for a homogeneous isotropic plasma]. This equivalent dielectric is then employed in the usual way in the Maxwell equations. When the plasma is immersed in a uniform steady magnetic field, this dielectric constant becomes anisotropic as well as frequency-dependent. For the homogeneous isotropic cold plasma, the uniform plane-wave solution of Maxwell's equations leads to a dispersion relation

$$\beta^2 = \omega^2 \mu_0 \epsilon_0 \left(1 - \frac{\omega_p^2}{\omega^2}\right) \tag{8.4}$$

This dispersion relation illustrates that electromagnetic waves do not propagate in an ionized gas below the plasma frequency and that the plasma has little influence on the waves at frequencies much greater than the plasma frequency. The results above assume that the collision frequency is much less than the wave frequency so that a given electron can oscillate for many periods of the wave before it suffers a collision. The macroscopic effects of collisions on the damping of the wave can be taken into account by a phenomenological approach which assumes the frequency to be complex $[\omega^2$ is replaced by $\omega(\omega - i\nu_c)]$. The collision frequency is then in turn related to some rate process in the plasma (e.g., electron-neutral collisions, etc.).

The properties of these electromagnetic waves are complicated considerably if a steady uniform magnetic field is superposed on the plasma, thus resulting in a medium in which its anisotropic nature plays an important role. For the purpose of plasma diagnostics, it is necessary to know the phase velocity in the direction of propagation in order to compare the measured phase shift over some known path with the theoretical phase shift over the same path. Some of these results are given in a later section of the chapter for a few specific cases which occur in typical plasma diagnostic experiments.

For the purpose of classification of the various wave types that a magnetized plasma can propagate, it is better to use a system devised by P. C. Clemmow, R. F. Mullaly, and W. P. Allis. This method of categorization, which will not be reproduced here, is described in detail in the books by Stix,[3] by Allis, Buchsbaum, and Bers,[4] and by Heald and

Wharton[13] and is summarized in the second edition of the monograph by Spitzer.[5]

In addition to electromagnetic and electrostatic waves, an ionized gas can propagate magnetohydrodynamic waves at frequencies much less than the ion cyclotron frequency (cf. Chap. 6). The simplest type of hydromagnetic wave is nondispersive and is generally known as an Alfvén wave. This wave propagates in the direction of the steady magnetic field as a transverse wave with no preferred polarization. It can be regarded as a propagation of a disturbance along a magnetic field line which acts as a stretched string loaded by the plasma mass per unit length that is in effect attached to the line of flux. It has the interesting property that it is incompressible and no density fluctuations occur. The phase velocity for these hydromagnetic waves in the limit where the displacement current is negligible is known as the *Alfvén speed*,

$$v_A = \frac{B}{\sqrt{\mu_0 \rho}} \tag{8.5}$$

where ρ = mass density

Some of the various conditions and situations in which magnetohydrodynamic waves propagate are given by Spitzer[5] and by Cowling.[6] The complete mode spectrum of waves in the MHD limit is complicated and includes the electrical resistivity and fluid viscosity as well as the pressure.[7]

A plasma of finite ion and electron temperature can also propagate longitudinal compression waves which are essentially acoustic waves. In an ionized gas these waves are usually called *positive ion* or *ion acoustic waves*, and they propagate with speed

$$v = \left(\frac{Z \gamma_e k T_e + \gamma_i k T_i}{m_i} \right)^{1/2} \tag{8.6}$$

The symbols γ_e and γ_i denote the usual ratios of specific heats, and Z is the ionic charge expressed in electron units.

In the limit where the perturbation frequency is much higher than the collision frequency these waves are strongly damped (just like sound waves in an ordinary gas) unless the electron temperature is much higher than the ion temperature. In this case the longitudinal wave propagates as an acoustic wave corresponding to the electron pressure and the ion mass. The positive ion or acoustic waves in an ionized gas are not changed by a magnetic field in the direction of the propagation vector. Perpendicular to the magnetic field the wave travels at a different speed and is known as the *magnetosonic wave*. As the angle between the propagation vector and the magnetic field direction comes together, the magnetosonic wave becomes either the ion acoustic or the Alfvén wave depending on the ratio of sound speed to Alfvén speed being greater or less than unity (see Chap. 6).

Whenever the plasma is of finite transverse extent, the appropriate boundary conditions must be supplied to account for the wave properties in such configurations.   For the infinite one-dimensional plasmas previously considered, some of the solutions involved waves of infinite phase velocity.   As might be expected, such solutions would be greatly modified for a plasma of finite extent, since the high phase velocity corresponds to wavelengths larger than the dimensions of the system under consideration.

The simplest problem involving a finite plasma is that of a perfectly conducting waveguide of cylindrical cross section filled with "cold" plasma immersed in an infinitely strong magnetic field parallel to the waveguide axis.   The transverse electric modes of the empty waveguide are not affected by the plasma, because the electrons are unable to respond to the transverse electric field.   However, electrons are free to move along the dc magnetic field lines, and the TM modes are now given by the following dispersion relation,

$$(\beta a)^2 = \left(\frac{\omega a}{c}\right)^2 + \frac{P_{n\nu}{}^2}{(\omega_p/\omega)^2 - 1} \tag{8.7}$$

where $P_{n\nu} = \nu$th zero of nth-order Bessel function of first kind
  a = waveguide radius
This dispersion relation shows that the empty-waveguide cutoff frequency has been shifted to a higher frequency ($\omega^2 = \omega_p{}^2 + \omega_{\text{cutoff}}^2$) and that the one-dimensional plasma resonance ("electrostatic" wave) has become a propagating wave which has a passband extending from $\omega = 0$ to $\omega = \omega_p$ and which has a phase velocity much less than the velocity of light over its entire passband.

If the magnetic field strength is reduced to a value such that the electron gyrofrequency is comparable with the plasma frequency, then there is an additional passband beginning at the plasma frequency, or the cyclotron frequency, whichever is smaller, and extending upward to $(\omega_p{}^2 + \omega_c{}^2)^{1/2}$.

If the magnetic field is further reduced to zero, then space-charge waves no longer propagate in a plasma-filled waveguide; however, surface waves of an incompressible nature can propagate if the plasma only partially fills the waveguide or if we are dealing with a plasma column in free space.   These waves also propagate at velocities less than the velocity of light and have a passband which extends from $\omega = 0$ to $\omega = \omega_p/\sqrt{2}$.[8]

The usual complexities of plasma properties are also present in finite plasmas, but the finite-plasma problems are further complicated by the variety of geometrical configurations which are possible.   A good source of details on the wave properties of finite plasmas is part II of "Waves in Anisotropic Plasmas" by Allis, Buchsbaum, and Bers.[4]

8.2 PLASMA - WAVE EXPERIMENTS

The propagation characteristics of the plasma waves described in the previous section are frequently represented by so called ω-β diagrams, which are plots of frequency vs. wave number for a given set of parameters. To compare theory and experiment, the ω-β diagram is obtained for the experimental configuration used.

The wavelength is measured as a function of frequency, and the experimental points are compared with the corresponding theoretical ω-β diagram.   In this way it is possible to determine some property of the plasma, usually charge density.

Because of the absence of a time-varying magnetic field, space-charge waves are nonradiating into space.   To couple them, then, we must use probes, slow-wave structures, waveguides, etc., whose external fields couple to the fringing electric fields of the electron wave, or we must rely on mode coupling, ordinarily a very weak effect.

8.2.1. Waves in a Plasma Column

The propagation of plasma waves along a plasma column can be studied by means of the experiment sketched in Fig. 8.1.   A simple wire probe

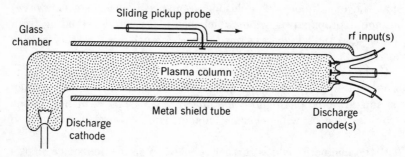

Fig. 8.1   Plasma-wave experiment, with rf input coupling to various space-charge and electromagnetic wave modes.

would excite nearly all the possible modes, space-charge and electromagnetic.   Properly grouped launching probes, helices, or grids can excite only one mode, while discriminating against others.   The electromagnetic modes will propagate only at high enough frequencies so that the metal container (as a waveguide enclosing the plasma) is not cut off ($f > f_{co}$).

If an external magnetic field is present, the electromagnetic modes as well as space-charge modes may propagate at $f < f_{co}$, especially at frequencies near gyroresonance, where the index of refraction is large.

Ordinarily, in the ω-β diagrams sketched for a plasma column it is assumed that the plasma density is uniform across the column, falling abruptly to zero at the boundary. The diagram in Fig. 8.2 is for such a

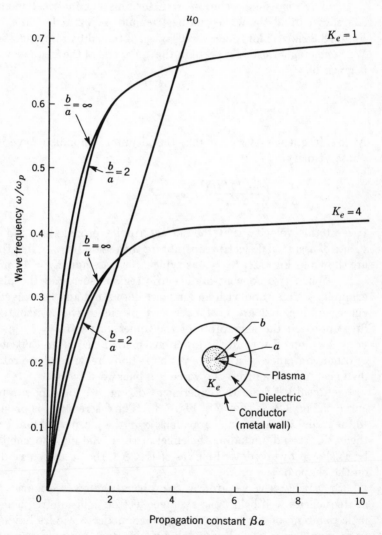

Fig. 8.2 Phase characteristics for axially symmetrical surface wave mode. K_e is the dielectric constant of the space between the plasma and the metal wall. $K_e = 4$ is typical for glass.

plasma column, where a is the plasma radius and b is the radius of the conducting container. When the plasma boundary does not reach to the metal walls of the container, various surface wave modes are possible. These modes have very slow phase velocities and propagate at frequencies well below the plasma frequency. To begin, let us discuss a simple mode, the $m = 0$ mode, having circular symmetry. The geometry is shown in Fig. 8.2.

The dispersion relation even for this symmetrical mode is very complicated and the wave number β cannot be extracted in closed form. Figure 8.2 shows plots of βa versus ω/ω_p, obtained by numerical solutions.[9] When no magnetic field is present, the upper end of the frequency passband is given by

$$\frac{\omega_u}{\omega_p} = \frac{1}{(1 + K_e)^{1/2}} \tag{8.8}$$

At low frequencies ($\omega/\omega_p \gtrsim 0.2$) the medium is nondispersive and has a phase velocity

$$v_p = \frac{\omega}{\beta} = \left(\frac{\log (b/a)}{2K_e}\right)^{1/2} \omega_p a \tag{8.9}$$

It is at this velocity that the surface ripples propagate. Plots for two values of assumed dielectric constant for the region surrounding the column are shown in Fig. 8.2; $K_e = 4$ is typical for glass and $K_e = 1$ for vacuum.

When two concentric dielectric regions surround the plasma, for example, a glass tube with an air space between it and the waveguide, the curves of Fig. 8.2 are no longer asymptotic to the horizontal, but at first approach the asymptote of the upper curve ($K_e = 1$) and then at $\beta a \sim 1$ droop off toward the lower curve, for K_e equal to that of the glass or other container. In the transition region the wave phase retards with distance, indicating that the wave is a backward wave.

Results[4,9] from the experiment of Fig. 8.1 yield points below the curve of Fig. 8.2, as shown in Fig. 8.3. The discrepancy is presumed due to the radial variation in charge density in the plasma column. The true theoretical trend, including the effect of the radial inhomogeneity, should be as shown by the dotted curves of Fig. 8.3, the actual curve depending on the shape of the profile.

Space-charge waves are seen to propagate at low frequencies, and even at frequencies well below the average electron collision frequency there is little damping of the lower-order surface modes. Modes having several variations across the plasma column are more strongly damped. The phase velocities of higher-order modes, especially ones having angular dependence, may or may not go to zero at $\omega \to 0$.

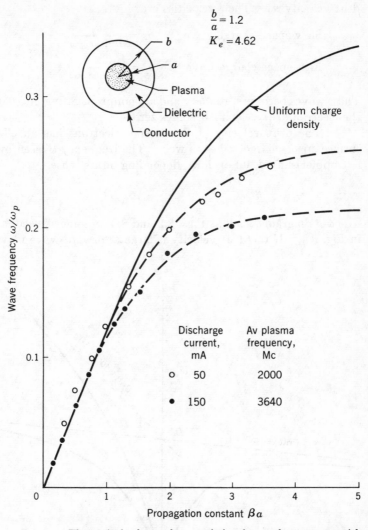

$$\frac{b}{a} = 1.2$$
$$K_e = 4.62$$

Fig. 8.3 Theoretical phase characteristics for surface waves with experimental points, demonstrating the effect of variation of charge density with radius. (*After Trivelpiece, Slow Wave Propagation in Plasma Waveguides, Calif. Inst. Technol. Tech. Rept. 7, May,* 1958.)

8.2.2. Space-charge Waves in a Drifting Electron Stream

In a plasma column (or infinite plasma) whose electrons are drifting, a disturbance can drift along with the stream. A periodic disturbance thus excites a wave, or rather a pair of waves grouped about the electron drift velocity u_0.   Their velocities are

Slow space-charge wave: $v_s = \dfrac{u_0}{1 + \omega_p/\omega}$ $\hspace{2cm}$ (8.10a)

Fast space-charge wave: $v_f = \dfrac{u_0}{1 - \omega_p/\omega}$ $\hspace{2cm}$ (8 10b)

These waves become stationary and are nonpropagating fluctuations when the electron drift velocity becomes zero.

The drift velocity of the plasma electrons has an effect on the surface space-charge waves as well.   The frequency ω' seen by the wave is Doppler-shifted, up or down depending on its sense.

$$\omega' = \omega \pm \beta u_0 \hspace{4cm} (8.11)$$

The ω-β diagrams, such as Figs. 8.2 and 8.3, become skewed, as sketched in Fig. 8.4.   If the drift velocity is large a forward wave may become a

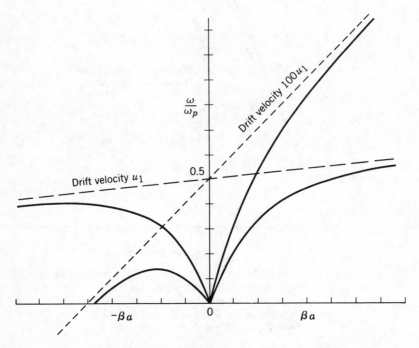

Fig. 8.4   The ω-β diagram for a plasma column containing drifting electrons having velocities of u_1 and $100u_1$.

backward wave, as can be seen for the $100u_1$ case.   In addition, if the drift velocity is large enough so that v_s or v_f coincides with the phase velocity of one of the plasma modes or with a cyclotron wave at some frequency, an interaction of the wave will occur.[10]

8.2.3. Interaction of an Electron Beam with Plasma Waves

An electron beam is capable of carrying space-charge waves by itself. A modulated beam entering a plasma then can lead to an interaction very much like that in a traveling-wave tube; i.e., an amplification of the modulation at certain frequencies can occur.   Referring again to the curve of Fig. 8.2, let the line labeled u_0 represent the electron-beam velocity. The beam plasma frequency is very much smaller than the electron plasma frequency, since the beam electron density is small.   The space-charge wave velocities given in Eqs. (8.10) are then approximately u_0.   In this example, at the frequency $\omega/\omega_p = 0.65$, where the velocities of the two waves are equal, we would expect coupling to occur and an interchange of energy between the waves to proceed,

When the coupling of the beam is with a backward plasma wave (one whose phase retards with distance), a situation analogous to a backward wave oscillator results; i.e., spontaneous oscillations occur. These oscillations originate from the amplification of noise fluctuations on the beam and build up only if the two-stream gain is large enough to overcome the wave damping.   There is thus a start-oscillation threshold in the beam current density, the value depending on the losses due to plasma collision frequency and phase mixing or Landau damping of plasma waves (cf. Chap. 3).   For many plasma experiments the beam current threshold is around 1 mA.   This spontaneous oscillation is often spoken of as a *two-stream instability* (cf. Chap. 5).   As with other instabilities, if the interaction is strong, the wave fluctuations may grow to a large enough amplitude to alter the electron beam or plasma energy distributions.   The interactions then saturate, and the growth rate levels off for further interpenetration of the beam.   A monoenergetic beam may be completely thermalized in a few centimeters of penetration; i.e., the thermal fluctuations may have grown to a level such that the beam's energy has been spread into a Maxwellian energy distribution.   The energies of the faster electrons in the final distribution often are greater than the energy of the initial electron beam, an effect first noted by Langmuir[1] and others in early probe studies on plasmas.[2]   The gain in energy of some of the electrons (at the expense of energy of others) appears to come from the electric fields arising from space-charge fluctuations.[11,12]

Figure 8.5 shows an example of an experiment in which the longitudinal space-charge waves on an electron beam, launched by a slow-wave helix, couple to the plasma electron oscillations in a mercury discharge. The drift energy of the electrons is converted into wave energy, giving a

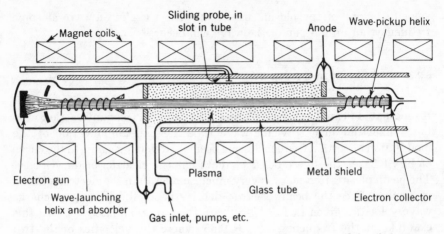

Fig. 8.5 Experiment to investigate the interactions between a modulated electron beam and a plasma. (*After Bogdanov et al., Interaction between an Electron Stream and Plasma, Proc. Symposium on Millimeter Waves Polytechnic Institute of Brooklyn, March, 1959.*)

wave amplitude which grows exponentially with distance. The sliding probe, extending through a slot in the metal shield, detects the fringing fields of the wave, to measure the growth rate. If the waves were terminated in a reflecting plate or if counterflowing waves are present, the probe could be used to pick up the standing waves and thus measure the wavelength and the wave velocity. The wave velocity can also be found by using a phase bridge or interferometer,[13] having one input from the pickup probe and the other from a sample of the input signal. The maxima and minima observed as the probe is moved along correspond to half wavelengths of the wave. The effects of a magnetic field may be studied by placing the experiment inside a solenoid.

8.2.4. Wave Coupling

Some coupling between plasma waves and electromagnetic waves is to be expected.[14,15] The coupling problem is a complicated one and is difficult to attack experimentally. Intrinsic coupling, i.e., without the presence of material walls, probes, beams, etc., can occur at steep density gradients or in certain directions in anisotropic media. An example of the steep-gradient coupling is the experiment by Dattner,[16] in which resonances due to space-charge oscillations were detected. A plasma column contained in a 5-mm-diameter glass tube was inserted across a waveguide, as shown in Fig. 8.6. The reflection and absorption of 6-cm waves by the column showed a series of peaks as the discharge current was varied. The major

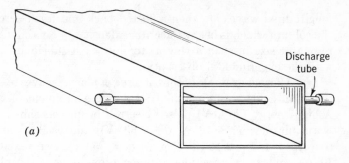

(a)

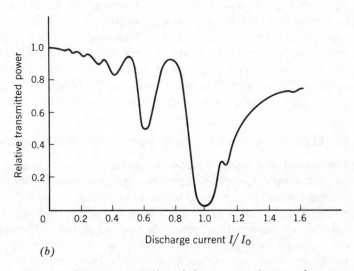

(b)

Fig. 8.6   Resonant scattering of electromagnetic waves by space-
charge waves. (*a*) Waveguide containing a discharge
tube; (*b*) transmission coefficient in the waveguide past
the discharge tube. [*After Dattner, Ericsson Technics
(Stockholm), no. 2, p. 310, 1957; Fifth Intern. Conf. on
Ionization Phenomena in Gases, Munich, 1961, p. 1447.*]

peak, associated with plasma frequency resonance, was the only one
expected; the minor peaks occurred at densities lower than cutoff.

An explanation for the subsidiary peaks has been given by Nickel
et al.[17]   They found that, whereas finite temperature effects will introduce
additional resonances associated with the reflection of longitudinal waves
back and forth across the column, the frequency separation predicted
theoretically did not agree with the experimental observations.   In order
to achieve agreement between theory and experiment, it was necessary to
include the effect of a nonuniform distribution of charge density.   The

longitudinal waves are then reflected back and forth across that part of the column which is below the critical density. This changes the effective resonator size in such a way as to shift each subsidiary peak consistent with the experimental observations.

Another example of the coupling of plasma waves to electromagnetic radiation fields is the instability-coupling experiment of Gardner et al.[18] A high-energy (20 keV) pulsed electron beam was shot into a column of warm plasma ($T_i \sim 10$ eV, $T_e \sim 50$ eV), and oscillations at the plasma frequency ($n_e = 2 \times 10^{13}/\text{cm}^3 \rightarrow f_p = 35$ Gc) were excited during the 10-μsec pulse. A receiving antenna brought near the plasma column's edge picked up a strong signal. As the antenna was pulled back, the signal fell off exponentially with distance out to about two column diameters and then the dependence became approximately $1/r^2$ out to the experimental limit of five column diameters. When a small Langmuir probe was inserted into the interaction region in front of the horn antenna, the radiation intensity increased tenfold, presumably owing to dipole radiation from currents induced in the probe by space-charge waves.

Mode coupling due to anisotropy will be treated in the next section.

8.3 ELECTROMAGNETIC WAVES IN PLASMA: BRIEF THEORY

Electromagnetic waves propagate through a plasma region as if it were an anisotropic dielectric material. An equivalent complex dielectric constant κ can be specified in terms of electron density, magnetic field strength, and average electron collision frequency, from which a refractive index N and attenuation index A can be calculated. The expressions are quite complicated and, since in general the quantities are tensors, ordinarily are written in matrix form.[13]   The tensor dielectric constant is

$$\kappa = \kappa_r + i\kappa_i = \frac{\varepsilon}{\epsilon_0} = \frac{1}{\epsilon_0} \begin{pmatrix} \epsilon_{11} & i\epsilon_{12} & 0 \\ -i\epsilon_{12} & \epsilon_{11} & 0 \\ 0 & 0 & \epsilon_{33} \end{pmatrix} \tag{8.12}$$

The components of the matrix are given as

Parallel permittivity: $\epsilon_{33} = \epsilon_0 \left(1 - \frac{\omega_p^2}{\omega} \frac{1}{\omega + i\nu} \right)$ (8.13)

Perpendicular permittivity: $\epsilon_{11} = \epsilon_{22}$

$$= \epsilon_0 \left[\frac{\omega_p^2}{\omega} \frac{\omega + i\nu}{(\omega + i\nu)^2 - \omega_b^2} \right] \tag{8.14}$$

Hall permittivity: $\epsilon_{12} = -\epsilon_{21}$

$$= \epsilon_0 \left[\frac{\omega_p^2}{\omega} \frac{\omega_b}{(\omega + i\nu)^2 - \omega_b^2} \right] \tag{8.15}$$

where $\omega_p = (ne^2/m\epsilon_0)^{1/2} = 2\pi \times 8.984\sqrt{n}$, plasma radian frequency

$\quad\quad \omega_b = (e/m)B = 1.76 \times 10^{11}B$, electron gyrofrequency

$\quad\quad \nu$ = collision frequency for momentum transfer

The complex wave propagation coefficient γ is

$$\gamma = \alpha + i\beta = i\frac{\omega}{c}\mu \quad\quad\quad (8.16)$$

where α = attenuation constant

$\quad\quad \beta$ = phase constant

$\quad\quad \mu$ = complex refractive index

The attenuation index A and real refractive index N are, respectively,

$$A = -\operatorname{Im}\kappa^{1/2} = \left[-\frac{\kappa_r}{2} + \frac{\kappa_r}{2}\left(1 + \frac{\kappa_i^2}{\kappa_r^2}\right)^{1/2}\right]^{1/2} \quad\quad (8.17)$$

$$N = \operatorname{Re}\kappa^{1/2} = \left[+\frac{\kappa_r}{2} + \frac{\kappa_r}{2}\left(1 + \frac{\kappa_i^2}{\kappa_r^2}\right)^{1/2}\right]^{1/2} \quad\quad (8.18)$$

The values of A and N for various cases of propagation along and across magnetic field lines will be discussed next.

8.3.1. Waves in a Plasma without Magnetic Field

When no magnetic field is present the permittivities ϵ_{11} and ϵ_{12} vanish, leaving the refractive index and attenuation index in fairly simple form,

$$N = \left\{\frac{1}{2}\left(1 - \frac{\omega_p^2}{\omega^2 + \nu^2}\right) + \frac{1}{2}\left[\left(1 - \frac{\omega_p^2}{\omega^2 + \nu^2}\right)^2\right.\right.$$
$$\left.\left. + \left(\frac{\omega_p^2}{\omega^2 + \nu^2}\frac{\nu}{\omega}\right)^2\right]^{1/2}\right\}^{1/2} \quad (8.19)$$

$$A = \left\{-\frac{1}{2}\left(1 - \frac{\omega_p^2}{\omega^2 + \nu^2}\right) + \frac{1}{2}\left[\left(1 - \frac{\omega_p^2}{\omega^2 + \nu^2}\right)^2\right.\right.$$
$$\left.\left. + \left(\frac{\omega_p^2}{\omega^2 + \nu^2}\frac{\nu}{\omega}\right)^2\right]^{1/2}\right\}^{1/2} \quad (8.20)$$

Plots of these expressions are shown in Fig. 8.7 for two different values of collision frequency ν. The refractive index is seen to fall and the attenuation to rise at a value of electron density such that $\omega_p/\omega \to 1$. This value of density will be called the *cutoff density* n_{co}. At high collision rates the onset of attenuation is at densities well below n_{co}; at low collision rates μ is either real or imaginary and little power absorption from the wave by the plasma is possible.[†] The wave attenuation then is due to reflection and scattering.

[†] That is, in the simple Lorentz-model plasma discussed here. When the plasma has a finite temperature, however, as real laboratory plasmas do, wave absorption due to Landau damping can occur.[13]

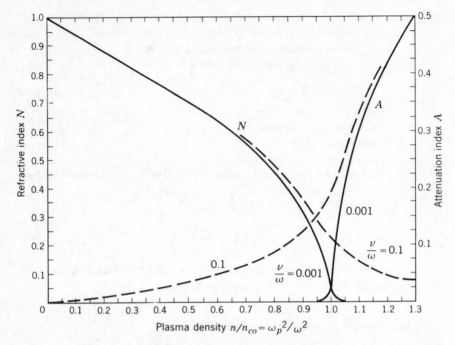

Fig. 8.7 Refractive index N and attenuation index A for propagation of microwaves through a plasma, in the absence of a B field, or with $E_{rf}||B_0$, as a function of plasma density.

8.3.2. Waves in a Plasma along Magnetic Field Lines

For wave propagation along the magnetic field the complex refractive index is given by

$$\mu^2 = \mu_R{}^2 = \kappa_R \quad \text{or} \quad \mu^2 = \mu_L{}^2 = \kappa_L \tag{8.21}$$

where κ_R = "right-hand" dielectric coefficient = ϵ_R/ϵ_0
$\quad\quad \kappa_L$ = "left-hand" dielectric coefficient = ϵ_L/ϵ_0
κ_R and κ_L are related to the permittivities of Eqs. (8.13), (8.14), and (8.15) by considering the propagating waves in rotating coordinates, so that

$$\mathbf{\mu}^2 = \begin{pmatrix} \kappa_L & 0 & 0 \\ 0 & \kappa_R & 0 \\ 0 & 0 & \kappa_{33} \end{pmatrix} \tag{8.22}$$

From these and other relationships we find that

$$\kappa_{11} = \frac{\kappa_L + \kappa_R}{2} \quad\quad \kappa_L = \kappa_{11} + \kappa_{12} \tag{8.23a}$$

$$\kappa_{12} = \frac{\kappa_L - \kappa_R}{2} \qquad \kappa_R = \kappa_{11} - \kappa_{12} \tag{8.23b}$$

and the refractive index [Eq. (8.21)] is

$$\mu_{L,R}^2 = 1 - \frac{\omega_p^2}{\omega} \frac{1}{\omega \pm \omega_b + i\nu}$$

$$= 1 - \frac{\omega_p^2}{\omega} \frac{\omega \pm \omega_b - i\nu}{(\omega \pm \omega_b)^2 + \nu^2} \tag{8.24}$$

The plus sign, coming from κ_L, is associated with a left-hand circularly polarized wave; the minus-sign solution applies to the right-hand circularly polarized wave. For a given plasma condition the indices, and thus the wave velocities, of the two wave types are different. A plane, linearly polarized wave incident on a plasma along B is decomposed into two contrarotating circularly polarized waves that travel with different velocities. After traveling through the plasma, if the waves reemerge and are reconstituted into a linearly polarized wave, we notice that its E vector has been rotated about the z axis (the direction of the B field) by an angle given by

$$\Psi = \tfrac{1}{2} \int_0^d [\beta_L(z) - \beta_R(z)]\, dz \tag{8.25}$$

where Ψ = Faraday rotation angle

d = path length in plasma

If there is damping present, the two wave types in general suffer differently so that the emerging wave is elliptically polarized. The phenomenon of Faraday rotation is also present in microwave transmission through ferrites and in light transmission through optically active substances.

The curves in Fig. 8.8a and b show the real refractive index for propagation along B as functions of plasma density and of magnetic field, respectively. In Fig. 8.8b, a resonance is noted at $\omega_b/\omega = 1$, called the *cyclotron frequency resonance* (or *gyroresonance*). It is at this frequency that the electrons spiral about the magnetic lines. Large power absorption (and by Kirchhoff's law "blackbody" emission) occurs at this frequency.

8.3.3. Waves in a Plasma at 90° to Magnetic Field Lines

For wave propagation across the field lines the complex refractive index μ again has two values,

$$\mu^2 = \mu_O^2 = \kappa_{33} \tag{8.26a}$$

$$\mu^2 = \mu_E^2 = \frac{\kappa_R \kappa_L}{\kappa_{11}} = \frac{\kappa_{11}^2 - \kappa_{12}^2}{\kappa_{11}} \tag{8.26b}$$

where the subscript O denotes ordinary waves and E denotes extraordinary

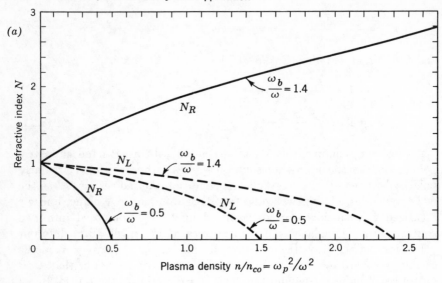

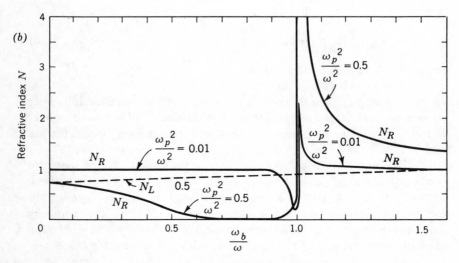

Fig. 8.8 (a) Plasma refractive index for propagation along B_0, as a function of electron density. N_R is the index for electron cyclotron waves (solid curves). N_L is the index for ion cyclotron waves (dashed curves). $\nu = 0$ is assumed. (b) Plasma refractive index for propagation along B_0, as a function of magnetic field, at fixed densities.

waves.   The first case is the same as that of Sec. 8.3.1, i.e., with no magnetic field present.   Equation (8.26a) is said to describe propagation of the *ordinary wave*, that having its electric vector parallel to the magnetic field and thus unaffected by it.   Equation (8.26b) describes propagation of the *extraordinary wave*, that having its electric vector perpendicular to the magnetic lines.   Substitution into Eq. (8.26b) leads to

$$\mu_E^2 = 1 - \frac{\omega_p^2}{\omega^2} \frac{1}{1 + i\frac{\nu}{\omega} - \dfrac{\omega_b^2/\omega^2}{1 + i\dfrac{\nu}{\omega} - \dfrac{\omega_p^2}{\omega^2}}} \tag{8.26c}$$

If collisional damping is small ($\nu \to 0$) this becomes

$$\mu_E \to N_E \approx \left(\frac{(1 - \omega_p^2/\omega^2)^2 - \omega_b^2/\omega^2}{1 - \omega_p^2/\omega^2 - \omega_b^2/\omega^2} \right)^{1/2} \tag{8.26d}$$

The wave cutoff and resonance are seen to involve both ω_b and ω_p, instead of ω_b alone as in the cyclotron waves of Sec. 8.3.2

$$\text{Cutoff:} \qquad \frac{\omega_p^2}{\omega^2} = 1 \pm \frac{\omega_b}{\omega} \tag{8.27a}$$

$$\text{Resonance:} \qquad \frac{\omega_p^2}{\omega^2} = 1 - \frac{\omega_b^2}{\omega^2} \tag{8.27b}$$

A plot of the refractive index vs. plasma density is shown in Fig. 8.9 for two different gyrofrequencies.   The resonance occurs only for $\omega_b/\omega \leq 1$.

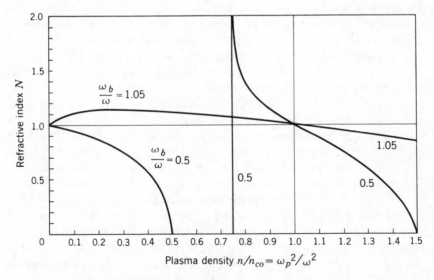

Fig. 8.9   Refractive index N for propagation of a wave through a plasma, with $E_{rf} \perp B_0$, and propagation across B_0, as a function of plasma density.

8.3.4. Waves in a Plasma at Angle θ to the Magnetic Field Lines

The refractive index for this case is somewhat complicated but can be written for characteristic waves† in the form of the Appleton equation[19]

$$\mu^2 = 1 - \cfrac{\omega_p{}^2/\omega^2}{\left(1 + i\dfrac{\nu}{\omega} - \dfrac{(\omega_b{}^2 \sin^2 \theta)/\omega^2}{2\left(1 + i\dfrac{\nu}{\omega} - \dfrac{\omega_p{}^2}{\omega^2}\right)} \pm \dfrac{(\omega_b{}^4 \sin^4 \theta)/\omega^4}{4\left(1 + i\dfrac{\nu}{\omega} - \dfrac{\omega_p{}^2}{\omega^2}\right)^2} + \dfrac{\omega_b{}^2 \cos^2 \theta}{\omega^2}\right)^{1/2}}$$

(8.28)

The cutoffs and resonances are given by

Cutoffs: $\qquad \dfrac{\omega_p{}^2}{\omega^2} = 1 \pm \dfrac{\omega_b}{\omega}$ $\qquad\qquad$ (8.29a)

Resonances: $\qquad \dfrac{\omega_p{}^2}{\omega^2} = \dfrac{1 - \omega_b{}^2/\omega^2}{1 - (\omega_b{}^2 \cos^2 \theta)/\omega^2}$ $\qquad$ (8.29b)

The cutoffs (or zeros) are thus independent of the angle θ, but the positions of resonances depend on the angle, the magnetic field, and the electron density.

8.3.5. Nonuniform Plasma in a B Field

An inhomogeneous anisotropic plasma causes the wave propagation quantities to be functions of both position and direction. The expressions derived above apply at each point, but the gross transmission characteristics are obtained only by summing over the entire path of the wave.

Inhomogeneities that change gradually compared with wavelength generally do not alter the wave types but do change the reflection and transmission properties. The depth at which cutoffs and resonances occur varies with plasma properties and frequency. The effective reflection coefficient of a nonuniform plasma is thus smaller than that of a uniform plasma, since the wave must penetrate the lossy plasma to the cutoff depth and then retrace its path again through the lossy region after reflection. Dense plasmas thus often appear surprisingly "black" to waves.

Since the electron density falls to zero at the boundaries of a plasma chamber, some of the hybrid resonances and windows cannot be seen from the outside; the propagating region is surrounded by a non-propagating cutoff region. If the cutoff region is physically thin, some evanescent waves may fringe through, coupling two propagating regions

---

† Waves whose polarizations do not change as they propagate, that is, $P_x/E_x = P_y/E_y$, where P is the polarization.

on either side. This effect is especially evident for inhomogeneities transverse to the magnetic field. Inhomogeneities transverse to the direction of propagation lead to rotation or bending of the wave front, causing refraction and scattering. A characteristic wave traversing a steep gradient in an anisotropic plasma may suffer conversion to another wave type or polarization, especially if the velocity of the other wave type is the same as that of the original wave at some point along its path. Coupling between electromagnetic waves and plasma waves also may be expected to occur in this fashion, as will coupling between right-circular and left-circular polarization cyclotron waves. For example, coupling and conversion are often seen in ionospheric propagation studies.[19] Some of the resonances hidden behind cutoff regions may become detectable through double conversion in highly nonuniform plasmas.

Scattering and refraction from time-varying gradients are also to be expected. Such gradients may be due to plasma oscillations or space-charge waves, in which case the scattered wave will have modulation corresponding to the frequency of the scattering fluctuations. If the fluctuations have a broad spectrum, an analysis of the frequency spectrum of scattered waves should give information about the plasma turbulence. The scattering cross section for turbulence scattering is many orders of magnitude larger than that for Thomson scattering, so that Doppler-shifted scattering from individual charges is negligibly small in turbulent plasmas. Further extensive discussion can be found in Ref. 13.

8.4 ELECTROMAGNETIC - WAVE EXPERIMENTS IN PLASMAS

An electromagnetic wave may be launched into a plasma, propagated through it, and received on the opposite side from the sending antenna under certain plasma conditions. For illustration let us suppose that we have an isotropic plasma slab many wavelengths thick, having a uniform electron density between sharp vacuum boundaries and no magnetic field. We see from Fig. 8.7 that the frequency must be high enough so that $\omega/\omega_p > 1$ for propagation to be possible. Waves are easily launched by electromagnetic horns, which give them directivity. The plasma interface is an impedance discontinuity, leading to both reflection and transmission. If there is a finite electron collision rate or Laudau damping within the plasma, the wave will suffer absorption. As the wave emerges, it will have part of its energy reflected internally from the backside interface. The transmitted power is then seen to be a function of the plasma reflection, transmission, and absorption coefficients and not the absorption alone.

Similar reasoning applies to the total phase shift of a transmitted wave. Since the wave impedance of the plasma interface in general is complex, there will be a phase shift due to the wave's penetration of the interface in addition to the phase shift due to the changing refractive index within the plasma.   The aggregate phase shift will be the sum of all the phase shifts along the path.

8.4.1. Measurements of Attenuation and Phase Shift

The microwave bridge or interferometer is commonly used to measure changes in the complex refractive index of a plasma region.[13]   A basic microwave interferometer, typically used to study transient plasmas as in shock waves or controlled-fusion containment experiments, is sketched in Fig. 8.10.

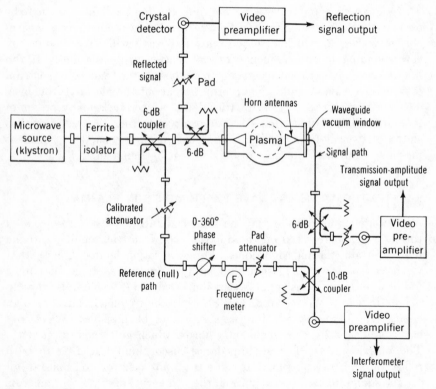

Fig. 8.10   Microwave interferometer bridge.

The attenuation caused by the plasma is measured by observing the changes in amplitude of the *transmission-signal output*.   The 6-dB directional coupler samples only the signal transmitted through the plasma. The magnitude of the reflected signal is sampled by another 6-dB directional coupler and observed in the *reflection-signal output*.   The reflected-signal amplitude is normalized to that caused by reflection from a surface of known reflectivity placed at the location of the plasma.   The transmission phase (and also the amplitude) is determined by observing the interferences of the waves coming into the 10-dB coupler from the reference path and from the plasma path.   The magnitude of the interference signal appears at the *interferometer-signal output*.

If the bridge is initially balanced, i.e., if the amplitudes and phases of the two waves arriving at the comparison point exactly cancel when no plasma is present, its response to a transient plasma without B field will be as sketched at the bottom of Fig. 8.11.   The density is shown to rise above n_{co}, the value at which $\omega_p = \omega$, and then later to fall back through this value and to decay slowly to zero.   The total phase shift over the path d is shown to be $5 \times 2\pi$, giving five interference "fringes" (by analogy with an optical interferometer).   The decrease in amplitude near the cutoff density represents a combination of absorption and scattering. When n is above cutoff, the plasma path offers no transmission (the detected power is then $\frac{1}{2}$) and the detected signal out of the bridge rests at one-fourth the peak value, since the microwave detector (ordinarily a silicon crystal diode) has a square-law response.

The reflected signal will depend upon the absorption and the steepness of the gradient, since the wave must penetrate the lossy layer of plasma twice (as pointed out in Sec. 8.3.5) to be reflected back and out.   A large reflection coefficient, then, indicates steep gradients and/or very low wave absorption.   The reflection signal sketched in Fig. 8.11 shows no fluctuations due to multiple internal reflections.   Often these fluctuations are present and may cause 5 to 10 percent oscillations of the skirts of the curve.

Multiple internal and external reflections would also lead to fluctuations on the skirts of the transmission-amplitude signal.   In addition, if the walls of the confining chamber reflect back some of the spurious radiation signals (such as that from antenna side lobes), these stray signals may interfere with the signals transmitted through the plasma, yielding completely erroneous phase and amplitude results.   It is usually a wise precaution to coat the chamber walls with a reflectionless material if they are closer than about 10 horn diameters to the measuring region.

8.4.2. Frequency Diversity

Additional information is gained by using two frequencies simultaneously. The waves are cut off at different times, and there are a different number of interference fringes in received data, since the higher frequency has more

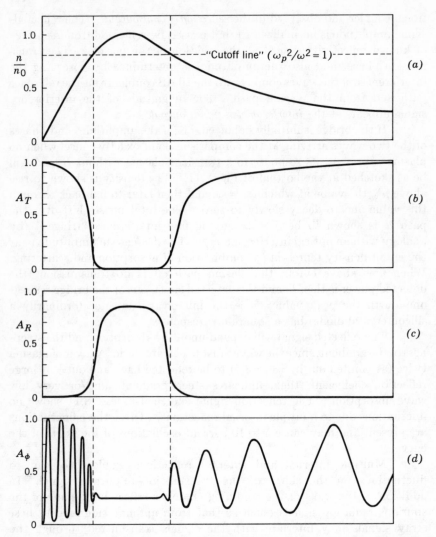

Fig. 8.11 Microwave transmission, reflection, and interferometer-bridge response to
an assumed transient plasma of depth $d/\lambda = 10$, using square-law (silicon-
diode) detectors. (*a*) Plasma density as function of time. Dashed line
shows the density at which the wave is cut off. (*b*) Transmission signal.
(*c*) Reflection signal. (*d*) Interferometer signal.

wavelengths than the lower one in a given path length. The two frequencies can be transmitted simultaneously with two sets of antenna horns
or even with one set by the use of polarization or frequency diplexers.
If both frequencies use one set of horns, there is good assurance that they
traverse the same path length even if the plasma is not symmetrical.

Two transmission-amplitude records, taken simultaneously at 70 Gc
and 90 Gc, are shown in Fig. 8.12.   The density rises rapidly and falls

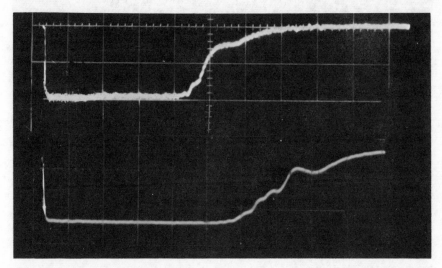

Fig. 8.12   Transmission attenuation signals for microwave propagation through a
transient plasma.   Frequencies of 90 Gc (top) and 70 Gc (bottom) were
used, leading to wave cutoff at different times.   (*Photograph courtesy of
Lawrence Radiation Laboratory, Livermore, Calif.*)

more slowly.   Fluctuations due to internal reflections are evident.   The
corresponding interferometer responses, also at 70 and 90 Gc, are shown
in Fig. 8.13.   Different numbers of interference fringes are seen at the
two frequencies.   This allows us to estimate the density profile of a
plasma column through an appropriate data analysis.[20,21]

8.4.3. The Fringe-shift Interferometer

A more sophisticated interferometer is the *fringe-shift*, or *zebra-stripe*,
interferometer.   In this type of data presentation the instantaneous phase
shift is plotted directly on the oscilloscope, and the effects of amplitude
variations are discriminated against.   The circuit of Fig. 8.14 shows the
comparison, or null, path to be very much shorter than the plasma path,
so that, when the frequency of the klystron is swept back and forth, by
varying the repeller voltage, the bridge will generate several maxima and
minima, the number depending upon the difference in length of the two
paths and the frequency excursion.   For example the 20-m path difference
of Fig. 8.14, with a frequency excursion of 50 Mc about a 35-Gc center

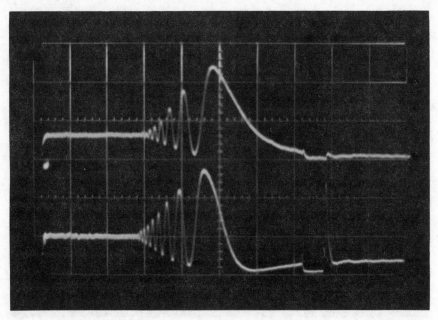

Fig. 8.13 Microwave interferometer response to a transient plasma. Frequencies of 70 Gc (top) and 90 Gc (bottom) were used. Small calibration pulses at 750 μsec indicate that the bridges were slightly off null at the time the traces were made. Sweep time was 100 μsec/cm. (*Photograph courtesy of Lawrence Radiation Laboratory, Livermore, Calif.*)

frequency, will produce about four interferences, or fringes, due to the difference in phase vs. frequency characteristics between the short and long paths. The attenuator in the short path adjusts the power level to be equal to that from the long path. The 0 to 360° phase shifter adjusts the position of the fringes on the frequency scale. Figure 8.15 shows how the fringes, when amplified, clipped flat, and applied to the oscilloscope intensity grid, produce parallel rows of bars on the screen. When the horizontal sweep is slow, the bars coalesce into zebra stripes.

If, during the horizontal sweep, the plasma density varies, the phase of the wave traveling through the plasma path varies and the frequency at which the fringe maxima occur is shifted. The stripes shift vertically, as indicated in Fig. 8.15. This figure represents the response of the fringe-shift interferometer to the same transient plasma as in Fig. 8.11. Five fringe shifts are observed, the same number as in the previous example. The deflection is shown downward, since the effective index *decreases* (from 1 down to 0 at cutoff).

Since the fringe-shift method discriminates against amplitude variations, separate reflection-transmission amplitude measurements must

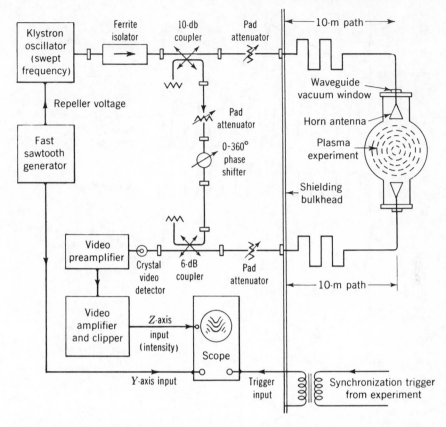

Fig. 8.14 Circuit diagram of fringe-shift or zebra-stripe microwave interferometer applied to a plasma experiment.

be made.   Two directional couplers (one to look at reflections, one to look at transmission) can be inserted on either side of the plasma experiment, as in Fig. 8.10, allowing all three measurements to be made simultaneously.   This can be done only if the components in the plasma path are well enough matched so that no deleterious effect due to the 50-Mc frequency swing is noticeable.

To determine the density distribution, either of the interferometer circuits can be used, with measurements made at various frequencies (over a 50 percent spectrum) to obtain various values of phase shift vs. time.

8.4.4. Magnetic Field Effects

When a magnetic field is present in the plasma, we have additional quantities to measure.   A wave propagated across the field lines sees a

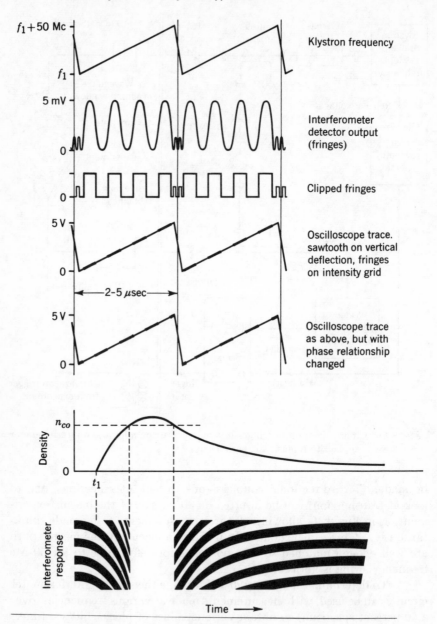

Fig. 8.15 Component operation of the fringe-shift microwave interferometer.

refractive index described by Eq. (8.26a) or (8.26b) for $\theta = 90°$. By using a square horn and waveguide and separating the two waveguide waves (polarized at right angles with each other), for example, with a polarization diplexer, we can observe both the ordinary and extraordinary wave measurements simultaneously on a dual-beam oscilloscope.

When we propagate waves along the field lines ($\theta = 0°$, Fig. 8.8), we can measure the Faraday rotation, either by rotating the receiving horn or again by using the polarization diplexer to compare the relative magnitudes of the two (x and y) components. When the phase shift and Faraday rotation are combined, the bridge output shows amplitude fluctuations superimposed on the phase fringes. A 180° phase reversal also occurs with each half rotation. When one of the circular polarizations goes to cutoff, the Faraday rotation ceases and both the x and y components behave alike. If circularly polarized antennas are used to study these waves, the two counterrotating waves can be studied independently.

If the walls of the discharge chamber are close to the plasma, some ellipticity will result and true right-hand and left-hand waves do not exist. Also when density and B-field gradients are present, the modes are no longer clean.

8.5 MICROWAVE RADIATION FROM A PLASMA

A plasma emits radiations at various wavelengths, from the frequency of radio waves, through infrared, optical, and ultraviolet to the x-ray end of the spectrum. The radiant energy comes, in general, from acceleration or from changes of quantum levels of electrons, but by a variety of mechanisms. Two broad categories of mechanisms can be distinguished: (1) the random (stochastic), incoherent emission of photons, causing bremsstrahlung and blackbody radiation; (2) coherent emission such as stimulated emission in a resonant system (lasers and masers) and "nonthermal" emission due to plasma oscillations, instabilities, and other cooperative effects.

Incoherent emission in the microwave region of the spectrum has been studied for some time and in considerable detail. It is this type of radiation that the radio astronomer usually views and that issues from gas-discharge random-noise sources. The limiting emission power density within a frequency interval is the blackbody emission level. The emission intensity at a distance from a source is given by Planck's law.

Coherent emission has only recently received detailed study, and few significant relationships between radiation intensity and plasma properties have been derived. Not only are the radiation mechanisms poorly understood, but the coherent or cooperative interactions themselves are of many possible origins. One common origin of disturbances having phase coherence over a relatively large area is plasma oscillations. These oscillations have been discussed in Sec. 8.2. Usually some kind of instability, such as two-streaming, is suspected of creating the space-charge or potential fluctuations, but the coupling mechanisms to electromagnetic radiation fields are not well understood.

8.5.1. Blackbody Radiation

A region having radiation equilibrium in a particular part of the frequency spectrum—i.e., one having sufficiently large absorption so that microwave power is neither propagated nor reflected—may be compared with a blackbody. Strictly speaking, a blackbody is in *thermodynamic equilibrium*. In a plasma, however, it is mainly the electrons that contribute to the radiation; if the electrons are in equilibrium, we may discuss blackbody equivalence even if the ions are non-Maxwellian or at a different temperature. Electron equilibrium is fairly common in low-temperature plasmas.

The energy density of radiation in equilibrium with the electrons of temperature T in the frequency interval df is given by Planck's radiation law,[22]

$$W \, df = 8\pi h \frac{f^3}{c^3} \frac{1}{\exp(hf/kT) - 1} df \qquad \text{J/m}^3 \qquad (8.30)$$

where $h = 6.6 \times 10^{-34}$ J-sec (Planck's constant)
$k = 1.38 \times 10^{-23}$ J/°K (Boltzmann's constant)
$c = 3 \times 10^8$ m/sec (velocity of light)
$f = $ frequency, cps
$T = $ electron temperature, °K

At low frequencies or high temperatures such that $hf \ll kT$, the Rayleigh-Jeans approximation applies and Eq. (8.30) may be written

$$W \, df \approx 8\pi \frac{f^2}{c^3} kT \, df \qquad (8.30a)$$

After the radiation escapes the plasma, it travels through space at velocity c. The isotropic power density flowing in the solid angle $d\Omega$ is expressed as the Planck function B_ω,

$$B_\omega = dP \, d\Omega = cW \, df \frac{d\Omega}{4\pi} \qquad \text{W/m}^2$$

$$= 8\pi \frac{f^2}{c^2} kT \, df \frac{d\Omega}{4\pi} \qquad (8.31)$$

Plasma microwave radiation usually is received by a polarized directional antenna. The directivity of an antenna can be specified in terms of its intercept area S (a kind of cross section for radiation pickup),[23] which is proportional to λ^2.   The increment of power received in a solid angle Ω in one polarization is then†

$$P_\omega \, df = \tfrac{1}{2} \int_\Omega B_\omega S \, d\Omega \, df \tag{8.32}$$

The power received in a bandwidth Δf by an antenna that sees only a blackbody at temperature T is then

$$P = kT \, \Delta f \quad \text{W} \tag{8.33}$$

This is seen to be a one-dimensional problem, and so long as the source is an extended one, i.e., its width is much greater than the receiving antenna's beamwidth, the received power is independent of distance.   If the source diameter is less than the intercept area of the antenna, an integration over the beam must be made to yield the received power.

8.5.2. Incomplete Opacity

The radiation equilibrium assumed above will be obtained only if the plasma is many *skin depths* in extent.   The skin depth δ of a medium—i.e., the distance in which a wave is attenuated to $1/\epsilon$ of its initial value—is given by

$$\delta = \frac{1}{\alpha} = \frac{c}{\omega A} = \frac{\lambda}{2\pi A} \tag{8.34}$$

where $\alpha = \text{Re } \gamma$, the wave attenuation constant
$A = -\text{Im } \kappa^{1/2}$, the attenuation index [Eq. (8.17)]
If the medium is not *optically thick* at the wavelength in question, there will not be complete self-absorption and part of the radiation that would have been removed according to Eq. (8.31) or (8.33) escapes elsewhere.‡ The ratio of power P_R received to that received from a blackbody is obtained by integrating α through the medium, of extent d,

$$\frac{P_R}{P_{BB}} = \int_0^d e^{-2\alpha z} d(2\alpha z) \tag{8.35}$$

The factor of 2 is present since we are talking about power here and α was previously defined in terms of E and H fields.   If the medium is uniform, combining Eq. (8.35) with (8.33) gives the received power

$$P_R = kT(1 - e^{-2\alpha d}) \, \Delta f \tag{8.36}$$

† If both polarizations were received, the factor of $\tfrac{1}{2}$ would vanish.
‡ Such a medium is often called a *gray body*.

8.5.3. Bremsstrahlung

The opposite extreme from the blackbody case of total self-absorption is the case of complete transparency. Equation (8.36) breaks down when α tends toward zero; the correct radiation intensity is then given by bremsstrahlung equations. Bremsstrahlung arises from free-free transitions, i.e., Coulomb encounters between electrons and ions. If the medium is transparent, the radiation passes out without further interaction, the total intensity being simply the sum of the uncorrelated events.

Bremsstrahlung may be considered in terms of quanta shaken off during electron acceleration or in terms of a cross section for bremsstrahlung. The radiant energy due to an electron-ion encounter is[24]

$$W = \int_{-\infty}^{\infty} \frac{dW}{dt}\, dt = \frac{e^2 N}{6\pi\epsilon_0 c^3} \int_{-\infty}^{\infty} a^2\, dt \quad \text{J} \tag{8.37}$$

where a = instantaneous acceleration
dW/dt = energy rate due to a
N = refractive index of medium
The frequency spectrum is obtained from a Fourier analysis of Eq. (8.37),

$$\int_{-\infty}^{\infty} [a(t)]^2\, dt \rightarrow \int_0^{\infty} \pi [a_\omega(\omega)]^2\, d\omega \tag{8.38}$$

For a Maxwellian energy distribution of electrons and parabolic collision orbits (low temperatures), the microwave part of the emission spectrum is almost frequency-independent. The power radiated in one polarization per cubic meter per unit solid angle and per frequency interval[25] [cf. also Eq. (9.21)] is

$$P\, df = \frac{8\sqrt{2\pi}}{3\sqrt{3}} \left(\frac{e^2}{4\pi\epsilon_0 c}\right)^3 \frac{n_e n_i Z^2}{m^{3/2}(kT)^{1/2}} N\bar{G}\, df \tag{8.39}$$

where n_e, n_i = electron and ion densities, respectively
Z = ion charge number
N = refractive index (≈ 1)
$\bar{G}$ = Gaunt factor
The Gaunt factor is close to unity for low temperatures except at low frequencies, for which it has logarithmic dependence. For a dilute plasma, that is, $\omega_p \ll \omega$,

$$\bar{G} = \frac{\sqrt{3}}{\pi} \ln\left[\frac{1}{3}\left(\frac{2}{\gamma}\right)^{5/2} \frac{\omega_p}{\omega}\Lambda\right] \tag{8.40}$$

where $\gamma = 1.781$, Euler's constant
$\Lambda = \lambda_D/b_{90°}$ = plasma parameter (cf. Chap. 1)
$\lambda_D = (kT_e/m\omega_p^2)^{1/2}$, Debye length
$b_{90°}$ = 90°-deflection impact parameter

In a dense plasma, that is, $\omega_p > \omega$, the radiators are shielded by the Coulomb fields of neighboring particles, the shielding distance being λ_D, the Debye length.   The Gaunt factor, when shielding is present, is then[26]

$$\bar{G} = \frac{\sqrt{3}}{\pi} \ln\left(\frac{4}{3\gamma^2 e^{1/2}} \Lambda\right) \tag{8.41}$$

Alternatively, the radiation may be calculated in terms of a cross section for bremsstrahlung, σ_i.[22]   The number of quanta lying in the energy range $d(hf)$, radiated per electron incident on an ion cloud, is

$$dQ = n_i \sigma_i v_e \, d(hf) \qquad \text{quanta/(electron)(m}^3) \tag{8.42}$$

where n_i = ion density, number/m^3
σ_i = cross section for bremsstrahlung, m^2
v_e = electron velocity, m/sec
The power flow associated with these quanta is

$$dp = hf \, dQ = n_i \sigma_i v_e hf \, d(hf) \tag{8.43}$$

If the electrons are plasma electrons and as a result of their collisions, etc., have assumed a Maxwellian energy distribution, their average velocity is given by

$$v_{th} = \bar{v}_e = \left(\frac{2kT_e}{m_e}\right)^{1/2} \qquad \text{m/sec} \tag{8.44}$$

The total power radiated in the range $d(hf)$ by n_e electrons is then

$$P \, df = n_e n_i \sigma_i \left(\frac{2kT}{m}\right)^{1/2} hf \, d(hf) \qquad \text{W/m}^3 \tag{8.45}$$

The Heitler cross section for nonrelativistic electrons, when converted to our units and terminology, is

$$\sigma_i \, df = \frac{8}{3} \bar{\Phi} \frac{mc^2}{kT_e} F(f, T_e, \lambda_D) \frac{df}{f} \tag{8.46}$$

where $\bar{\Phi}$ = unit of cross section,

$$\bar{\Phi} = \frac{Z^2 r_0{}^2}{a} = \frac{Z^2 r_0{}^2}{137} \qquad \text{m}^2$$

with Z = ion charge
$r_0 = e^2/4\pi\epsilon_0 mc^2$, m
   = classical electron radius
$a = 4\pi\epsilon_0 hc/e^2 = 137$
   = fine-structure constant
$k = 1.38 \times 10^{-23}$ J/°K
$m = 9.1 \times 10^{-31}$ kg = electron mass

The factor F suggested by Heitler for nonrelativistic electrons is

$$F = \log \left[\frac{(\sqrt{kT_e} + \sqrt{kT_e - \hbar\omega})^2}{\hbar\omega} \right] \tag{8.47a}$$

where $\hbar = \dfrac{h}{2\pi} = \dfrac{6.6}{2\pi} \times 10^{-34}$ J-sec

At low temperatures and for pure Coulomb fields, that is, $\omega \ll \omega_p$,

$$F \approx \log \frac{4kT_e}{hf} \tag{8.47b}$$

However, in a plasma the electrons tend to screen the ions and a pure Coulomb field does not exist.

The factor F then becomes[26]

$$F = \frac{1}{2} \left[\log \frac{(4/b_0{}^2)(2kT/m)}{e^{2C}\omega^2} \frac{1 + (b_0/\lambda_D)^2}{1 + 2kT/m\omega^2\lambda_D{}^2} \right.$$
$$\left. + \frac{(b_0/\lambda_D)^2}{1 + (b_0/\lambda_D)^2} - \frac{2kT/m\omega^2\lambda_D{}^2}{1 + 2kT/m\omega^2\lambda_D{}^2} \right] \tag{8.47c}$$

where $b_0 = \dfrac{Ze^2}{mv^{-2}} = \dfrac{Ze^2}{2kT_e}$

= impact parameter or classical distance of closest approach

In a low-temperature plasma ($T_e \lesssim 1$ keV) at microwave frequencies of the order of ω_p, the wavelength can become fairly long, so that b_0/λ_D is small. The factor then becomes

$$F = e^{(\hbar\omega/2kT)} K_0 \left(\frac{\hbar\omega}{2kT} \right) \tag{8.47d}$$

where K_0 = modified Bessel function of second kind

By neglecting induced emissions (complete transparency) but including the effect of Debye screening, the total radiated microwave power density in $d(hf)$ is

$$P\,df = n_e n_i \frac{8}{3} \Phi \frac{mc^2}{kT} \exp\left(\frac{\hbar\omega}{2kT}\right) K_0\left(\frac{\hbar\omega}{2kT}\right) \left(\frac{2kT}{m}\right)^{1/2} hf \frac{df}{f}$$
$$= \frac{\sqrt{2}}{12} \frac{n^2 Z^2 e^6}{\pi^2 \epsilon_0{}^3 m^2 c^3 h} \left(\frac{m}{kT}\right)^{1/2} \exp\left(\frac{hf}{2kT}\right) K_0\left(\frac{hf}{2kT}\right) hf \frac{df}{f} \quad \text{W/m}^3$$

$$\tag{8.48}$$

The total power received is obtained by a volume integration over the receiving-antenna radiation pattern. The numerical evaluation of this volume integral is difficult because of the presence of absorbing and reflecting walls in most plasma experiments. As compared with blackbody radiation the bremsstrahlung power is small and, unless a very good receiver is available, may not even be detectable above background noise.

8.5.4. Cyclotron Radiation

When a magnetic field is present, the electrons spiral about the lines with angular frequency ω_b. Their nonrelativistic centripetal acceleration is given by

$$a_b = \frac{e}{m} \, |(\mathbf{v} \times \mathbf{B})| = \omega_b v_\perp \tag{8.49}$$

Substitution of Eq. (8.49) into Eq. (8.37) yields the radiation rate per electron in a plasma of refractive index N,

$$P = \frac{e^2 N a^2}{6\pi\epsilon_0 c^3} = \frac{e^2 \omega_b^2 v_\perp^2 N}{6\pi\epsilon_0 c^3} \tag{8.50}$$

If the plasma electrons have a Maxwellian velocity distribution, taking $v_\perp$ as the thermal velocity, $\bar{v} = (2kT/m)^{1/2}$, gives the cyclotron radiation due to low-temperature electrons in a dilute plasma (i.e., no self-absorption) of density n and depth L,

$$P = \frac{e^2 \omega_b^2}{3\pi\epsilon_0 c^3} \frac{kT}{m} \, nL \qquad \text{W/m}^2 \tag{8.51}$$

For nL of 10^{14} or so, the power density would far exceed the blackbody level. This cannot occur, however, since self-absorption (leading to stimulated emission)[28] is also very strong at the gyrofrequency. In the absence of cooperative effects, then, the cyclotron radiation intensity of an equilibrium plasma becomes the blackbody limit.

Electron collisions interrupt the phase of the cyclotron oscillations, leading to frequency broadening about ω_b. Transit-time effects (due to streaming), field inhomogeneities, and Doppler shifts may also lead to broadening. If the radiation is viewed along the magnetic field lines and there is a drift velocity along the lines, the observed frequency will be Doppler-shifted either up or down depending upon the sense of the drift. A drifting plasma column thus has different observed gyroradiation frequencies, depending upon which end of the column is viewed, leading to a method of measuring drift velocities.

If the electrons have high energies so that they are relativistic, their instantaneous radiation intensities have a maximum in the direction of their motion. An observer viewing the electrons across the magnetic field lines then sees pulses of radiation, one for each cycle, leading to gyrofrequency harmonics. In dense high-temperature plasmas the total power loss due to radiation in the higher harmonics can be severe.[30]

Within the microwave part of the spectrum, however, the harmonics have much smaller power densities than the blackbody level and, in general, are difficult to detect in uniform equilibrium plasmas.[31]

8.5.5. Cooperative Effects

Many plasmas have sharp boundaries, sheaths, electron streams, non-equilibrium velocity distributions, etc., that may lead to coupling between space-charge oscillations and electromagnetic waves.   Occasionally microwave radiation levels several orders of magnitude above the blackbody level are observed emanating from plasmas.[32]   The frequencies may be related to the plasma frequency or to the gyrofrequency and its harmonics. Ordinarily such plasmas are suspected of being unstable or at least of containing plasma oscillations.   The coupling between plasma waves and electromagnetic waves ordinarily is small.   However, in regions where both waves happen to have similar velocities and electric-field directions the coupling may become appreciable.[14]   Also at sharp boundaries where charge neutrality cannot be conserved or where currents are induced in probes or walls, an electric dipole may be set up that radiates rather strongly.   In any given plasma experiment the possible sources of radiation must be carefully scrutinized before far-reaching conclusions are reached.

8.6   RADIATION EXPERIMENTS

8.6.1. Radiometers

Microwave radiometers can be made which possess great sensitivity. When steady-state or slowly varying phenomena are being studied, coherent detection with very long averaging times permits temperatures of only a few degrees Kelvin to be measured.[33]   However, for measurements on transient plasmas a much shorter averaging time is required, and bandwidths of several megacycles are necessary.   A radiometer circuit which allows either method of operation is shown in Fig. 8.16.   A photograph is shown in Fig. 8.18.

Such radiometers for millimeter wavelengths having 12- to 15-dB noise figures are commonly used.   The unit sketched in Fig. 8.16 provides absolute calibration against a thermal source of known temperature or approximate calibration against a standard gas-tube noise source.   In the chopped operation the input signal is compared with the standard signal, and the calibrated attenuator is adjusted to give a null output, the reading on the attenuator being calibrated in terms of temperature.   For transient work the ferrite switch is left connected to the antenna until after the

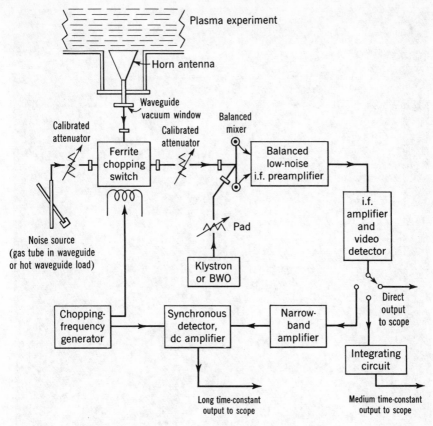

Fig. 8.16   Microwave radiometer for plasma radiation studies.

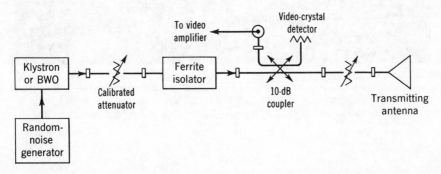

Fig. 8.17   Noise-modulated transmitter for plasma absorption measurements.

Fig. 8.18 Photograph of a microwave radiometer for operation in the 3-mm band (88 to 92 Gc). The input is at the top, through the ferrite isolator. (*Photograph courtesy of Lawrence Radiation Laboratory, Livermore, Calif.*)

plasma event is over and then is switched to the noise source to give a calibration signal at the end of the oscilloscope trace.   Transient plasma temperatures in the range 0.2 eV to 20 keV have been measured with such a radiometer.[31,34]

8.6.2. Absorption-Radiation Experiments

In order to draw conclusions regarding the "blackness" of the radiating region, we must perform transmission and reflection measurements in conjunction with the radiation measurements.   Figure 8.17 shows a noise-modulated transmitter used as a signal source for these measurements.[31] The noise spectrum is some 200 Mc wide, which looks to the receiver like white noise, and which averages out any effects of high-order modes or internal reflections in the discharge chamber.   Having a modulated signal also permits the use of ac-coupled video amplifiers instead of dc.

In order to eliminate the plasma-generated noise from the receiver (of order micromicrowatts) an attenuator of $\sim$100 dB must be placed between the receiving antenna and the input.   The transmitter level is then adjusted by its attenuator until a received noise signal of convenient amplitude is obtained.   When the plasma experiment is operated, the reflected and transmitted signals are recorded simultaneously to obtain a power balance.   The absorption coefficient A is then

$$A = 1 - R - T \tag{8.52}$$

where R = reflection coefficient
T = transmission coefficient

When A is very large, the plasma looks "black" and the radiation is then also maximum.   Even when low-level plasma oscillations and other cooperative effects are present, if A is large and the plasma depth is several wavelengths, the radiation intensities that escape the plasma are not expected to be much different from blackbody levels.

REFERENCES

1. Tonks, L., and I. Langmuir: *Phys. Rev.*, **33**, 195 (1929).
2. Drummond (ed.), J. E.: "Plasma Physics," chap. 1, McGraw-Hill, New York, 1961.
3. Stix, T. H.: "The Theory of Plasma Waves," McGraw-Hill, New York, 1962.
4. Allis, W. P., S. J. Buchsbaum, and A. Bers: "Waves in Anisotropic Plasmas," M.I.T., Cambridge, Mass., 1963.
5. Spitzer, L., Jr.: "Physics of Fully Ionized Gases," 2d ed., Interscience, New York, 1962.

6. Cowling, T. G.: "Magnetohydrodynamics," Interscience, New York, 1957.
7. Baños, A., Jr.: In L. N. Ridenour and W. A. Nierenberg (eds.), "Modern Physics for the Engineer," chap. 8, McGraw-Hill, New York, 1961.
8. Ref. 4, p. 218.
9. Trivelpiece, A. W.: Slow Wave Propagation in Plasma Waveguides, *Calif. Inst. Technol. Tech. Rept.* 7, May, 1958.
10. Harris, E.: Plasma Instabilities Associated with Anisotropic Velocity Distribution, *J. Nuclear Energy*, **C2**, 138 (1961).
11. Emeleus, K. G., D. Mahaffy, and G. McCullagh: Beam-Plasma Interaction, *Queen's Univ. Rept.* (Belfast, North Ireland), July, 1958; *Phys. Rev.*, **112**, 1052 (1958).
12. Bogdanov, R. V., V. J. Kislov, and Z. S. Tchernov: Interaction between an Electron Stream and Plasma, *Proc. Symposium on Millimeter Waves*, Polytechnic Institute of Brooklyn, March, 1959.
13. Heald, M. A., and C. B. Wharton: "Plasma Diagnostic with Microwaves," Wiley, New York, 1965.
14. Trivelpiece, A. W.: Space-charge Waves and Plasma Diagnostics, *Proc. Second Symposium on Eng. Aspects of Magnetohydrodynamics*, 1962, p. 419.
15. Boyd, T. J. M.: Emission of Radio Noise by Plasmas, *Phys. Fluids*, **7**, 59 (1964).
16. Dattner, A.: *Ericsson Technics* (Stockholm), no. 2, p. 310, 1957; *Proc. Fifth Intern. Conf. on Ionization Phenomena in Gases*, Munich, 1961, p. 1477.
17. Nickel, J. C., J. V. Parker, and R. W. Gould: Resonance Oscillations in a Non-uniform Plasma Column, *Phys. Rev. Letters*, **11**, 183 (1963).
18. Gardner, A. L.: Diagnostic Measurements of a Highly Ionized, Steady-state Plasma, *Proc. Second Sumposium on Eng. Aspects of Magnetohydrodynamics*, 1962, p. 438.
19. Ratcliffe, J. A.: "The Magneto-ionic Theory," Cambridge, New York, 1959.
20. Wharton, C. B., and D. Slager: Microwave Determination of Plasma Density Profiles, *J. Appl. Phys.*, **31**, 428 (February, 1960).
21. Motley, R., and M. A. Heald: Use of Multiple Polarizations for Electron Density Profile Measurements in High Temperature Plasmas, *Proc. Symposium on Millimeter Waves*, Polytechnic Institute of Brooklyn, 1959.
22. Planck, M.: "Theory of Heat Radiation," Dover, New York, 1959.
23. Schelkunoff, S. A., and H. T. Friis: "Antennas: Theory and Practice," Wiley, New York, 1952.
24. Panofsky, W. K. H., and M. Phillips: "Classical Electricity and Magnetism," Addison-Wesley, Reading, Mass., 1955.
25. Kramers, H. A.: On the Theory of X-ray Absorption and of the Continuous X-ray Spectrum, *Phil. Mag.*, **46**, 836 (1923).
26. DeWitt, H.: Free-free Absorption Coefficient in Ionized Gases, *Univ. Calif. Radiation Lab. Rept.* UCRL5377, October, 1958.
27. Heitler, W.: "Quantum Theory of Radiation," Oxford University Press, New York, 1936.
28. Cillié, G.: The Hydrogen Emission in Gaseous Nebulae, *Mon. Not. R. astr. Soc.*, **92**, 820 (1932).
29. Trubnikov, B. A., and V. S. Kudryavtsev: Plasma Radiation in Magnetic

Field, *Proc. Second UN Conf. on Peaceful Uses of Atomic Energy*, 1958, A/Conf. 15/P/2213, vol. 31, p. 93.

30. Drummond, W. E., and M. N. Rosenbluth: Cyclotron Radiation from a Hot Plasma, *Phys. Fluids*, **6** (2), 276 (February, 1963).

31. Wharton, C. B.: Microwave Radiation Measurements of Very Hot Plasmas, *Proc. Fourth Intern. Conf. on Ionization Phenomena in Gases*, Uppsala, 1959, vol. 2, p. 707.

32. Wharton, C. B., F. R. Scott, and T. H. Jensen: Microwave Radiation from an Unstable Plasma, *Bull. Am. Phys. Soc.*, **9** (5), 541 (1964).

33. Dicke, R. H.: Measurements of Thermal Radiation at Microwave Frequencies, *Rev. Sci. Instr.*, **17,** 268 (1946).

34. Dellis, A. N.: The Measurement of Electron Temperatures by Microwave Methods, *Atomic Energy Research Establishment Rept.* GP/R 2265, Harwell, England, 1957.

9

OPTICAL RADIATION

GRIEM

HANS R. GRIEM, *Professor of Physics, University of Maryland, College Park, Maryland*

9

9.1 INTRODUCTION

Electromagnetic radiation from plasmas is of interest for two main reasons: it provides in many instances almost the only information concerning the physical properties of a plasma, and it may also be a very important cause of energy losses. The first aspect has been exploited to a large extent by astrophysicists, and a major fraction of some texts on astrophysics[1,2] is devoted to quantitative spectroscopical methods for the analysis of stellar atmospheres and the solar corona.

Similar methods have been applied to discharges in gases for many years. There exist some review articles on spectroscopic methods for the determination of temperatures and densities in such cases.[3,4] Descriptions of techniques developed for the spectroscopic diagnostics of plasmas produced in more recent devices can be found in the original literature or in certain conference proceedings,[5-7] which contain some relevant review papers.[8-10] Theory and applications of spectroscopic methods to plasmas are also summarized in a recent book.[11]

The quantitative understanding of radiative energy losses is in a more rudimentary stage. However, for special situations theoretical estimates of losses due to spectral-line radiation[12,13] or synchrotron radiation[14] have been made. That the theory for the diagnostic methods can usually be carried through with a larger precision than that for radiative-

278

loss rates is not surprising, because in the first case one has a choice of various methods, whereas in the second case one can hardly control the relative importance of specific loss mechanisms.

A detailed discussion of the applications of spectroscopic methods must be preceded by a review of the basic processes producing electromagnetic radiation. If one excludes both nuclear γ rays and also all molecular processes, there will be line radiation caused by transitions between electronic states of atoms or ions and continuum radiation due to collisions of free electrons with atoms and ions or due to recombination of electrons with such systems In plasmas containing magnetic fields, cyclotron radiation is another process that can be understood in terms of individual particles. But electromagnetic radiation may also be produced by collective motions of a large number of electrons in certain types of plasma oscillations. This radiation is usually important only in the microwave range and requires, therefore, besides a very different theoretical treatment, experimental techniques that have little in common with conventional spectroscopy.

Before a system radiates, it obviously must be excited. This can occur by single-particle impacts, by the absorption of radiation, but in case of plasma oscillations also by collective phenomena connected with instabilities. Again, the latter mechanism is vastly different from the others, and it will not be treated in this chapter.

If reabsorption of radiation within a given plasma is negligible, knowledge of excitation (and deexcitation) rates and of the radiation-producing processes proper suffices for a complete description of the radiation. Otherwise it becomes necessary simultaneously to treat the problem of radiative energy transfer within the plasma. It turns out that very often most of the radiative energy transport takes place in the strongest lines, which implies that besides cross sections and transition probabilities one has to know rather precisely the wavelength dependence of absorption coefficients in the spectral range covered by any given line. (The problem of entrapment of resonance-line radiation in finite systems has been treated by Holstein.[15])

Even though normally the radiation intensity in laboratory plasmas is far below the blackbody limit, velocity distributions of electrons, atoms, and ions and the degrees of excitation and ionization may still be very close to those pertaining to a system in complete thermodynamic equilibrium. Such systems are said to be in *local thermal equilibrium*, which can exist if collisional processes dominate radiative processes. If this is the case, no detailed knowledge of cross sections is required, because then excitation and ionization equilibria are solely determined by statistical relations (Boltzmann factors and Saha equations). This simplifies the energy-transfer problem considerably.

There is another important class of stationary or quasistationary systems in which equilibrium is obtained by balancing electron impact excitation or ionization rates with radiative decay or recombination rates. Such a situation prevails in the solar corona[16] and also in some high-energy laboratory plasmas.   (It is more usual here, however, that no stationary situation whatever is attained.)

In conclusion, a complete theory of electromagnetic radiation from plasmas must be based on atomic physics (quantum mechanics) combined with transport theory and statistical mechanics.   Experimentally, this close relationship between plasma physics, especially plasma spectroscopy, with other fields is extremely fruitful, and plasma physics can already repay some of the debts incurred, for example, to astrophysics.

9.2 RADIATION FROM SMALL VOLUMES

9.2.1. Line Intensities

The power dP radiated in a spectral line by a small-volume element $d\tau$ is proportional to the number density N' of atoms in that excited state from which the line originates owing to a transition to the proper lower state. The difference of the excitation energies equals $h\nu$, h being Planck's constant, ν the frequency of the line.   If m and e are mass and charge of the electron and c is the velocity of light, the expression for dP (in Gaussian units) is

$$dP = \frac{8\pi^2 h\nu^3 e^2}{mc^3} f_e N' \, d\tau \tag{9.1}$$

Numerically this becomes

$$dP = 4.3 \times 10^7 (h\nu)^3 f_e N' \, d\tau \qquad \text{eV/sec}$$

if $h\nu$ is expressed in electron volts.   Besides the frequency, the only quantity characteristic of the states involved is the (dimensionless) *emission* oscillator strength f_e, which must be calculated from quantum mechanics and is of order 1 for stronger lines.

In the language of quantum mechanics, one would say that the transition rate is given by the product of transition probability (for spontaneous emission) A and the number of atoms available, $N' \, d\tau$, and that the power radiated away follows from this rate by multiplying with $h\nu$,

$$dP = h\nu A N' \, d\tau \tag{9.2}$$

Astrophysicists prefer still another way of writing this equation. They use *absorption* oscillator strengths f_a, in terms of which the power can be written as

$$dP = \frac{8\pi^2 h\nu^3 e^2}{mc^3} \frac{g}{g'} f_a N' \, d\tau \tag{9.3}$$

Here g and g' are the statistical weights of lower and upper states of the transition giving rise to the line, i.e., the number of degenerate states having the appropriate energies.

Most widely tabulated are the absorption oscillator strengths.[11,17] The conversion into emission oscillator strengths and transition probabilities for spontaneous emission follows readily from Eqs. (9.1) to (9.3) as

$$f_e = \frac{g}{g'} f_a = \frac{mc^3}{8\pi^2 \nu^2 e^2} A = 2.3 \times 10^{-8} (h\nu)^{-2} A \tag{9.4}$$

with $h\nu$ again expressed in electron volts. Besides spontaneous emission, there are induced emission and absorption processes and the corresponding transition probabilities. These processes are especially important in connection with the problem of radiation from finite volumes of plasma, i.e., from a large number of volume elements. Then, in general, one cannot simply add the contributions of the individual elements to the radiated power.

9.2.2. Line Shapes

Infinitely sharp spectral lines do not exist. Even if atoms radiate without appreciable interactions with neighboring particles, the emitted energies (or frequencies) will be spread over a band whose width is determined by Heisenberg's uncertainty principle as

$$\Delta E_n \equiv 2h \, \Delta\nu_n = h(2\pi \, \Delta t)^{-1}$$

where Δt is the lifetime of the excited state, which is determined by spontaneous decay. The inverse of this lifetime is equal to the sum of the spontaneous transition probabilities for *all* lines originating from a given upper state. Actually this yields only the contribution to the line width from the upper state, to which that from the lower state must be added, unless the latter is stable against radiative decay and not too much affected by absorption processes. (Occasionally, induced emission and absorption processes contribute to the radiation width as well.)

If no other broadening mechanism is active besides this "natural" line broadening, then the frequency distribution within the spectral line follows a dispersion profile

$$S_n(\nu) = \frac{\Delta\nu_n/\pi}{(\nu - \nu_0)^2 + (\Delta\nu_n)^2} \tag{9.5}$$

where ν_0 is the frequency of the ideally sharp line.   Since

$$\int_{-\infty}^{+\infty} S_n(\nu)\, d\nu = 1$$

the radiated power per unit frequency interval can be expressed as

$$dP_\nu = S_n(\nu)\, dP \tag{9.6}$$

In the same way a line of any shape can be specified by a spectral distribution $S(\nu)$.

Another source of broadening is the Doppler effect.   It causes the frequencies to shift by $\nu - \nu_0 = \nu_0(v/c)$, where v is the relative velocity component of the emitting atom with respect to the observer in the direction of the line of sight.   If the statistical distribution of these velocity components is described by some (normalized) function $W(v)$ and if no other important broadening process complicates the situation, the resulting line shape, again properly normalized, is

$$S_d(\nu) = \left(\frac{c}{\nu_0}\right) W\left(\frac{\nu - \nu_0}{\nu_0}c\right) \tag{9.7}$$

If the emitting atoms are in kinetic equilibrium, this leads to Gaussian line shapes

$$S_d(\nu) = \frac{1}{\Delta\nu_d \sqrt{\pi}} \exp\left[-\left(\frac{\nu - \nu_0}{\Delta\nu_d}\right)^2\right] \tag{9.8}$$

with the Doppler $(1/e)$ width given by

$$\Delta\nu_d = \nu_0 \sqrt{\frac{2kT}{Mc^2}} \tag{9.9}$$

in terms of Boltzmann's constant k, kinetic temperature T, and mass M of the emitting atoms or ions.

Usually emitting atoms are perturbed to a significant extent by collisions with other particles.   In cases where the duration of the collisions is short compared with the average time between collisions, or where the combined effect of collisions overlapping in time is small, this leads to impact profiles.   They have dispersion shapes (in nondegenerate cases) with impact widths $\Delta\nu_i$ and shifts $\Delta\nu_s$,

$$S_i(\nu) = \frac{\Delta\nu_i/\pi}{(\nu - \nu_0 - \Delta\nu_s)^2 + (\Delta\nu_i)^2} \tag{9.10}$$

Impact widths and shifts are proportional to the density of the perturbing particles and are, in general, quite weak functions of the kinetic temperature.   Before these widths and shifts can be calculated, the exact nature of the interaction between perturbing particles and the emitting atoms (or ions) must be specified.

In plasmas, the strongest interaction is due to electric fields produced by charged particles, i.e., electrons and ions, which cause Stark broadening. For stronger lines, the impact approximation is practically always valid for electrons as perturbers, but complications arise from the degeneracy in case of hydrogen and hydrogenlike lines.[18,19] The perturbing electrons can be treated as classical particles, even though for them the adiabatic approximation is almost never valid, not even for lines normally subject to the quadratic Stark effect like typical neutral helium lines.[20] Ions do not contribute very much to the impact broadening as long as temperatures are not extremely high. But in high-energy plasmas they will eventually override the electron contribution.[9]

Other causes of impact broadening may be resonance interactions with atoms in the ground state (resonance broadening[21,22]). This mechanism is quite important for resonance lines in partially ionized gases. It is a contributing factor for nonresonance lines if the lower state connects with the ground state via an allowed transition.

Nonresonance interactions with other atoms or the same atoms (if none of the states involved in the line under consideration have an allowed transition to the ground state) can give rise to some broadening, due to van der Waals forces. This broadening mechanism is almost always negligible in plasmas but must be considered in cases where the degree of ionization is very small, because it may then dominate the Stark broadening.

The impact approximation cannot be used in all cases to calculate the pressure broadening, i.e., the broadening other than natural and Doppler broadening. Often the opposite extreme approximation is applicable. In these quasistatic theories one first calculates the frequency shifts corresponding to some configuration of perturbing particles and then averages over the configurations to obtain the actual line profiles.

The most important example of quasistatic broadening is the ion broadening of lines subject to the linear Stark effect. Here the frequency shift is given by the product of the Stark coefficient C and the electric-field strength F,

$$\nu - \nu_0 = CF \tag{9.11}$$

Each value of the field strength has a certain probability $W(F)$ corresponding to the probability of finding perturbing ions in a configuration giving rise to this field. The line profile is then simply given by

$$S_S(\nu) = \frac{1}{C} W \left(\frac{\nu - \nu_0}{C} \right) \tag{9.12}$$

Normally several Stark components exist, and a sum must be taken using the partial oscillator strengths as weighing factors.[23]

As long as interactions between perturbing (and emitting) ions and Debye shielding by electrons are small, $W(F)$ is a universal function of the two-thirds power of the electron density (Holtsmark function). Otherwise the situation is more complicated, and it is also necessary to know the temperature. A consistent theory exists only for cases where deviations from the Holtsmark result are not too large and then only for atoms or singly ionized ions as emitting systems.[24,25]

Very often more than one line-broadening mechanism is effective for a given line. As long as these processes are statistically independent, the actual profile can be obtained by "folding" the parent line shapes. The best known examples for this are the Voigt profiles,[26] which are a superposition of Doppler profiles on dispersion profiles (from impact or natural broadening), and the Stark profiles of hydrogen and ionized helium lines produced by quasistatic ion and electron impact broadening.[18,19] (Detailed formulas and tables of line-broadening parameters are contained in Ref. 11.)

9.2.3. Continuum Intensities

The most important sources of continuous radiation in plasmas are transitions of free electrons into lower-energy free states or into bound states. The intensity is proportional to the product of ion (N_i) and electron density (N_e), as often are the line intensities. This can be seen from (9.3) if one invokes the appropriate Saha equation (9.32), described in Sec. 9.4 on Stationary States, and introduces $\Delta E'$ as the energy separation of the upper level from the ionization limit (counted positive),

$$dP = \frac{8\pi^2 h \nu^3 e^2}{mc^3} \frac{g'}{2g_i} \left(\frac{h^2}{2\pi mkT} \right)^{3/2} f_a \exp\left(\frac{\Delta E'}{kT} \right) \times N_i N_e \, d\tau \tag{9.13}$$

This relation can easily be generalized to describe the free-bound continuum of hydrogen, ionized helium, etc., by "smearing out" a line from a high-lying state with principal quantum number n' over a frequency range corresponding to a change in n' by one unit, $\Delta n' = 1$. The power radiated in the free-bound continuum into a unit frequency interval thus becomes

$$dP_\nu{}^{bf} \approx \frac{dP}{\nu_{n'+1} - \nu_{n'}} \approx \frac{dP}{|d\nu/dn'|} \tag{9.14}$$

One now introduces the Bohr formula

$$h\nu = E_\infty \left(\frac{1}{n^2} - \frac{1}{n'^2} \right) \approx \frac{E_\infty}{n^2} \tag{9.15}$$

differentiates,

$$h \left| \frac{d\nu}{dn'} \right| = \frac{2E_\infty}{n'^3} \tag{9.16}$$

uses the asymptotic formula for the absorption oscillator strengths[27]

$$f_a \approx \frac{G(n)}{n'^3} \approx \frac{32n}{3\pi \sqrt{3}\, n'^3} \qquad (9.17)$$

and the statistical weights

$$g = 2n^2 \qquad g' = 2n'^2 \qquad g_i = 1 \qquad (9.18)$$

and finally substitutes (9.13) and (9.15) to (9.18) into (9.14),

$$dP_\nu{}^{bf} = \frac{128\pi e^2 E_\infty{}^2}{3\sqrt{3}\, hmc^3 n^3} \left(\frac{h^2}{2\pi mkT}\right)^{3/2} \exp\left(\frac{E_\infty}{n^2 kT}\right)$$
$$\times\, N_i N_e \exp\left(\frac{-h\nu}{kT}\right) d\tau \qquad (9.19)$$

noting that $\Delta E' = \Delta E - h\nu$ and $\Delta E = E_\infty/n^2$.

The total bound-free continuum is the sum of such expressions over all n for which $h\nu \geq \Delta E$, and the free-free continuum can be considered as an extension of this sum (which becomes an integral) to positive energy states ($\Delta E < 0$). By using $E_\infty/n^2 = \Delta E$,

$$-\left(\frac{2E_\infty}{n^3}\right) dn = d(\Delta E) \qquad (9.20)$$

the free-free continuum derived in this way is

$$dP_\nu{}^{ff} \approx -\frac{64\pi e^2 E_\infty}{3\sqrt{3}\, hmc^3} \left(\frac{h^2}{2\pi mkT}\right)^{3/2} N_i N_e$$
$$\exp\left(-\frac{h\nu}{kT}\right) \int_0^{-\infty} \exp\left(\frac{\Delta E}{kT}\right) d(\Delta E)\, d\tau$$
$$= \frac{64\pi e^2 E_\infty}{3\sqrt{3}\, hmc^3} \left(\frac{h^2}{2\pi mkT}\right)^{3/2} N_i N_e \exp\left(-\frac{h\nu}{kT}\right) kT\, d\tau \qquad (9.21)$$

It is more convenient to include formally all states for which ΔE is smaller than some suitable $\Delta E''$ in the integral. Then the complete continuum formula can be written as

$$dP_\nu = dP_\nu{}^{ff} + dP_\nu{}^{bf}$$
$$= \frac{32\sqrt{\pi}\, he^4}{3\sqrt{3}\, m^2 c^3} z \sqrt{\frac{E_\infty}{kT}} \left[\exp\left(\frac{E_\infty}{n''^2 kT}\right) + \frac{2E_\infty}{kT}\sum_{n*}^{n''} \frac{1}{n^3}\exp\left(\frac{E_\infty}{n^2 kT}\right)\right]$$
$$N_i N_e \exp\left(-\frac{h\nu}{kT}\right) d\tau \qquad (9.22)$$

if one also introduces $E_\infty = 2\pi^2 e^4 mz^2/h^2$, the quantum theoretical result for the ionization energy of hydrogen and hydrogenic ions.

The actual choice of n'' is obviously not critical, but it must be remembered that the sum includes only such n for which a free-bound transition is possible for a given value of $h\nu$, that is, $h\nu \geq E_\infty/n^{*2}$ or

$$n^* \geq \sqrt{\frac{E_\infty}{h\nu}} \tag{9.23}$$

In cases where the series limits advance by $\Delta E(N_e)$ (merging of lines), e.g., owing to Stark effects, n'' should be chosen to correspond just to $\Delta E(N_e)$, that is,

$$n'' \approx \sqrt{\frac{E_\infty}{\Delta E(N_e)}} \tag{9.24}$$

Also, if the reduction ΔE_∞ of the ionization potential is important, (9.22) must be multiplied with $\exp(-\Delta E_\infty/kT)$, and if $n^* \geq n''$, there will be no bound-free contribution. Then E_∞/n''^2 in (9.22) should be replaced by $h\nu + \Delta E_\infty$.

The charge dependence of the continuum intensity is quite strong (hydrogen has $z = 1$, ionized helium $z = 2$, etc.). The bremsstrahlung continuum (free-free transitions) scales with z^2, but the recombination continuum (free-bound transitions) with z^4. The latter is therefore very important if highly ionized species are present, especially because the exponentials in (9.22) can be quite large.

The Kramers-Unsöld formulas for the continuum intensities described in this section follow essentially from an application of Bohr's correspondence principle. The exactly known (Gaunt) correction is normally quite small,[10,11] except in high-energy plasmas. More serious errors may be incurred if one wants to use these formulas for continua from nonhydrogenic atoms or ions. But even then a good approximation is often obtained, as long as only high-lying states, which are quite hydrogenlike, contribute.

Another source of continuous emission in hydrogen should at least be mentioned, namely, negative hydrogen ions, which are, e.g., very important in the solar atmosphere. In laboratory plasmas, the H^- continuum is, however, mostly negligible because of the higher fractional ionization. (The H^- continuum is proportional to the product of the densities of *neutrals* and electrons, as compared with the normal H continuum, which is proportional to N_e^2, on the assumption that $N_i = N_e$.)

9.2.4. Magnetic Radiation

All high-energy plasmas must be confined by magnetic fields, in which the electrons of the plasma will, in general, describe helical orbits with angular (cyclotron) frequencies

$$\omega = \frac{eB}{mc} \tag{9.25}$$

where B = magnetic field strength

The radius of these orbits is for nonrelativistic electron velocities, $v_\perp$ being the velocity component at right angles to the field,

$$r = \frac{mcv_\perp}{eB} \tag{9.26}$$

The total radiated power can be estimated from that of a dipole whose dipole moment is

$$\mu = \sqrt{2}\, er \cos \omega t \tag{9.27}$$

That is, according to the classical theory for $N_e\, d\tau$ incoherent radiators,

$$d\bar{P} \approx \frac{\omega^4 \mu^2}{3c^3}\, N_e\, d\tau = \frac{2e^4 kTB^2}{3m^3 c^5}\, N_e\, d\tau \tag{9.28}$$

with $\overline{v_\perp{}^2} = kT/m$ used.

The shape of this cyclotron resonance line may be determined by collision broadening, which gives rise to dispersion profiles whose half width is essentially given by the electron-ion collision frequency or by Doppler broadening due to motions along the field lines. The radiation is not at all isotropic, because all equivalent dipoles are in the plane at right angles to the field.

An important modification occurs at electron velocities approaching the velocity of light. Then the wavelength of the cyclotron radiation is no longer much larger than the gyroradius, and higher harmonics (synchrotron radiation) come in with increasing intensities, which are not as effectively reabsorbed as the first harmonic. It is this radiation which constitutes an important energy-loss process in plasmas required for eventual thermonuclear reactors.[14]

There is also, of course, some radiation at the ion cyclotron frequency. But the radiated power, because of the large mass, is so much smaller that it is of no practical interest. However, the inverse process, i.e., absorption of electromagnetic radiation near this frequency, can be useful as a selective heating mechanism for ions.[28]

9.2.5. Plasma Radiation

The processes discussed until now produce electromagnetic radiation in ways that are not at all restricted to plasmas, which does not mean that they would not be useful to analyze a plasma or could not influence its energy balance.   But none of the typical aspects of plasma physics are involved, except in phenomena like the Stark broadening of spectral lines.

The only type of electromagnetic radiation that is really characteristic of plasmas is that due to various plasma oscillations which are sometimes capable of radiating away a certain fraction of their energy. These phenomena have already been discussed in Chap. 8.

9.3 RATE PROCESSES

9.3.1. Collisions

In plasmas, electrons are responsible for most collisional processes because of their high velocities, and because electron-impact cross sections tend to be larger or at least equal to those of ions, not to mention neutrals.   This can be understood in terms of the classical picture, since usually only colliding particles in resonance with orbital electrons can have a strong perturbing effect, i.e., only electrons for which the characteristic frequency, velocity over impact parameter, is of the order of typical orbital frequencies of bound electrons.

To fully describe all important collisional rate processes, the cross sections for the following reactions are required: excitation and deexcitation from any bound state to any other bound state, ionization from any bound state into any free-electron state and collisional (three-body) recombination from any free state into any bound state, and energy and momentum transfer between free electrons, between electrons and ions, and between electrons or ions and neutrals.

The quantum-mechanical theory[29] of most of these cross sections is much more complicated than the theory of radiative process, excluding line profiles.   This is true even for hydrogen, etc., where exact wave functions exist.   For the energy and momentum-transfer cross sections between charged particles, classical theory suffices, but the long-range-nature of the Coulomb interaction causes some difficulties that are characteristic of plasmas.[30]   One has to realize that most of these transfer processes involve more than two particles at a time.   If one wants to maintain the binary-collision assumption, a cutoff must be introduced at impact parameters of the order of the Debye radius, the exact value of which will depend on the type of reaction that one wants to calculate.

Fortunately, for many purposes it is quite adequate to use rough estimates for the cross sections, because it is frequently sufficient to know the order of magnitude of various reaction rates.   And if the cross section for some reaction is known, that of the inverse reaction can be obtained from the principle of detailed balancing.   This says that any reaction and the corresponding inverse reaction will lead to a steady state which is equal to the thermodynamic equilibrium state, if the active partners in this reaction, e.g., electrons or photons, possess an equilibrium energy distribution. It is therefore sufficient to have excitation and ionization cross sections. For the former, a reasonable approximation is often

$$\sigma \approx \frac{2\pi e^4}{mv^2 \, \Delta E} f_a \qquad v^2 \geq \frac{2 \, \Delta E}{m} \tag{9.29}$$

where ΔE = difference in excitation energies of initial and final states
f_a = corresponding absorption oscillator strength
Expressions valid for ionization follow from (9.29) by summing (integrating) over final states.[11]

Even if all cross sections are known and if radiative excitation and ionization are negligible, the simultaneous solution of all rate equations still poses a difficult problem. But for atoms like hydrogen, again an important simplification can be made. Here all excited states are so close in energy that collisional processes between them are much more rapid than those also involving the ground state. Therefore the populations of excited bound states and free states can often be considered to be quasi-stationary, which reduces the problem essentially to one rate equation for the ground-state population.[31]

9.3.2. Radiation

Besides spontaneous emission, which is characterized by the transition probabilities mentioned in Sec. 9.2.1 on Line Intensities, there are two more radiative processes, absorption and induced (stimulated) emission, which can again be described by transition probabilities (Einstein coefficients). These coefficients, however, are proportional to the amount of radiation present in the relevant frequency interval and are related to each other by the principle of detailed balancing.

It is accordingly sufficient to consider only one of these processes. The quantitative relation for the absorption cross section is

$$\sigma_\nu = \frac{\pi e^2}{mc} f_a S(\nu) \tag{9.30}$$

where f_a = absorption oscillator strength
$S(\nu)$ = normalized line shape
This is the cross section of an atom in some state, which is the lower state for the line under consideration. The same formula can be applied to the absorption of continuous radiation by an atom in some bound state, if $f_a S(\nu)$ is replaced by f_ν, the oscillator strength per unit frequency interval. The generalization to free-free absorption poses no serious difficulty either.[1]

From the cross sections, photoexcitation or -ionization rates can be obtained by multiplying with the velocity of light and the energy density (per unit frequency interval) of the radiation divided by $h\nu$ and finally integrating over the frequency range covered by the line or the continuum.

The rates for deexcitation by stimulated emission can be obtained in exactly the same manner, only that now emission oscillator strengths must be used in the equation corresponding to (9.30), which would give the cross section for stimulated deexcitation from some upper to some lower state. Obviously, each stimulated emission adds one photon to the radiation field.

9.4   STATIONARY STATES

9.4.1. Local Thermal Equilibrium

In many experiments, excitation rates, etc., are so high that the plasma as a whole and also small parts of it adjust practically instantaneously to changes in their environment, net energy and particle losses or gains. Then it is no longer necessary actually to solve the rate equations, but only to find their appropriate stationary solution.   In principle, this also requires the knowledge of all cross sections, but often important simplifications are possible, which will be described below.

If population and depopulation rates of different energy states are dominated by processes that are inverse of each other, and if the coupling agents, usually free electrons, have an energy distribution that corresponds to that in a thermodynamic equilibrium system at some temperature, one needs no cross sections. Detailed balancing applies, and the relative populations in the states coupled in this manner will be equal to those in a system in thermodynamic equilibrium at the same temperature. Such systems are said to be in *local thermal equilibrium*, the "temperature" being in general a function of both space and time.

As a definition of local thermal equilibrium one can therefore say that it prevails if distributions over excited states, the degree of ionization, and the various velocity distributions can be characterized by one temperature.   But the radiation, except in the resonance lines, will usually be far from that of a blackbody at this temperature.   A necessary condition for the existence of local thermal equilibrium is then obviously that radiative processes are negligible against their collisional counterparts, i.e., essentially against the corresponding electron impact processes.   Based on estimates for the ratios of the relevant rates, relatively simple validity criteria for local thermal equilibrium can be derived.[11,32,33]

Since electron impacts are so much more important than heavy particle impacts, it does not matter much if the velocity distributions of atoms and ions are not in equilibrium with that of the electrons, which is practically always Maxwellian. Excitation and ionization equilibria will accordingly reflect only the electron temperature. This is the reason why most spectroscopic techniques applicable to local-thermal-equilibrium plasmas will primarily yield the value of the kinetic electron temperature.

If local thermal equilibrium can be assumed, the relative populations in different energy states are related by statistical weights and Boltzmann factors,

$$\frac{N'}{N} = \frac{g'}{g} \exp\left(-\frac{E' - E}{kT}\right) \tag{9.31}$$

and the ionization equilibrium is determined by the Saha equation (cf. Chap. 3),

$$\frac{N_e N_i}{N'} = \frac{2g_i}{g'}\left(\frac{2\pi mkT}{h^2}\right)^{3/2} \exp\left(-\frac{E_\infty - E'}{kT}\right) \tag{9.32}$$

where E_∞ = ionization energy
E' = excitation energy of that state to whose population N' one wants to relate the electron and ion density

That these two statistical relations are of common origin becomes obvious when one realizes that $2g_i(2\pi mkT/h^2)^{3/2}/N_i$ is the number of free-electron states available per ion, i.e., in fact the quantity that corresponds to the statistical weight g' in (9.31).

At high electron densities it is necessary to use reduced ionization energies because the ionization energy of a free atom is lowered by an amount corresponding to the energy set free when an electron-ion pair is embedded into the plasma. At the same time one must take into account that the higher-lying excited states may cease to exist for the same reason. Uncertainties in these corrections usually cause no significant errors if one is consistent in their application.[10,11,34] [Note that the reduction ΔE_∞ in the ionization potential does not necessarily correspond to the advance $\Delta E(N_e)$ of the series limits mentioned in Sec. 9.2.3 on Continuum Intensities. ΔE_∞ can normally be calculated[11,34] by using the Debye theory and almost never exceeds a fraction of an electron volt in laboratory plasmas. $\Delta E(N_e)$ is usually larger and must be estimated from the line-broadening theory.]

9.4.2. Corona Equilibrium

With decreasing electron densities and increasing charges of the ions, radiative processes gain in relative importance. Radiative decay of excited states and radiative recombination take the place of deexciting electron collisions and three-body recombination, but photoexcitation and -ionization remain normally negligible in laboratory plasmas. Therefore the stationary state is obtained by balancing collisional excitation and ionization rates with the radiative decay and recombination rates. This leads to relations that are applicable, e.g., to the excitation and ionization equilibrium in the solar corona and to some extent also to high-energy plasmas produced in the laboratory, even though in the latter case confinement times are often too short for the formation of a stationary state.

The transition from local thermal equilibrium to the corona equilibrium is a gradual one. Radiative processes become important first for processes that populate the ground state, whereas excited states and free electrons remain in local thermal equilibrium down to surprisingly low electron densities[11,31-33] for atoms in which the energy gaps between excited states are relatively small, as in hydrogen. This is a fortunate circumstance, because most spectroscopic temperature-measurement techniques require only the assumption of local thermal equilibrium to be valid for these states.

Actually to calculate the corona equilibrium, electron impact excitation and ionization cross sections, transition probabilities for spontaneous transitions, and radiative recombination coefficients must be known and an assumption be made concerning the electron velocity distribution, which will normally be Maxwellian. Then electron density and temperature and the chemical composition of the plasma are the only independent variables.

Because excitation rates are proportional to the density N_0 of atoms in the ground state and to the electron density, whereas deexcitation rates are proportional to the density in the excited state, the excitation equilibrium is now described by equations of the type

$$\frac{N'}{N_0} = N_e f'(T) \tag{9.33}$$

This is true if secondary (cascading) processes involving other states are negligible, which is the usual situation when coronalike conditions prevail. The equation for the ionization equilibrium is

$$\frac{N_i}{N_0} = f_i(T) \tag{9.34}$$

because ionziation rates scale with the product of neutral and electron densities and radiative recombination rates with the product of electron and ion densities. The functions $f'(T)$, $f_i(T)$ depend on the temperature, cross sections, and radiative coefficients. They have been calculated approximately for astrophysical applications[16,35] and also for conditions occurring in high-energy laboratory plasmas,[12,31] including multiple ionization (cf. also Fig. 10.2).

It is interesting to note that relative excited-state populations are again independent of the electron density, as can be seen by applying (9.33) twice and taking the ratio.   But the temperature dependence is not quite the same as that of the corresponding Boltzmann factor.   As to the question where local thermal equilibrium or corona equilibrium applies in stationary or quasistationary plasmas, the simplest answer is that one should always use the relations yielding the lowest degree of ionization and excitation for a given electron temperature.

9.5 RADIATION FROM LARGE VOLUMES

9.5.1. Systems in Local Thermal Equilibrium

The calculation of radiation from large volumes is in general an extremely complicated problem.   Besides the various local rate processes, one must consider all transport phenomena, including radiative energy transfer within the plasma, and a simultaneous solution of rate and transport equations must be found.   As pointed out before, this requires knowledge of all relevant data from atomic physics, cross sections, transition probabilities, line-broadening parameters, and also transport coefficients. Besides this, the choice of the proper radiative-transport equation is quite difficult in itself and depends on the amount to which absorbed and reemitted photons are correlated.

However, for local-thermal-equilibrium systems, there are no such difficulties, because now absorption and emission in small-volume elements can be calculated from local temperatures and densities.   The radiative-transfer problem can be solved separately and is simplified considerably, since Kirchhoff's law applies, which relates absorption and emission in such systems by the Planck function $B_\nu(T)$.

Consider the change of the intensity $j_\nu(x)$ of radiation in the x direction (per unit frequency interval, unit area, and unit solid angle). If x is changed by an infinitesimal amount, we have

$$dj_\nu = k'_\nu B_\nu \, dx - k'_\nu j_\nu \, dx \tag{9.35}$$

where k'_ν is the effective absorption coefficient, i.e., the actual absorption coefficient minus the coefficient of stimulated emission.   The first term on the right-hand side gives the emission in dx according to Kirchhoff's law, the second term the absorption.   A general solution can be found if one introduces the "optical path" $\tau \equiv \int_0^x k'_\nu \, dx$ (not to be confused with the volume element $d\tau$ in Sec. 9.3)

$$j_\nu(\tau) = e^{-\tau} \int_0^\tau B_\nu(\tau') e^{\tau'} \, d\tau' \tag{9.36}$$

The absorption coefficient is related to the absorption cross section (9.30) by

$$k'_\nu = N\sigma_\nu \left[1 - \exp\left(-\frac{h\nu}{kT}\right) \right] \tag{9.37}$$

where N = number density of absorbing systems

[. . .] = correction factor for induced emission

In the limit of large optical paths $\left(\int_0^x k'_\nu \, dx \gg 1 \right)$, one properly obtains (by partial integration) the blackbody formula from (9.36), provided that the fractional change of B_ν over a unit optical path near the surface is small. At the opposite extreme one has for optically thin layers ($e^{-\tau} \approx e^{+\tau} \approx 1$)

$$j_\nu \approx \int_0^x B_\nu k_\nu \, dx \tag{9.38}$$

which becomes with (9.37) and $B_\nu = (2h\nu^3/c^2)/[\exp (h\nu/kT) - 1]$

$$j_\nu \approx \frac{2h\nu^3}{c^2} \int_0^x \exp\left(\frac{-h\nu}{kT}\right) N\sigma_\nu \, dx \tag{9.39}$$

This can be brought into a more familiar form by using the general expression for the absorption cross section (9.30) and by introducing the number density in the upper state N' via (9.31), it being noted that $E' - E = h\nu$. This yields

$$j_\nu \approx \frac{2\pi h\nu^3 e^2}{mc^3} \frac{g}{g'} f_a \int_0^x S(\nu)N' \, dx \tag{9.40}$$

which is, as expected, precisely what one obtains from (9.3). This may be shown by writing the volume element as a product of element of area and dx and by recognizing that (9.3) gives the power radiated into the whole solid angle 4π in the total frequency range of a line. Note also that

$$\int_{-\infty}^{+\infty} S(\nu) \, d\nu = 1$$

9.5.2. Systems in Corona Equilibrium

In this case there is no radiative-transfer problem, because absorption and induced emission are both negligible. Therefore the expressions for line emission become extremely simple, especially for resonance lines, where the rate of photons emitted is just equal to the rate at which electron collisions populate the upper state, if cascading is negligible. For nonresonance lines, this rate has to be multiplied by the branching ratio, i.e., the ratio of the transition probability A_k for the line whose intensity one wishes to calculate to the sum of the probabilities A_i for all spontaneous transitions that originate from the upper state of this line.   The total line intensity is accordingly

$$\int_{-\infty}^{+\infty} j_\nu \, d\nu = \frac{h\nu}{4\pi} \frac{A_k}{\Sigma A_i} \int_0^x \langle \sigma v \rangle N_e N_0 \, dx, \tag{9.41}$$

where $\langle \sigma v \rangle$ is the average of the product of excitation cross section as taken, e.g., from (9.29) and electron velocity.   Again the factor $1/4\pi$ enters because of the conversion to the unit solid angle and dx instead of $d\tau$ because intensities are defined per unit area.

Equation (9.41) differs considerably from the corresponding formula for optically thin plasmas in local thermal equilibrium.   Not only is the intensity now proportional to the density in the ground state N_0, but it also scales with the electron density.   This has rather drastic consequences for dense high-energy plasmas containing impurity ions of some charge z, which are not completely stripped.   If (9.41) is applicable to the resonance line of these impurities, one has for the relative rate at which electrons lose their energy by exciting this line

$$\frac{1}{E} \frac{dE}{dt} = \frac{1}{T} \frac{dT}{d\tau} = - \frac{h\nu \langle \sigma v \rangle N_0}{\frac{3}{2}kT} \tag{9.42}$$

That is, the cooling rate is proportional to the *absolute* concentration of impurities.

There is no need to derive special formulas for continuum emission due to free-free and free-bound transitions because Eq. (9.22), after converting to intensities in the usual way, or in case of large deviations from hydrogenlikeness a properly modified relation can always be used when the electron velocity distribution is Maxwellian and when absorption and induced emission are negligible in the continuum.   Both these assumptions are almost always fulfilled in the laboratory.

For stationary plasmas, whether they are in local thermal or in corona equilibrium, we are therefore able to calculate, in principle, the intensity distribution of the electromagnetic radiation in the whole wavelength range covered by conventional spectroscopy as function of number density and electron temperature.   This allows us to calculate energy losses and, by inverting the procedure, to determine from spectroscopic measurements many of the quantities that characterize a plasma. Besides the parameters mentioned above, atom and ion kinetic temperatures can be obtained from Doppler widths and magnetic fields from Zeeman splittings of suitable spectral lines.   In transient plasmas, the appropriate rate equations must be solved for each case.   From their solutions the time characteristics of the spectrum can be found without further difficulty and can be compared with experiment.

REFERENCES

1. Unsöld, A.: "Physik der Sternatmosphären," 2d ed., Springer, Berlin, 1955.
2. Aller, L. H.: "Atmospheres of the Sun and Stars," 2d ed., Ronald, New York, 1964; R. v. d. R. Woolley and D. W. N. Stibbs: "The Outer Layers of a Star," Oxford University Press, New York, 1953.
3. Finkelnburg, W., and H. Maecker: "Handbuch der Physik," vol. XXII/II, Springer, Berlin, 1956.
4. Lochte-Holtgreven, W.: *Rept. Prog. Phys.*, **21**, 312 (1958).
5. Dickerman, P. J.: "Optical Spectrometric Measurements of High Temperatures," University of Chicago Press, Chicago, 1961.
6. *Proc. Fourth Symposium on Temperature, Its Measurement and Control Sci. Ind.*, 1961.
7. *Proc. Fifth Intern. Conf. on Ionization Phenomena in Gases*, 1962.
8. Stratton, T. F.: In Ref. 6.
9. Griem, H. R.: In Ref. 6.
10. Griem, H. R.: In Ref. 7, p. 1857.
11. Griem, H. R.: "Plasma Spectroscopy," McGraw-Hill, New York, 1964. See also R. H. Huddlestone and S. L. Leonard (eds.), "Plasma Diagnostic Techniques," Academic, New York, 1965.
12. Knorr, G.: *Z. Naturforsch.*, **13a**, 941 (1958); V. I. Kogan: *Soviet Phys. Doklady*, **4**, 1057 (1960).
13. Post, R. F.: *Ann. Rev. Nuclear Sci.*, **9**, 378 (1959); *J. Nuclear Energy*, **C3**, 273 (1961).
14. Trubnikov, B. A.: *Phys. Fluids*, **4**, 195 (1961).
15. Holstein, T.: *Phys. Rev.*, **72**, 1212 (1947), **83**, 1159 (1951).
16. Elwert, G.: *Z. Naturforsch.*, **7a**, 432 (1952).
17. Allen, C. W.: "Astrophysical Quantities," 2d ed., Athlone Press, London, 1963; B. M. Glennon and W. L. Wiese: *Natl. Bur. Standards (U.S.) Monograph* 50, 1962.
18. Griem, H. R., A. C. Kolb, and K. Y. Shen: *Phys. Rev.*, **116**, 4 (1959).
19. Griem, H. R., and K. Y. Shen: *Phys. Rev.*, **122**, 1490 (1961).
20. Griem, H. R., M. Baranger, A. C. Kolb, and G. Oertel: *Phys. Rev.*, **125**, 177 (1962).
21. Weisskopf, V.: *Physik. Z.*, **34**, 1 (1933).
22. Wlassow, A., and W. Furssow: *Physik. Z. Sowjetunion*, **10**, 378 (1936).
23. Underhill, A., and J. Waddell: *Bur. Standards Circ.* 604, 1959.
24. Baranger, M., and B. Mozer: *Phys. Rev.*, **115**, 521 (1959).
25. Mozer, B., and M. Baranger: *Phys. Rev.*, **118**, 626 (1960).
26. Traving, G.: "Über die Theorie der Druckverbreiterung von Spektrallinien," Braun, Karlsruhe, 1960.
27. Bethe, H. A., and E. E. Salpeter: "Handbuch der Physik," vol. XXXV/I, Springer, Berlin, 1957. Their eq. (63.11) is for the sum of all oscillator strengths $n \rightarrow n'$, and (9.17) follows from it by dividing with $g = 2n^2$ in the limit $n^2 \ll n'^2$.
28. Stix, T. H.: *Phys. Rev.*, **106**, 1146 (1957); *Phys. Fluids*, **1**, 308 (1958).

29. Seaton, M. J.: In D. R. Bates (ed.), "Atomic and Molecular Processes," chap. 11, Academic, New York, 1962.

30. Spitzer, L.: "Physics of Fully Ionized Gases," 2d ed., chap. 5, Interscience, 1962.

31. McWhirter, R. W. P.: *Nature*, **190**, 902 (1961); D. R. Bates and A. E. Kingston: *Planetary Space Sci.*, **11**, 1 (1963); D. R. Bates, A. E. Kingston, and R. W. P. McWhirter, *Proc. Roy. Soc. (London)*, **A267**, 297 (1962), **270**, 155 (1962); R. W. P. McWhirter and A. G. Hearn: *Proc. Phys. Soc.*, **82**, 641 (1963).

32. Wilson, R.: *J. Quant. Spectroscopy and Radiative Transfer*, **2**, 477 (1962).

33. Griem, H. R.: *Phys. Rev.*, **131**, 1170 (1963).

34. Griem, H. R.: *Phys. Rev.*, **128**, 997, (1962).

35. Seaton, M. J.: *Planetary Space Sci.*, **12**, 55 (1964).

10

LABORATORY PLASMA PRODUCTION

KUNKEL

WULF B. KUNKEL, *Senior Physicist, Lawrence Radiation Laboratory, and Lecturer in Physics, University of California, Berkeley, California*

10

In this chapter some of the properties of real plasmas are discussed, with emphasis on the methods of production of ionized gases in the laboratory. No claim for completeness can be made, and it is not attempted to provide great detail. The subject is large enough to fill several volumes. The purpose here is to provide merely a concise summary of the field as an introduction for the newcomer and a survey for the outsider. The first four sections deal with some general considerations, and the remainder contains brief descriptions and discussions of the most important or most basic types of laboratory plasmas.

10.1 NONEQUILIBRIUM

Experimental research on plasma is difficult, because under laboratory conditions it is not possible to maintain common gases in an interesting state of ionization if they are to remain in complete thermodynamic equilibrium. This statement follows directly from Saha's equation (3.40), which, for a single state of ionization, takes the approximate form

$$\frac{(n^e)^2}{n^0} \approx G(T) T^{3/2} \times 10^{21-0.43 V_i/T} \qquad \text{cm}^{-3} \tag{10.1}$$

where n^e = electron density (assumed equal to ion density), (cm^{-3})

n^0 = neutral particle density, (cm^{-3})

T = temperature, V

V_i = ionization potential, V

$G(T)$ is a factor (of order unity) that normally varies only slowly with temperature; it depends on the nature of the gas, of course, and usually lies between 2 and 6. For atomic hydrogen it is nearly equal to 3 if $T < 2.5$ V. Substitution of the ionization potential V_i into Eq. (10.1) shows that at the maximum tolerable surface temperatures of solid materials, i.e., at $T < 0.3$ V, the equilibrium degree of ionization of most substances is extremely small. In consequence practically all experimental plasmas either are ionized out of equilibrium or have to be much hotter in their interior than the surrounding walls; i.e., they must be nonuniform in space, at least near their boundaries. In the latter case they transfer energy at a rather rapid rate toward their boundaries, and very often they can be created only as brief transients (pulsed discharges, shocks, etc.).

A discussion of laboratory plasmas must therefore be largely concerned with rate processes such as ionization, recombination, heat transfer, diffusion, and lifetimes in general. Analytic work becomes exceedingly complicated, especially because the neutral species and the many states of excitation must be taken into consideration. More specifically, the equations of change introduced in Chap. 4 must include the effects of inelastic collisions, and, in particular, they must be augmented by source terms representing the results of chemical "reactions," such as creation and destruction of the free charged particles and the accompanying absorption or liberation of binding energy. In addition, in many cases it is essential that radiative processes be taken into account, both in the equations for energy transport and in the continuity equations of the various species.

The energy equation (4.58) can be formally retained without modification if the following redefinitions are introduced: (1) All free particles in a given state of excitation are considered as a separate species and identified by a separate superscript a. (2) The internal energy density $\mathcal{E}^a$ is taken to include the energy stored in excitation, dissociation, and ionization. This is most systematically accomplished by counting all electronic binding energies as negative quantities; i.e., the zero energy refers to the fully stripped state where all nuclei and electrons are separated. Nuclear reactions are not considered here, but all possible chemical compounds are readily included in this way. We thus have for the total internal energy density of a species

$$\mathcal{E}^a = \tfrac{3}{2} p^a + \rho^a w^a \tag{10.2}$$

where $-w_a > 0 =$ "energy of formation" per unit mass of species. (3) The energy-transfer-rate density R^{ab} must now be replaced by a sum over all types of interactions and must include binding energy transferred by inelastic processes which create or destroy particles belonging to species a. To be exact, even radiative processes must not be omitted.[1]

If a sum over all species is then performed, the general relations for the change of energy (4.71) is recovered, of course. We write it in the form

$$\frac{\partial}{\partial t} (\mathcal{E}' + \tfrac{1}{2}\rho u^2) + \nabla \cdot (\mathcal{E}' + \tfrac{1}{2}\rho u^2)\mathbf{u}$$
$$= \mathbf{j} \cdot \mathbf{E} - \nabla \cdot (\mathbf{Q}' + \mathbf{p}' \cdot \mathbf{u}) \quad (10.3)$$

where the symbols have the same meaning as in Chap. 4 except that $\mathbf{Q}'$ must be understood now to include the rate of radiative energy transfer as well as of kinetic transport of internal energy through the fluid. In that case, however, Eq. (10.3) is not exact, because the rate of change of the energy stored in the electromagnetic-radiation field, which should appear on the left, has been neglected.[2] Fortunately, in laboratory plasmas, this is always justified. Of course, the point can be avoided if $\mathbf{j} \cdot \mathbf{E}$ is eliminated with the help of Poynting's theorem and if the electromagnetic energy density $\mathcal{E}_M$ is interpreted to extend over all the frequencies. Denoting the Poynting vector by $\mathbf{S}$ the equation can be written as a true conservation law:

$$\frac{\partial}{\partial t} (\mathcal{E}_M + \mathcal{E}' + \tfrac{1}{2}\rho u^2) + \nabla \cdot (\mathbf{S} + \mathbf{Q}' + \mathbf{p}' \cdot \mathbf{u}$$
$$+ \mathcal{E}'\mathbf{u} + \tfrac{1}{2}\rho u^2 u) = 0 \quad (10.3a)$$

A thorough treatment of laboratory plasmas using Eq. (10.3) or (10.3a) is clearly beyond the scope of this text. We restrict this introduction therefore to some rather general remarks. Likewise, the subsequent discussion of specific examples will in many cases have to be limited to qualitative descriptions.

10.2 CONTINUITY

In addition to the energy flow the most important consideration in the present context is the conservation of charge. In terms of the macroscopic equations derived in Chap. 4, this is expressed by the equations of continuity for the charged-particle species, i.e., for ions and electrons. For simplicity we shall discuss formally only the case for a single species of singly charged ions. Extension to a mixture of different and possibly multiply charged ions is straightforward but lengthy and does not add particularly to an understanding of the physical processes involved.

Nevertheless, there are in general several source terms that have to be added to Eq. (4.40), because there are a number of different processes by which the state of ionization of atoms and molecules can be affected. The principal ones are interactions with electrons, with photons, and with ions and with electronically excited neutral particles.   Extensive discussions of these processes can be found in the literature.[3-7]   The frequency of ionizing collisions between electrons and neutrals per unit volume can be expressed as $n^e \nu_i$.   Here the ionizing frequency per electron is given by

$$\nu_i = \sum_k n^k \langle \sigma_{ik} v' \rangle = \int_0^\infty \int_0^\infty \sum_k \sigma_{ik}(v') v' f^k(\mathbf{v}^k) f^e(\mathbf{v}^e) \, d\mathbf{v}^k \, d\mathbf{v}^e \qquad (10.4)$$

where n^k and $f^k(\mathbf{v}^k)$ are the number density and normalized distribution in velocity of neutral particles in state k, so that $\sum_k n^k = n^0$ is the total

density of neutrals, and $\sigma_{ik}(v')$ is, of course, the cross section for ionization from state k by electrons whose speed relative to the neutral is

$$v' = |\mathbf{v}^k - \mathbf{v}^e|$$

The function $f^e(\mathbf{v}^e)$ is as usual the distribution of electron velocities. Since usually $|\mathbf{v}^k| \ll |\mathbf{v}^e|$, we frequently are justified in the approximation $v' = |\mathbf{v}^e|$ so that (10.4) reduces to a summation over the populations n^k, each term involving a single integral over $\mathbf{v}^e$.

For the other production processes (such as photoionization, ionization by ion-neutral and by neutral-neutral collisions) similar expressions can be written down.   Since we are not going to make explicit use of them, we shall simply sum them up by a single source term s_i.   For each of these ionization processes the inverse exists as a type of recombination process.   The deionization rate can then also be expressed in terms of integrals over the distribution functions of the particles, but the calculations become increasingly more complex.   For the purpose of our discussion we again simply denote the entire recombination rate, as is usually done, by a term of the form $\alpha n^i n^e$, where it must be remembered that the "coefficient" α will itself in general depend on the electron and neutral-gas density as well as on the distribution in speeds.   Obviously these rates are functions of position and time if the f^a are.

The continuity equations (4.40) for ions and electrons thus take the form

$$\frac{\partial n^a}{\partial t} + \nabla \cdot n^a \mathbf{u}^a = n^e \nu_i + s_i - \alpha n^e n^i \qquad (10.5)$$

When Eq. (10.5) is multiplied by the charge q^a and summed over all species, the terms on the right always cancel, of course, so that the continuity of the total charge [Eq. (4.46)] is never affected by these source terms.   If in Eq. (10.5) all terms on the right are negligible for all charged particle species when compared with those on the left, we may speak of *frozen ionization,* particularly when the interdiffusion of ions and neutrals is also negligible.   This is usually the assumption made in the idealized theories of pure plasma dynamics.   If, on the other hand, the terms on the left are negligible compared with those on the right, we are dealing with what might be called *local ionization equilibrium.*   Let us first discuss the latter case in a little more detail.

In the description of local ionization equilibrium, the various neutral-particle species of type k with number densities n^k must be included in the considerations and the degree of ionization must be calculated from the various rates involved, i.e., from

$$n^e \nu_i + s_i = \alpha n^e n^i \qquad (10.6)$$

where ν_i and s_i are sums of terms which are proportional to the n^k.   Condition (10.6) as it stands does not necessarily imply thermodynamic equilibrium.   Specific nonequilibrium situations will be briefly discussed later on in this chapter.   By far the most important cases, however, are those in which the electrons have a Maxwellian velocity distribution that can be characterized by a temperature T^e.   Strictly speaking this can occur only in complete thermodynamic equilibrium, in which case detailed balancing holds.   For example, in equilibrium, ionization by electron impact is exactly balanced by its inverse, three-body recombination.   These reactions may be represented symbolically by equations such as

$$A + e \rightleftharpoons A^+ + 2e \qquad (10.7)$$

The same would be true for photoionizing reactions and their inverse, radiative recombination, schematically written as

$$A + h\nu \rightleftharpoons A^+ + e \qquad (10.8)$$

In that case, of course, Saha's equation (10.1) is obtained, and a detailed knowledge of the individual rates is not needed.   Now, because of the very rapid equilibration rate of the electrons among themselves, at sufficiently high densities the electrons remain rather well thermalized even in the absence of detailed balancing of all the processes.   In particular, the velocity distribution of the ions is relatively unimportant as long as their speeds are small compared with those of the electrons.   It follows that calculations of the rates assuming Maxwellian distributions of the electron speeds yield good approximations provided the reaction cross sections are known.   Such calculations have recently been carried out for hydrogenlike systems by Bates and coworkers.[8]   These authors considered both optically thin (transparent) plasmas and optically thick (opaque to resonance-line radiation) plasmas.   Some of their results are summarized in Fig. 10.1, where the recombination coefficient α is shown for atomic

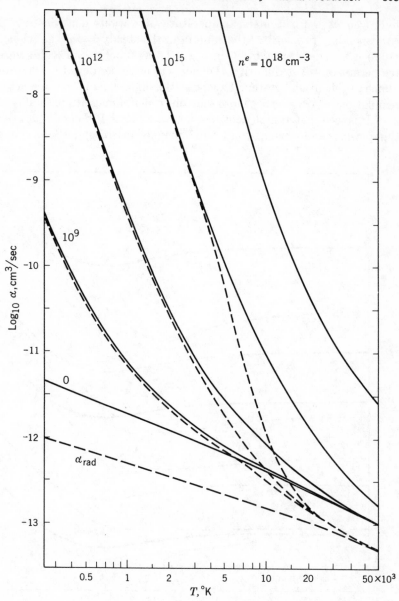

Fig. 10.1 Collisional-radiative recombination coefficient for electrons and atomic hydrogen ions as a function of electron temperature at various electron densities. Solid curves: plasma transparent to all radiation. Dashed curves: plasma transparent except to Lyman-line radiation, which is completely reabsorbed. For comparison the coefficient for radiative recombination into the ground state alone, α_{rad}, is also shown.

hydrogen as a function of temperature at various densities n^e. The corresponding ratios n^1/n^e that obtain in the steady state, i.e., when Eq. (10.6) is satisfied, are plotted in Fig. 10.2 as functions of n^e for various temperatures. The quantity n^1 denotes the number density of hydrogen atoms in the ground state. Except at the highest temperatures here the populations n^k ($k > 1$) are more than one order of magnitude smaller than n^1. It is seen that at high densities, i.e., when $n^e > 10^{17}$ cm^{-3} for optically thin plasmas and when $n^e > 10^{15}$ for "opaque" plasmas, the Saha equation

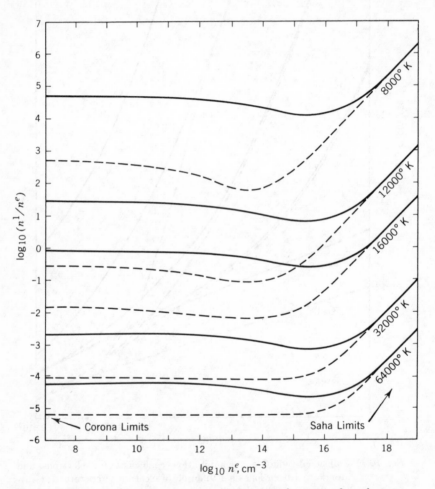

Fig. 10.2 The ratio of hydrogen atoms in the ground state to free electrons, n^1/n^e, in local ionization equilibrium [i.e., when Eq. (10.6) is satisfied] in transparent plasmas ($s_i = 0$) as a function of n^e at various electron temperatures. Solid curves: plasma transparent to all radiation. Dashed curves: plasma opaque to Lyman-line radiation only.

(10.1) with the value of $G(T) = 3$ gives a good approximation, although complete equilibrium is not established.   The reason is that under these conditions the reactions (10.7) dominate the steady state and are nearly balancing in detail for all values of k.   In other words, the Saha equation is a good approximation if the rate at which energy is transported by radiation and by general heat conduction is much smaller than the rate at which energy is exchanged by collisional ionization and its inverse.

Figure 10.2 demonstrates clearly that the degree of ionization of laboratory plasmas in ionization equilibrium is a very strong function of the temperature, particularly in the neighborhood of $T \approx 0.1 V_i$.   Moreover, it is seen that the degree of ionization depends also explicitly on n^e (or n^i).   It follows that in equilibrium the right sides of Eq. (10.5) can safely be neglected only when the degree of ionization is guaranteed to be very high.   Finally, we note that in the steady state as well as during build-up Saha's equation overestimates the degree of ionization if the electron temperature is known, sometimes by many orders of magnitude. Conversely, when the degree of ionization is known, or when a certain degree of ionization is desired, Saha's equation invariably underestimates the required electron temperature except in rapidly decaying plasmas.

It is also clear that at the maximum attainable temperatures of solid enclosures Eq. (10.6), just like Eq. (10.1), can yield appreciable degrees of ionization only for the vapors of the alkali metals and of barium.   Moreover, collisions with solid or liquid surfaces tend to be completely inelastic.   Thus ions have a high probability of recombining upon striking a solid surface whenever $V_i > \varphi_w + T_w$, where φ_w is the work function and T_w is the temperature of the surface, all expressed in the same units.   This means that walls must usually be considered as efficient sinks for plasmas.[†]   It therefore follows that, near boundaries at least, the transport of charge represented by the second terms on the left in Eq. (10.5) can never be neglected, and in very tenuous plasmas these terms may be important throughout the volume whether the gas is highly ionized or not.   In our discussion of the various methods of production of ionized gases in the laboratory we thus always have to assess the relative magnitude of charge removal by transport and by recombination.

10.3 LIFETIMES

Without going into any detail we can make several statements concerning the rates that yield very crude but easily made estimates of plasma lifetimes, power requirements, and heat loading of the surrounding surfaces. We can for instance define an *energy-containment* time τ_w for any given

[†] As will be discussed in more detail later, the exceptions are again some of the metal vapors mentioned above.

plasma by the ratio

$$\tau_w = -\frac{W}{\left(\dfrac{dW}{dt}\right)_0} \tag{10.9}$$

where W = total energy content

$\quad(dW/dt)_0$ = rate of change of W when no energy is supplied from external sources

The quantity W in this expression must include the energy of motion and the total internal energy of all particles, as well as the electromagnetic energy caused by plasma currents and charges in the volume considered,

$$W = \int(\tfrac{1}{2}\rho u^2 + \mathcal{E}' + \mathcal{E}'_M)\,d\mathbf{r}$$

Since the electrostatic energy density is usually negligible, the electromagnetic energy density $\mathcal{E}'_M$ is given by $\mathcal{E}'_M = (B^2 - B_0{}^2)/8\pi$, where B_0 is the magnetic induction field due to external sources; i.e., we have subtracted out the energy in the magnetic field produced by external sources. If the random motion of the ions and electrons can be described by the temperatures T^e and T^i, the internal energy density of the particles is conveniently approximated by

$$\mathcal{E}' = \mathcal{E}^{0\prime} + (p^i + p^e)\left[1.5 + \frac{\bar{\mathcal{E}}_i}{k(T^i + \bar{z}T^e)}\right] \tag{10.10}$$

where $\mathcal{E}^0$ = energy density of unionized components

$\quad p^i,\ p^e$ = partial pressures of the ions and electrons

$\quad\bar{\mathcal{E}}_i$ = mean "chemical" energy stored per ion (including dissociation, excitation, and ionization)

$\quad\bar{z}$ = the average number of free electrons per ion

The above form is useful for quick estimates of energy contents for highly ionized plasmas (where $\mathcal{E}^0$ is negligible), because we frequently have some information about the pressures involved and according to Eqs. (10.1) and (10.6) $\bar{\mathcal{E}}_i/kT^e$ can be a large number only in rapidly recombining or weakly ionized plasmas.   Since 1 atm is about 0.1 J/cm³, it is seen that except at extreme pressures even very hot, highly ionized plasmas do not generally contain more than a few joules per cubic centimeter in thermal and internal energy.   In this connection it is well to remember that $\mathcal{E}'_M \approx 0.1$ J/cm³ at about $B = 5{,}000$ G.

No such simple estimates can be made for the energy-loss rate $-(dW/dt)_0$.   Depending on the nature of the mechanisms as well as on the composition, energy content, structure, and size of the plasma the transport rates can vary over a very wide range indeed.   In fact, it happens not infrequently that observed transport rates deviate markedly from those predicted from various theoretical analyses.   For the purpose of the present discussion we shall usually consider $(dW/dt)_0$ as an experimentally determined rate, and the value of τ_w thus arrived at we shall call the *actual energy-containment time*.   For a steady-state plasma this time is of course simply given by

$$\tau_w = \frac{W}{P} \tag{10.11}$$

where P = power input

It is then interesting to compare τ_w with some other relevant time characteristic of the particular plasma.   The most appropriate scaling time to be introduced for this purpose is the ion transit time τ_t defined by

$$\tau_t = \frac{L}{\bar{v}^i} \approx 10^{-6} L \left(\frac{A}{w^i}\right)^{1/2} \qquad \text{sec} \tag{10.12}$$

where L = characteristic length such as diameter of plasma
    $\bar{v}^i$ = ion speed averaged over all the ions
The quantitative expression results if the mean ion energy w^i is expressed in electron volts and the mass A is expressed in atomic units, while L is measured in centimeters.   In general we expect, of course, that $\tau_w > \tau_t$ unless the dominant energy loss is by radiation.   In this sense τ_t represents a sort of minimum containment time.   Inasmuch as this minimum containment time is a fairly well-defined quantity, we can now define the actual *quality of confinement*, or *insulation factor*, Q_w as the ratio

$$Q_w \equiv \frac{\tau_w}{\tau_t} \tag{10.13}$$

It is occasionally more appropriate or more interesting to consider the actual confinement or lifetime of the charged particles rather than of the energy.   In that case we speak of an *ion residence time* τ_p, and we can introduce the quality of particle confinement,

$$Q_p \equiv \frac{\tau_p}{\tau_t} \tag{10.14}$$

Again, we expect $Q_p > 1$ unless the particles disappear primarily by recombination in the volume.   When a hot, highly ionized plasma cools primarily by radiation, we have $Q_w < Q_p$.   Conversely, when a cool, weakly ionized plasma decays primarily by recombination, we find $Q_p < Q_w$.   But when the dominant loss is by diffusion to the walls, be it a high-energy plasma insulated by a magnetic field or a low-energy plasma insulated by cool neutral gas—or a combination of the two—in general we have $Q_w \approx Q_p$.

It is very instructive, in this connection, to consider the fundamental difference between gas insulation of a low-temperature, very weakly ionized plasma and the magnetic confinement of a very high-temperature, highly ionized plasma. In the first case the ion is experiencing many gas collisions and migrates to the wall by diffusion; hence we have

$$Q_p \approx \frac{L}{\lambda} \propto n^0 L \tag{10.15}$$

where λ = mean-free path for ion-neutral collisions

n^0 = neutral density

In the second case the ion spirals in the field many times until it suffers a charge-exchange (electron-capture) collision. The resulting neutral particle carries the energy away, and a cold ion stays behind, so that now we find

$$Q_w \approx \frac{\lambda}{L} \propto (n^0 L)^{-1} \tag{10.16}$$

In the first case the total plasma energy is usually roughly proportional to the energy invested in ionization; in the second case, on the other hand, the cold ion is usually not confined. It follows therefore that indeed in both cases the energy and the particles leave the system essentially simultaneously. It is only the dependence on the scattering rates that is inverted.

The foregoing also indicates that if $\lambda \approx L$ neither the gas nor the field is likely to give good insulation, and $Q_w \approx Q_p \approx 1$. Relations (10.15) and (10.16) may be regarded as the idealized upper limits for the quality of confinement in the high and low gas-density cases and may be used for estimates of upper bounds on plasma lifetimes. It is noteworthy, however, that quite in general it is found very difficult to achieve, experimentally, insulation factors much in excess of about 100.[†] This means that, in view of Eq. (10.12), lifetimes of most laboratory plasmas turn out to be below 1 msecond. In other words, it usually takes powers in the kilowatt range to maintain plasmas with energy contents of the order of 1 J.

10.4 CLASSIFICATION OF LABORATORY PLASMAS

A glance at the literature reveals that there exist an enormous variety of experimental plasmas and a confusing diversity of names and terms used in their description and identification. Some systematic approach, in the nature of a classification, is required to afford an orderly and comprehensible presentation of the summary attempted here. For this purpose it appears most economical to classify laboratory plasmas with

---

[†] The only exceptions are to be found in some of the carefully designed containment experiments performed in connection with the controlled-fusion research (cf. Chap. 13).

respect to the character of the source terms of Eqs. (10.3) and (10.5), both in the volume and on the boundaries.   This means that we define the categories here according to the type of energy input and the method of production and removal of the charged particles which make up the plasma.   The result is summarized in the organizational chart shown in Fig. 10.3.   As will be seen, such a classification is useful because it correlates well with many other distinguishing features of the various types of plasmas that are encountered in experimental research.

In the individual discussions which are to follow only the relatively important or basic categories are included, and even there only the main characteristics of interest to the plasma physicist are mentioned.   For more complete information the reader is referred to the literature cited which, wherever possible, is chosen from standard texts and reference volumes rather than from original publications.   It is hoped that in this way this chapter may serve not only as a general introduction but also as a useful guide to further study for anyone desiring to familiarize himself with the great variety of experimental approaches to plasma physics.

10.5   ASSEMBLED PLASMAS

10.5.1. General Considerations

It is possible to form a plasma by bringing together positive and negative charges that are generated externally in separate ion and electron sources. In a certain sense this method may be regarded as the cleanest way of producing a plasma inasmuch as the "chemistry" is minimized and one would have, in principle, complete control over the injected velocity distribution of all constituents.   If one wishes to create a fully ionized plasma, for instance, one could think of a suitably disposed set of ion and electron guns aimed into a well-evacuated space.

In terms of our basic equations (10.3) and (10.5), assembled plasmas are characterized by $\mathbf{j} \cdot \mathbf{E} = 0$ and $\nu_i = s_i = 0$.   This says that no dissipation of electrical energy and no ionizing processes in the volume considered are contributing to the plasma generation.   Any build-up in density must be entirely due to negative values of the divergence of the individual flow terms.   The implied divergences of positive and negative charges must of course be sensibly equal in magnitude, since otherwise large unbalanced space charges would accumulate, electric fields would appear, and $\mathbf{j} \cdot \mathbf{E} \neq 0$ would result.†   Fortunately quasineutrality is easily preserved in prac-

† Injection of quasineutral plasma from plasma guns into a given space should not be included in this category.   Such methods must be listed under the appropriate types of electrical discharges or, better yet, under the heading Afterglows (cf. Sec. 10.8.4).

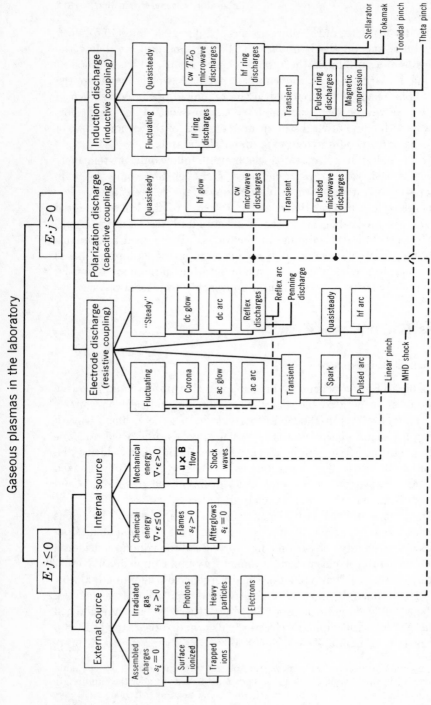

Fig. 10.3 The family of gaseous laboratory plasmas, showing their interrelation in regard to production of ionization. The symbol s_i here denotes all ionization processes, including ionization by electron impact. The electric field **E** is measured in the frame of the observer.

tice, since electrons are both very much more mobile than ions and readily emitted by metal surfaces, particularly when the latter are heated. It is therefore necessary only to provide a sufficient area of electron-emitting surfaces in the immediate neighborhood of the space to be filled with plasma. The positive space charge created by the ions from the ion source will then automatically attract the required amount of electrons to form the nearly neutral plasma.

In essence, this is also the principle used in the so-called ion-propulsion technique for advanced space-flight systems as discussed in Chap. 12. The most important considerations for laboratory applications are presented in the next two sections.

10.5.2. Surface-ionized Plasma

As pointed out in Sec. 10.2 many of the difficulties in the production of laboratory plasmas stem from the fact that physical boundaries usually act as rapid sinks for ions. Such wall recombination is expected, because, in general, the work function of surfaces is smaller than the ionization potential of most atoms or molecules. Whenever the reverse is the case, however, not only is wall recombination much reduced, but also the inverse process, i.e., ionization of the neutral particle on contact with the surface, becomes dominant. This phenomenon, commonly called *surface*, or *contact, ionization*, has long been known.[9] It has been put to good use in detectors of atomic beams of potassium, rubidium, and cesium, the ionization potentials of which are 4.34, 4.18, and 3.89 eV, respectively, with clean platinum or tungsten surfaces used whose work functions are above 4.5 V. In order to maintain these high work functions, it is essential that the surfaces stay clean. It is therefore usually necessary that they be kept at an elevated temperature, certainly to the point that condensation of the alkali vapor is avoided.

It is obvious that surface ionization affords an ideal method of assembly and maintenance of alkali-ion plasmas. Indeed, the method is being applied in the development of plasma diodes for the direct conversion of thermal into electrical energy (cf. Chap. 11) and in various ion-propulsion schemes (cf. Chap. 12). For the production of a steady-state plasma in the laboratory, it would seem most desirable if the volume were completely enclosed by such an ionizing surface. However, if no separate source of electrons is provided, i.e., if the plasma electrons have to come from thermionic emission by the same surface, complete equilibrium must be obtained, of course, even if collisions in the volume are scarce. The plasma potential simply adjusts itself in such a way that the net flow of charge across the boundary for all species is exactly zero. The resulting degree of ionization is readily shown to be precisely as given by Saha's equation (10.1) and is independent of the wall material, just as would be expected from thermodynamic considerations. Although enclosing the entire plasma by a heated metal wall is quite cumbersome in practice and the experiments that could be carried out are severely limited, some work along these lines has been reported.[10]   This equilibrium can be upset, of course, as is done in plasma thermoelectric elements, if a separate, more highly emitting electron source (a cathode) is provided and if currents are permitted to flow.

Considerably greater flexibility is achieved if surface ionization is used as an ion source covering only a portion of the plasma boundary. The resulting loss of ion containment and hence deviations from equilibrium can be minimized with the help of a strong magnetic field arranged so that the flow toward the quenching boundaries is significantly retarded. Of course, this latter ingredient can be added only in studies in which isotropy can be sacrificed or in which the effect of the magnetic field is just the feature under investigation.   It is clear, also, that at least two ionizing surfaces have to be provided, facing each other on a flux tube, if excessive plasma flow along the lines of force is to be prevented.†   Several interesting and very successful experiments based on these principles are in progress at various research laboratories,[12] and at least one of them has been described in some detail in the literature.[12a]   We shall therefore limit the discussion to the most important general remarks.

In order to avoid coating of the source area and hence loss of the ionizing efficiency, the surface has to be maintained at a temperature of at least 1400°K.   It is convenient in many cases to use the same surface as the electron emitter, in which case the temperature has to be high enough to supply an adequate electron current density.   If an adequate flow of electrons is available, by emission either from the surface itself or from auxiliary filaments, the plasma potential is negative and the rate at which ions are generated depends solely on the rate at which neutral atoms are striking the ionizing surface.   The density in the steady state then is given by the balance between this ionization rate and the removal rate caused by diffusion to the quenching surfaces and by recombination in the volume.

Three principal types of operation suggest themselves: the vapor-pressure method, the porous-plug method, and the atomic-beam method. In the first an excess of the alkali metal is introduced into the experimental chamber, and the walls are maintained at a given temperature.   At low pressures this presumably fixes both the neutral-particle density and temperature in most of the space and hence the rate at which neutrals

† None of these complications are needed if a streaming plasma is desired. The production of such a synthetic plasma "wind" has recently been described by Sellen et al.[11]

impinge on the ionizing surfaces.   The other methods require the use of separately heated ovens that contain the metal and produce predetermined flows of the alkali vapor.   In the second method the flow is forced through a heated porous tungsten plug, while the third requires formation of an atomic beam which is directed at one of the ionizing surfaces.   The chamber walls are kept cool to freeze out all particles striking them. Evidently the latter methods can yield higher degrees of ionization, but the first is simpler in operation.   When the mean-free path is long compared with the plasma dimensions, the ion and electron temperatures in both cases are very close to the temperatures of the ionizing and electron-emitting surfaces.   Electron densities exceeding 10^{12} cm^{-3} have been reached, and degrees of ionization approached 100 percent under favorable conditions.

The principal merits of such surface-ionized plasmas lie in the fact that they can be maintained in a steady state without the help of plasma currents and that their ion and electron velocity distributions presumably do not deviate much from Maxwellians.   In addition to the practical applications mentioned before, these plasmas seem admirably well suited for basic studies such as (1) recombination, (2) diffusion, (3) the effects of small currents on the transport phenomena, and (4) various small-amplitude waves in highly ionized plasmas.

Surface-ionized plasmas suffer from certain serious limitations, however: (1) they are restricted to species with a very low ionization energy; (2) the operating temperature is always very low; and (3) as the consequence of the low electron temperature the maximum permissible current density that can flow without heating the electrons or causing ion-wave instabilities is too low to produce magnetohydrodynamic effects.   If we wish to assemble plasmas with higher temperatures or plasmas made up of other species of ions, we have to abandon the surface-ionization effects both as ion sources and as ion-reflecting boundaries.

It is easily understood that surface-ionized plasmas are not found in nature.

10.5.3. Trapped-ion Plasma

Quite in general, if the method of ion injection into vacuum is used in the laboratory to fill a certain volume V in a steady state with a mean density n, the required ion current is given by

$$I = \frac{enV}{c\tau_p} \tag{10.17}$$

where τ_p = mean ion residence time introduced in Eq. (10.14)
Upon using the particle-confinement factor Q_p defined in Eq. (10.14) and

substituting Eq. (10.12), the number of ions per cubic centimeter becomes

$$n \approx 5 \times 10^{12} \left(\frac{A}{w_i}\right)^{1/2} \frac{LQ_p}{V} I \qquad (10.18)$$

where V/L is a surface area expressed in square centimeters and I is measured in amperes. In the case of a single transit of an ion beam we have $Q_p = 1$, and LI/V is just the ion current density. It is clear that such a beam represents a rather inefficient way of making a plasma, and this plasma with its very anisotropic ion velocity distribution may have very undesirable properties. In the alkali plasmas discussed in the previous section Q_p is enhanced by reflection on the ionizing surfaces, and the value of A/w_i is moreover very large.

In order to assemble a generally interesting plasma with a modest current, the value of Q_p must be raised in some other way, however. In other words, the ions must be reflected or trapped in the vacuum. This can be done only by coiling the ion trajectory into tight spirals with the help of a magnetic field in a specifically designed confinement configuration. A static field will not really solve the problem, however, since it is easily seen that any ion injected from the outside into such a field will leave again. In other words, although the path length of an ion trajectory can be lengthened enormously, we are still essentially dealing with a single transit of a coiled and twisted beam (except in certain limiting cases[13]). Real trapping of ions can presumably be achieved only by either using a time-varying magnetic field or causing an increase in the value of e/m somehow during the transit (a "stripping" or "breakup"; cf. Chap. 13).

Unfortunately, as indicated before, the values of Q_p for "trapped" ions are in practice always limited by scattering phenomena such as collisions with residual gas atoms (including charge exchange) and ion-electron and ion-ion collisions and also by instabilities which may be primarily driven by the anisotropic velocity distribution resulting from the beam-type injection (cf. Chap. 5). At low plasma temperatures the loss of ions by recombination may also contribute noticeably. Since most of the loss processes decrease with increasing energy, plasma production by magnetic trapping of ion beams is used only when extremely high temperatures are desired, as in controlled-fusion research. Further details of this and related methods will therefore be discussed in Chap. 13. It is clear from Eq. (10.18), however, that very large values of Q_p are required, i.e., a stable magnetic confinement system and extreme vacuum conditions, if interesting high-energy plasmas are to be assembled and maintained by the low beam currents that can be realized in practice (at most a fraction of 1 A). It should also be pointed out that such plasmas are likely to remain very far from any equilibrium in that the ion distribution cannot be expected to become isotropic, let alone Maxwellian, and the electrons are probably going to remain relatively cold unless they, too, are being injected from an electron gun with comparable energy.

Natural trapped-ion plasmas are presumably responsible for the radiation belts in the magnetospheres of planets.

10.6 IRRADIATED GASES

10.6.1. General Considerations

In this section we consider briefly the formation of plasmas by "irradiation" of a gas with photons or beams of energetic particles. This means that we allow $s_i > 0$ or $n^e \nu_i > 0$, or both, in Eqs. (10.5), but we still restrict ourselves to $\mathbf{j} \cdot \mathbf{E} = 0$ in Eq. (10.3). The energy needed for the ionization must then come from sources which should properly be included primarily in the term $\nabla \cdot \mathbf{Q}'$. The ionizing events in general do not impart any significant kinetic energy to the newly formed ion, but the liberated electron may, under certain circumstances, acquire a considerable speed. Plasmas formed in this way appear automatically whenever ionizing radiation of sufficient intensity penetrates a region containing gas. The glow surrounding very powerful radioactive sources is evidence for ionization, and beams from particle accelerators, when passing through a gas-filled space, create plasma columns. The best-known natural phenomenon is the ionosphere, which is caused by ultraviolet radiation from the sun.

10.6.2. Photons

Photoionization is not very suitable for plasma production in the laboratory, however, because the required photon sources are notoriously inefficient.[†] Moreover, the cross section for photoionization has its maximum value (usually of the order of 10^{-17} cm^2)[14] at the threshold and decreases rapidly with increasing energy (see Fig. 10.4). This means, for most gases, that the most effective radiation lies in the vacuum ultraviolet region so that windows and lenses cannot be used. In the x-ray region, on the other hand, where windows are again permissible, the ionizing free path becomes enormous so that absurd intensities would be needed for the creation of interesting plasma densities. It must also be borne in mind that on each photoionizing event a photon is removed completely. The excess energy is usually imparted to the liberated electron, which may cause further ionization by collisions with the gas if it has enough speed.

[†] Present laser-produced plasmas do not belong to this category but, more appropriately, to the electrodeless discharges since the energy is primarily absorbed by *already free* electrons. Nevertheless we note here that energetic plasmas have recently been generated by evaporation and ionization of small solid particles in vacuum using high-intensity laser pulses.[14a]

10.6.3. Heavy-particle Beams

Energetic heavy particles such as ions or neutral atoms are somewhat more practical ionizing agents for certain purposes since large power levels can be achieved (even fission products in a reactor have been used[14b]). In contrast to photoionization the cross section for ionization by heavy particles rises very gradually from zero at the threshold and passes through a broad maximum (usually of the order of 10^{-16} cm²) in the 10- or 100-kV range[5-7] (cf. Fig. 10.4). The rule seems to be that the maximum occurs

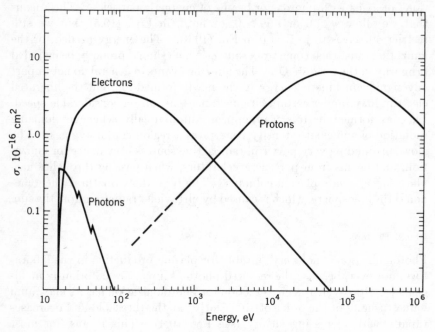

Fig. 10.4   Typical ionization cross sections for photons, electrons, and protons as a function of energy. The example shown here applies almost exactly to the case of argon.

when the relative speed between the colliding particles is approximately equal to the orbital speed of the valence electron. Another important difference lies in the fact that the fast particle is usually barely deflected by the ionizing collision and, except for relatively rare close encounters, gives up only slightly more than the energy needed to free the electron. On the average, including recoil and electronic excitation, a fast ion loses only between two and three times the ionization energy per ion pair formed. A beam of ions with tens of kilovolts of energy continues therefore unabated for many ionizing free paths. This property can be used if a well-defined column of a weakly but fairly uniformly ionized plasma is to be created.   The method is not very efficient, of course, and there exists the danger of oscillations excited by the streaming ions.   When energetic ions are trapped by a magnetic field, the above effect may contribute substantially to the ionization of the residual gas, thus reducing the charge exchange rate, and hence may be instrumental in improving the quality of confinement, Q_p.   Also, if the newly formed ions are not contained, the liberated electrons are available for the neutralization of the space charge of the fast ions so that no separate electron source is needed for that purpose.

10.6.4. Electron Beams

The simplest technique of generating plasma by ionizing beams makes use of electrons, however, because with these it is easiest to achieve high source intensities.   The ionization in this case is entirely due to the term $n^e \nu_i$ in Eq. (10.5), provided that the injected electrons do not give rise to x-rays, which in turn produce photoionization.   If, moreover, we neglect the motion of the neutral gas molecules and assume that all of them are in the ground state, Eq. (10.4) reduces to

$$\nu_i = n^0 \int_0^\infty \sigma_i(v^e) v^e f(v^e) \, dv^e = n^0 \langle \sigma_i v^e \rangle \tag{10.19}$$

The integral becomes trivial if the newly created electrons do not have sufficient energy to create secondary ionization and if the beam is and remains essentially monoenergetic.   In that case the source term can be written as

$$n^e \nu_i = n^0 n^b \sigma_i(v^b) v^b = \frac{n^0 \sigma_i j^b}{e} \tag{10.20}$$

where the superscript b refers to the beam electrons and j^b is the beam current density.   The cross section for ionization by electron impact for most gases rises rather steeply from zero at the threshold to a maximum (again of the order of 10^{-16} cm^2) at an energy between three and six times the ionization energy and then falls off approximately exponentially with an e-folding energy of about 500 eV (cf. Fig. 10.4).   The process of ionization by electrons of a few hundred electron volts of energy is therefore a rather efficient one.   The only possible disadvantage as compared with high-energy heavy-particle beams is the rather large spreading that results because electrons are more deflected.

When the source term (10.20) is inserted in Eq. (10.5) and when all other terms on the right of Eq. (10.5) are negligible, the resulting plasma can be readily calculated.[15]   However, as already pointed out in Sec.

10.6.3, the streaming charges are prone to interact with the stationary plasma, causing complications as discussed in Chaps. 5 and 8. In fact, conditions can be reached in which such violent oscillations are excited that the beam itself is destroyed by scattering from fluctuating potentials. In this way much of the beam energy is imparted to the plasma electrons, which heat up until they themselves are able to participate in the ionizing process should any neutral gas still be present. This process is indicated in Fig. 10.3 by a broken line linking electron injection with the microwave discharges. The vigorous interaction undoubtedly represents a very effective way of transferring energy to a plasma.[16] It has been claimed also that this phenomenon is responsible for the structure of certain low-pressure discharges.[17] The only drawback of the method is that the same fluctuating potentials that transfer the energy from the beam to the plasma probably also enhance the spatial transport rates and therefore reduce the quality of containment.

10.7 ELECTRICAL DISCHARGES

10.7.1. General Considerations

By far the most convenient and most popular techniques of preparing and maintaining ionized gases in the laboratory make use of the dissipation of electrical energy in the plasma itself. In contrast to the cases discussed before, these methods, which are usually identified by the generic term *electrical discharges* or *gas discharges*, thus depend primarily on the effects of $\mathbf{j} \cdot \mathbf{E} > 0$ in Eq. (10.3).

The electron beam mentioned in the previous section, of course, also represents an electric current. In the absence of an accelerating electric field, however, the energy of such a beam is attenuated by the dissipative processes along its path; and one of the disadvantages of the plasma so produced is the resulting nonuniformity in space. In principle the attenuation can be avoided if an electric field of the proper magnitude is maintained parallel to the beam so that the dissipation is exactly balanced by $\mathbf{j} \cdot \mathbf{E}$. This, then, is the basic ingredient of the electrical discharges. The scattering and randomization of the beam are irreversible, however, and cannot be compensated for by the addition of the electric field. The situation is further complicated by the fact that the newly created plasma electrons are also acted upon by the same field and contribute to the current. Indeed, in most discharges any distinction between the plasma electrons and those produced externally (i.e., emitted from a cathode) is not meaningful, and in the cases of electrodeless discharges no electrons whatever are introduced from the outside. At any rate, the function $f^e(\mathbf{v}^e)$ that appears in Eq. (10.4) or (10.19) must be calculated in a self-consistent manner, using the Boltzmann equation and including inelastic collisions, before ν_i can be computed.[18]   It is therefore not surprising that quantitative theoretical work on electrical discharges is not very highly developed.   Most analyses are based on the assumption that $f^e(\mathbf{v}^e)$ is controlled by local conditions; i.e., they exclude the existence of penetrating beams and of strong gradients and often make use of ionization rates which are empirically determined for given fields and given density and composition of the gas.   The theoretical picture is most satisfactory in the limits of high density and low electric fields or, equivalently, at very high frequencies, i.e., in high-pressure arcs and in microwave discharges. In these cases the electrons gain relatively little energy between collisions so that they tend to become and remain well randomized and the entire distribution $f^e(v^e)$ can be approximated by a Maxwellian with only small correction terms, which result both from the electric field and from the inelastic collisions.   Obviously, these cases no longer have any similarity with the ionizing beams of the previous section.

Most types of electrical discharges have been abundantly described in the literature during the past thirty years so that we shall limit this presentation to a simple classification of the principal categories, with only short résumés of their outstanding features.   A logical classification is again conveniently accomplished by a consideration of the various terms in our basic equations (10.3) and (10.5).   Except for dynamic-pinch discharges and other hydromagnetic or gas-dynamic acceleration problems the mass velocity $\mathbf{u}$ is unimportant, and Eq. (10.3) can be expressed in the frame of reference in which $\mathbf{u} = 0$.   The energy equation then reduces to the simple form

$$\frac{\partial \mathcal{E}'}{\partial t} = \mathbf{j} \cdot \mathbf{E} - \nabla \cdot \mathbf{Q}' \tag{10.21}$$

No simplification of this kind is, of course, possible in Eq. (10.5).

As indicated in Fig. 10.3, we may divide these types of plasma production into *steady* (S) discharges, characterized by $\partial/\partial t = 0$, and fluctuating (F) or *transient* (T) discharges, where $\partial/\partial t \neq 0$.   When the time derivatives in Eq. (10.5) and (10.21) are not zero but unimportant, the discharges may be called *quasisteady* (QS).   Another type of classification is derived from the nature of $\mathbf{j}$ and $\mathbf{E}$.   If the electric field is primarily caused by free charges within the plasma and on the plasma boundaries and if $\nabla \cdot \mathbf{j} \approx 0$ everywhere, we speak of *electrode* (E) discharges.   Conversely, if we still have $\nabla \cdot \mathbf{j} \approx 0$ everywhere but if the electric field is primarily generated by a changing magnetic flux, we are dealing with *induced* (I) discharges.   Finally, we denote as *polarization* (P) discharges all those transient or quasisteady discharges in which $\nabla \cdot \mathbf{j} \neq 0$ within the plasma or on the plasma surface.

10.7.2. Electrode Discharges

As indicated by the name, E discharges always require electrodes in conducting contact with the gas and a finite potential difference between them. The cathode always collects more ions than electrons and in addition always supplies extra electrons; the anode invariably collects more electrons than ions but rarely reemits any ions.   There exists a large variety of such discharges which are commonly known under such names as *glow discharges, electric arcs, Penning discharges*, etc.   The nomenclature has not been standardized, however, and one finds some confusion of terms in the literature.

At the risk of oversimplification, we shall here distinguish between the various kinds of E discharges primarily by the conditions at the cathode and only secondarily by the state of affairs in the body of the plasma.   We begin by noting that in the body of almost all E discharges most of the current is carried by the electrons because of their smaller mass, i.e., larger mobility.   Since the cathode performs a dual role as collector of ion current $I_c{}^i$ and emitter of electron current $I_c{}^e$, however, the current ratio at the cathode, $I_c{}^i/I_c{}^e$, can vary over a very wide range. This boundary condition has a strong influence on the properties of the discharge in the neighborhood of the cathode.   It is clearly very unlikely that complete charge neutrality exists in this region; in fact one must usually expect to find $n^i > n^e$.   This means that in general there is a marked potential difference between the cathode and the main body of the discharge which we shall call the cathode drop V_c.   Let us first confine our attention to the steady state, i.e., to SE discharges.   Roughly in accordance with common usage, we shall here divide the SE discharges into two categories depending on the value of V_c.   It should be noted that the distance over which V_c appears, i.e., the so-called cathode-drop thickness, is adjusted automatically by the discharge itself.   It follows that the currents $I_c{}^e$ and $I_c{}^i$ are in general not space-charge-limited but are determined by the emission properties of the cathode and of the discharge plasma.

GLOW DISCHARGES   We define arbitrarily as glows all those SE discharges in which $eV_c > \varepsilon_i$.   This situation arises whenever neither the cathode nor the discharge plasma is a copious emitter of current carriers.   The current continuity in the cathode region can in this case be accomplished only by a drastically enhanced ionization rate due to the electrons from the cathode that have gained energy in traversing the cathode drop. These electrons effectively form an ionizing beam somewhat as discussed in Sec. 10.6.4.   This connection is indicated by a broken line in Fig. 10.3. The plasma produced by this beam is usually called the *negative glow*.   It serves as the source of photons as well as of the ions which strike the cathode in turn after crossing the cathode drop in the reverse direction. In self-sustaining cold-cathode glow discharges the electron current emitted from the cathode is controlled primarily by the flux of ions and photons from the negative glow region. As a result these discharges have interesting distinctive characteristics which are discussed extensively in the pertinent literature.[19] Here we point out only those few features which are important for our general classification:

1. Since the ionization in the negative glow is caused primarily by dissipation of the energy of the electrons from the cathode, we must have $eV_c \gtrsim \varepsilon_i I_c{}^i / I_c{}^e$. This means that cold-cathode glow discharges are characterized by large cathode drops; i.e., they are high-impedance discharges because in these $I_c{}^e / I_c{}^i = \gamma$ and the secondary-emission coefficient γ is of the order of 0.1 or less.

2. The efficiency of the regenerative processes is maximized in the so-called hollow-cathode configuration in which the cathode surrounds the entire negative glow.[20]

3. Steady glows are limited to low currents or rather to low current densities at the cathode, because of an instability in the emission mechanism. The ion bombardment affects the characteristics of the cathode surface by heating and by sputtering.† Under heavy loading thermionic electron emission will contribute to $I_c{}^e$, and frequently, unless special precautions are taken, highly emitting cathode spots tend to appear in an uncontrolled fashion. If the external circuit does not limit the current to a low value, the discharge will then change its character automatically to one in which $I_c{}^e > I_c{}^i$; that is, it becomes an arc.

4. As a corollary of (3), glows are limited to low degrees of ionization, i.e., to well below 10^{-3} ion/neutral particle, and operate most readily at low gas densities.

5. Because of the low ion densities (of the order of 10^9 cm^{-3}) and relatively high electron temperatures (several electron volts) electron recombination in the volume plays a minor role in glow discharges.

6. It follows that the particle losses in these discharges are dominated by removal at the surfaces. This means that glow discharges are diffuse and tend to fill the volume available. The quality of charged-particle confinement Q_p simply increases with the gas density.

7. It also follows from the transport rates that neither the gas temperature nor the ion temperature exceeds the temperature of the walls

---

† The operating range can be extended drastically if a so-called *brush cathode* is used which consists of a multitude of needle-sharp points. The electron density of the negative glow in such discharges has been increased by orders of magnitude above the normal level and seems to be limited by recombination rather than by diffusion.[21]

by very much and only the electrons are energized far out of equilibrium by the electric power input $\mathbf{j} \cdot \mathbf{E}$.

8. If the separation between the anode and the cathode is sufficiently large, a region is created in which the conditions are independent of electrode effects. The conditions in this so-called positive column are then merely a function of the geometry, the current, and the transport properties of the ionized medium. The description of this region is therefore relatively simple, and positive columns have long been popular for certain plasma-physics experiments.

9. Owing to the complicated ionization phenomena long columns in glow discharges frequently are subject to density fluctuations in space and time (striations).[19,22] The visible light associated with these striations may vary too rapidly to be resolved by the unaided eye. Uniform appearance is therefore no guarantee that a column is truly stationary.

10. In cold-cathode glow discharges the ratio $I_c^e/I_c^i = \gamma$ is primarily a function of V_c and of the materials used and is sensitive to operating conditions only because V_c and the cathode surface properties will depend on the latter. The current I_c^e can be increased at will in relation to I_c^i, however, if a cathode with an independently controlled electron emission is provided. Simultaneously, of course, a reduction of the cathode drop will result, and the distance between the cathode and the positive column will shrink.

ARC DISCHARGES   All E discharges in which $eV_c \leq \varepsilon_i$ will here be classed as electric arcs regardless of specific conditions at the cathode or in the body of the plasma. This extension of the usual concept is recommended if discharges with independently controlled electron emission are to be identified as either arcs or glows. Thus, in contrast to the position taken in most of the earlier literature in the field, we do not make a fundamental distinction here between discharges which are self-sustaining and those which need electrons injected by an independently powered cathode. Obviously, in this way the category of arcs is very large, since no restrictions are placed on either gas pressure or current density. In fact, all thermionic diodes must be included here as long as sufficient gas ionization occurs to produce a substantial neutralization of the space charge, so that the term *discharge plasma* may still be considered appropriate. Depending on gas density, current density, and the conditions at the cathode, arc discharges can differ greatly in their characteristic properties. Again, many of these have been discussed at length in the literature, and we shall restrict the discussion to a few general remarks:[23,24]

1. As already indicated, one mode of operation of arcs involves electron emission at the cathode that is practically as large as the entire discharge current. If such a discharge is to be self-sustaining, the power

dissipated by the cathode must be supplied by the discharge current itself. It is easily seen that this requirement imposes certain restrictions on the current density. It is also possible that the current continuity at the cathode is provided by an enhanced emission of ions from a constricted high-density plasma region which forms at the cathode end of the arc column. These matters are also discussed extensively in the literature,[25] and we shall not go into them here. We are here primarily interested in considerations concerning the plasma columns of arcs.

2. At low gas densities and particularly at low current densities the columns of arcs do not differ in an essential way from those of glow discharges. The removal of charged particles and of energy is dominated by kinetic transport to the surfaces, and Q_p increases with pressure. An increase in the current density is simply reflected by a proportional increase in the degree of ionization.

3. As the degree of ionization is raised by increasing the current density, the heating of the gas will become more significant. Eventually a thermal instability may set in, resulting in a constriction of the entire column.[26]

4. As the gas pressure is increased, particle transport is reduced and eventually at high densities the deionization rate is controlled by recombination in the volume; i.e., Eq. (10.6) is a good approximation, and the concept of particle confinement loses its meaning. However, under such conditions the column is naturally constricted, and the energy insulation factor Q_w continues to increase with pressure. This may properly be called *gas insulation*.

5. The most striking consequence of the improved insulation in high-pressure arcs is seen in the fact that much lower electron temperatures are required than in low density discharges, and high degrees of ionization are readily achieved. In this sense it may be said that high-pressure arcs are more efficient in plasma production than low-pressure discharges.

6. Since the energy-exchange rates increase rapidly with the particle densities, at appreciable degrees of ionization, the entire gas in the core of high-pressure arcs becomes heated to almost the same temperature as the electrons. In other words, conditions approach those characteristic of thermodynamic equilibrium in spite of the energy source $j \cdot E$. This fact makes high-pressure high-current arc discharges suitable for the study of the properties of gases at high temperatures.[27]

7. It is, for instance, possible to maintain small amounts of fully ionized gas in a steady state by operating steady arc discharges of several hundred amperes of current at pressures of 1 atm or more. So far the highest temperatures that have been reached in this manner lie in the neighborhood of $50,000°K$.

8. The method of gas insulation is limited, however, inasmuch as the generalized heat flow Q' increases with the absolute temperature as well as with the temperature gradient. At very high values of p and T the losses are of course dominated by radiative transfer, and it is questionable that under such conditions steady arc discharges are suitable for the production of high-energy plasmas.

9. Moreover, as will be pointed out later, all constricted discharges have a tendency to be unstable, i.e., to develop fluctuations and macroscopic motion of the current channel. It has therefore proved desirable or necessary to stabilize high-current arc discharges in a more or less mechanical way by enclosing the channel with efficiently cooled solid walls (wall stabilization) or by surrounding it with a rapidly spinning circulation of gas (vortex stabilization). In the first case the losses due to heat conduction are increased, of course, and in the second, convection cooling becomes important. All these techniques are well described in Ref. 24.

10. Since plasma temperatures are high compared with the boiling point of most substances, it is easy to operate arcs in the vapors of the electrode materials. The best known examples are the carbon arc and the mercury-vapor arc.

11. The quality of confinement in low-pressure arc columns can be increased by superposition of a strong magnetic field parallel to the electric field. This reduces the transverse transport rates drastically. In this way it is possible to achieve rather high degrees of ionization,[28] and in long high-current arcs of this type impressive "temperatures" have been produced.[29] Unfortunately, the deviations from any kind of equilibrium tend to be appreciable in these cases.

12. It is obvious that magnetically insulated E discharges at low pressures have to be very long to be effective, because the lifetimes in these are ultimately limited by transport to the electrodes. In particular, it is clear that the spacing between the electrodes must be large compared with the electron mean-free path for ionization, since otherwise the electrons will deliver much of their energy to the anode.

REFLEX DISCHARGES   The magnetic field mentioned in the previous sections was introduced as a means of increasing the quality of confinement of the discharge plasma by reducing the rate of the *lateral* transport of charged particles and of energy. For this purpose it was sufficient that the magnetic field was everywhere parallel to the primary current, that is, $\mathbf{j} \times \mathbf{B} = 0$, and no zero-order force was exerted on the plasma. At low pressures the electron lifetime was then limited by the rapid drift toward the anode. In such cases the electron containment can be much improved, however, if the anode is so arranged that the electron current has to cross the magnetic field before it is collected.   This is the principle underlying all so-called $\mathbf{E} \times \mathbf{B}$ discharges.   The electron drift along a steady electric field that is crossed by a magnetic field is roughly proportional to $\nu_c/$ $[\nu_c{}^2 + (\omega_c{}^e)^2]$, where $\omega_c{}^e$ is the gyrofrequency of the electrons and ν_c is the frequency of their collisions for momentum exchange with particles of other species.   For fully ionized gases this has already been indicated in Eq. (4.127), but the expression has general validity.[18,30]   It is thus seen that this drift decreases with decreasing density as long as $\omega_c{}^e > \nu_c$, and in fact with properly designed cathodes $\mathbf{E} \times \mathbf{B}$ discharges can be operated at arbitrarily low gas pressures.   It is merely required that the cathode has an emitting surface which is not everywhere tangent to the magnetic field and that no obstruction is present blocking the electrons' spiraling paths toward the anode.   The best-known examples of plasma sources utilizing this principle have been given the name *reflex discharges*.

In reflex discharges two cathodes or one cathode and one repeller electrode face each other on opposite ends of a magnetic flux tube with axial symmetry.   The anode is a ring-shaped or cylindrical collector arranged coaxially with but not intercepting this flux tube.   The electrons that leave the cathode in the absence of collisions are constrained to move along magnetic flux lines and drift in the $\mathbf{E} \times \mathbf{B}$ direction (cf. Chap. 2).   Therefore they do not reach the anode but strike the other cathode or, after being reflected by the repeller electrode, return to the cathode from which they came.   This means that in the absence of collisions the net current is zero.   Any electron that suffers a collision is likely to have its velocity component parallel to the magnetic field reduced, however, which means that it can no longer return to the cathode.   Such an electron is temporarily trapped, and in general, as a result of subsequent collisions, it will migrate toward the anode and eventually it will be removed there.

It is seen that the residence time of a trapped primary electron (or the effective path length for its transit from cathode to anode) varies inversely with the scattering frequency.   This means that the number of collisions suffered by the electron during its transit should vary only a little with the density and should depend primarily on the field strength and on the geometry.   It is clear also that such an electron has a certain probability of causing ionization in the volume.   The newly created electrons are of course also trapped, but the ions are drawn promptly to the cathodes.   Negative space charge accumulates until the electric field is changed in such a way that the electron and ion removal rates just balance.   Unfortunately, this redistribution does not, in general, lead to a stable steady state.

At low gas pressures and very low current densities, e.g., with cold cathodes, the discharge seems relatively quiescent.   It operates with a pronounced electron sheath adjacent to the anode, and most of the voltage (in the kilovolt range) appears as an anode drop.[31,32]   This feature illustrates well the general rule that in E discharges the voltage drop is largest near those electrodes which collect the least mobile current carriers.   Such an $E \times B$ anode sheath suffers from the so-called diocotron, or slipping-stream, instability,[33] however, and space-charge flutes develop which enhance the transport rate of electrons across the magnetic field. At low levels, in particular when $\omega_p \ll \omega_c{}^e$ everywhere, these disturbances grow only slowly or remain limited in amplitude and the discharge is well behaved in its gross features.[34]   We may call this mode of operation the *Penning-magnetron discharge*.   The current in this case is determined by the ionization rate.   Thus, when the voltage is kept constant, the current is proportional to the gas density, and as a result Penning discharges have found application as pressure-measuring devices under the name of *Penning* or *Philips ionization gauges*.[35]   Hence the term *P.I.G. discharge* has become a popular synonym for reflex discharge.

When the gas density (or the current density from an emitting cathode) is increased beyond a certain critical value, the discharge changes its characteristics markedly.   The pronounced anode sheath disappears, the discharge impedance decreases, and the electrical "noise" increases considerably.   It is apparent that the charge transport across the magnetic field in this low-impedance mode of operation is facilitated by very rapid space-charge fluctuations which cause nonsteady electric fields.   This means, of course, that the electron residence time is drastically reduced, but not quite to the value that would obtain if the magnetic field were zero.   When poorly emitting cathodes are used, and when the current is low enough so that no strongly emitting cathode spots appear, the discharge now operates somewhat like a glow; i.e., it is characterized by a large cathode drop which serves to accelerate the electrons from the cathode.   The electron beam both causes the ionization which maintains the discharge and generates plasma fluctuations to help the current flow across the magnetic field in spite of the low gas density.   This is the usual reflex glow, or "hashy" Penning discharge, described in the earlier literature.[36] Because of its severe irregularities, it is not very practical for plasma studies, but it can be used as an ion source.

Just as in ordinary direct discharges, the cathode drop of reflex discharges can be much reduced if the cathode or cathodes are made to emit the major portion of the current thermionically.   In this case we may properly speak of a *reflex arc*.   Reflex arcs are, of course, also usually quite "noisy," but since they need lower voltages (in the 100-V range), the amplitude of the fluctuating potentials is expected to be lower than in cold-cathode nonsteady Penning discharges.   Such arcs have been operated in a continuous manner with tens of amperes, i.e., at power levels in the kilowatt range.[37,38]   Degrees of ionization of the order of 10 percent can be achieved, which means that n^e may be as high as 10^{13} when the gas pressure is only in the micron range.   Under these conditions rotation of the plasma as a whole is a pronounced feature; it is driven by the $\mathbf{j} \times \mathbf{B}$ body force, very much as in an electric motor.   In the steady state this torque must be balanced by the viscous losses that are characteristic of low-density gases.   Unfortunately again, the rotating plasmas hardly ever remain axisymmetric; the preferred mode of operation seems to involve one or more revolving flutes, and low-frequency oscillations are observed frequently.[39]   In pulsed operation currents of several hundred amperes amplitudes have been passed, and degrees of ionization as high as 80 percent have been achieved.[40]

Reflex arcs have one feature that makes them attractive as continuous plasma sources.   It is possible to extrude a portion of the plasma through a hole in one of the cathodes into an evacuated expansion chamber and in this way maintain a fairly highly ionized plasma column in an essentially current-free and electric-field-free space.[38]   The only disadvantage of such plasmas is the complex electron energy distribution, which always seems to contain some energetic primary electrons surviving from the discharge region.   The mechanisms operative in reflex arcs are not yet fully understood, but they are currently under investigation by various laboratories.

NONSTEADY E DISCHARGES   The electrode discharges discussed in the preceding sections are all powered by direct-current supplies.   Yet many of them are not true SE discharges because instabilities cause fluctuations to appear automatically.   Quite in general we may say that the majority of electrical discharges are not really steady.   Perhaps the best-known example of a gaseous conduction phenomenon that cannot be steady although it takes place between dc electrodes is the so-called brush, or corona, discharge observed in very nonuniform fields at high gas densities. In this case the fluctuations are caused by space-charge accumulation, which results in a quenching or choking effect.   Of course this means that $\nabla \cdot \mathbf{j} \neq 0$, and, according to our definitions given in Sec. 10.4 we are no longer dealing with a pure E discharge.   Coronas have found an application in Geiger counters and are very important phenomena in high-voltage technology.   Recently they have been thoroughly discussed in the literature,[41] and since they are not important as plasma sources, we shall not elaborate on them here.

Other well-known examples of nonsteady "dc discharges" at high pressure are the hissing and the spinning high-current arcs.[42]   The former involve motion of contracted anode spots, while the latter rotate probably because of an interaction between the self-magnetic field and the driving current.[43]   Low-pressure discharges may become nonsteady under certain conditions, even when a strong longitudinal magnetic field is added. Again, the configurations are unstable against deviations from axial symmetry because of an interaction with the external field. The best-known illustrations in this instance are the so-called pressure-gradient arc[44] and the helical glow-discharge column.[45] These phenomena are always very complicated, and in some cases they are not yet well understood. Therefore we cannot discuss these matters in this summarizing chapter.

On the other hand, there exist large classes of E discharges that are driven on purpose in a nonsteady manner by supplying time-varying electrical power. These may be divided into two major groups: continuously varying (periodic) discharges and transient (aperiodic) discharges. Inasmuch as both these are of considerable interest as methods of plasma production in the laboratory some remarks concerning them must be included in this section.

Clearly, when the time scale of the applied variations is long compared with the formative time τ_s of the corresponding fully developed steady discharge, then all the properties are essentially the same as those discussed in the three preceding sections. Since in most cases "steady" discharges are established in less than 1 millisec, alternating discharges up to about 1 kc and pulsed discharges of 1 millisec or more duration do not differ in an essential way from those which are completely steady. Only when the discharge completely extinguishes (deionizes) between pulses does each pulse, or half cycle, require a starting transient. Starting transients are in general characterized by high electric fields and low currents (high-impedance phase). It should be noted in this connection that partial or complete rectification may occur if the electrodes in an oscillatory discharge are not equivalent.[24] Pulsed discharges and alternating discharges that extinguish between half cycles should be classed as transient discharges, although the name is best limited to those phenomena which never reach steady-state conditions. Alternating discharges that do not require starting transients every half cycle may well be included among the type FE discharges.

Transient E Discharges If a high-pressure discharge is interrupted, or if its power supply is exhausted, during the starting transient we usually speak of it as a *spark*. The same phenomenon at low pressures is sometimes called a *flash*. During the early phase of a starting transient conditions are always quite different from those later on, i.e., when the steady state is being approached. In particular, $\mathbf{j}$ is low and usually $\mathbf{E}$ is high in the beginning and the electron gas deviates even more from equilibrium than later on. Hence sparks and flashes are used as light sources in spectroscopy when high levels of excitation are desired, but they are not very practical for most purposes of plasma production.

In the section on Arc Discharges it was stated that steady arcs have been operated in the kiloampere range and total dissipation rates in excess of 100 kW could be handled by special cooling techniques. When power levels beyond 1 MW are involved and particularly when currents above 10 kA are needed, one is generally limited to pulsed operation. This is typically the situation in all those plasma experiments in which the magnetic field created by the discharge current is performing a significant function. The best-known examples of these are the magnetic insulation by the self-field (pinch effect) and the magnetic acceleration of plasma (dynamic-pinch discharge, magnetic shock tubes, and plasma guns). The simplest and most popular method used for the generation of the large currents involves merely the discharging of high-voltage capacitors or of banks of such capacitors. As a result the conditions are hardly ever completely steady, and we may consider most of the ultrahigh current discharges as transients.

In recent years an impressive number of experiments have been performed with pulsed high-power discharges, mostly in connection with advanced propulsion concepts (cf. Chap. 12), or controlled-fusion research (cf. Chap. 13), or even what might be called *laboratory astrophysics*. Yet many features are still poorly understood, and the investigations are still in progress, so that only a few remarks can be made that have general validity.

As the current density usually exceeds by far the maximum that can be drawn from independently heated emitters, the discharges practically always operate with self-maintaining cathode spots. There seems to be no practical limit on the amount of current that can be drawn from metals in this manner. It is known that a single cathode spot may emit electrons with a current density in excess of $10^6 \, \mathrm{A/cm^2}$.[22] Thus, even at the highest recorded pinch-current densities, which are of the order of $10^6 \, \mathrm{A/cm^2}$, only a fraction of the cathode surface need be covered with emission spots. In this respect these plasmas may belong to the class called *cold-cathode arc discharges*. The matter has not yet been thoroughly investigated, however, and the question of the electrode contact of ultrahigh current discharges is still a subject for research. The practical limitations may well have to do with vaporization rates of the electrode material and, at very high currents, with plastic deformation due to the enormous magnetic forces.

Quite in general it would appear that phenomena such as outgassing, vaporization, or ablation of surface material result in poor control of plasma density and composition in high-power discharges. Only in very short pulses and at some distance from all surfaces, where insufficient time has elapsed for transport of material, can the plasma be expected to be uncontaminated.

Finally we note that discharges with appreciable magnetic interactions are subject to all the hydromagnetic instabilities (cf. Chap. 5), in addition to the microinstabilities mentioned in the preceding sections. It is not surprising, therefore, that a thorough discussion of their properties is beyond the scope of this chapter. It may suffice to state here merely that it has proved possible to produce fully ionized plasmas by pulsed E discharges with energy densities in the range of a few J per cubic centimeter and lifetimes of a few microseconds. Obviously, the power levels required are of the order of megawatt per cubic centimeter. If instabilities do not ruin the containment, the cooling process is invariably dominated by radiative heat transfer, even if the initial gas is hydrogen, because it has never been possible to avoid contamination. This means that the quality of energy containment has generally been quite disappointing, often with the values of Q_w near unity.

Alternating E discharges   The properties of E discharges that operate with a continuously alternating current will depend very much on the frequency ω of the oscillations. As is readily seen by inspection of Eqs. (10.3) to (10.5) we may distinguish between three major regimes. First, as long as $1/\tau_s < \omega < \nu_i$, the plasma density and electron temperature are fluctuating with the applied frequency; i.e., we are dealing with a true type FE discharge. In typical cases this regime may extend from 1 to 100 kc. If the electrodes are identical, the discharge may be quite symmetrical, since there is insufficient time to build up and remove fully developed cathode-drop regions on each half cycle.

Second, when $\nu_i < \omega < \nu_c$, where ν_c is the mean electron collision frequency for momentum transfer, the ion density no longer varies much with the applied frequency and the discharge will be called *quasisteady*. The characteristic is still largely resistive, however, and the mean electron energy is still slightly oscillatory. It is primarily the high-energy tail of the electron velocity distribution that has too long a relaxation time to respond to the variations in the electric field. These are the typical high-frequency discharges in the megacycle range. Since, at the same time, the electrons are often no longer able to cross the distance from the cathode to the anode in less than a half cycle, the importance of the electrodes is much reduced as regards both removal and replenishment of current carriers. This means also that in spite of the electrode contact we may be dealing more with type P than with type E discharges. As the relative loss rate of current carriers at the electrode decreases, the value of Q_p increases; i.e., high-frequency discharges tend to be more efficient than dc discharges of similar power densities and dimensions.

Finally, when $\omega > \nu_c$, none of the random energy of the electrons is able to follow the rapidly varying field. In fact, the current is now largely out of phase with the electric field; i.e., the discharge response is primarily reactive. This is usually in the kilomegacycle range or, as it is often called, the *microwave region*. It is clear that here the electrodes are utterly unimportant and that microwave glows belong more properly to the electrodeless discharges.

10.7.3. Polarization Discharges

The term *electrodeless* is widely used for all those discharges in which the plasma has no direct contact with any electrodes. This has caused some confusion, because, as indicated in Fig. 10.3, there are actually two distinct classes, which we have called *polarization discharges* and *induction discharges*, respectively. In the former, for short P discharges, the electric field E is basically electrostatic and can be derived from a potential. In the latter, for short I discharges, the electric field is induced by a changing magnetic field and cannot therefore be derived from a potential.† The separation is not always meaningful, because in certain high-frequency discharges the two types of fields exist simultaneously. The distinction is useful for purposes of clarification, however. In both cases the current cannot be completely steady, of course. Either it must be transient, or if it is continuous, it must be oscillatory.

The usual P discharges are oscillatory, and the current passes in the form of displacement current from the plasma boundary through a nonconducting region to what might be called "external" electrodes. In this sense, P discharges are not really electrodeless, but the coupling between the electrodes and the plasma is capacitive, whereas in E discharges it is resistive. Thus in P discharges the coupling increases with frequency and varies inversely with the distance between the plasma and the electrodes. It follows that in general, and particularly in the lower frequencies, a rigid insulating layer is needed to prevent the plasma surface from approaching the electrode until resistive contact dominates. At low frequencies the insulator surface facing the plasma may be regarded as a pseudoelectrode on which charges can accumulate and from which charges (electrons) can also be emitted. These matters, particularly in regard to starting conditions, are well discussed in the literature.[46,47] An obvious advantage of this arrangement lies in the fact that the current is unable to constrict at these pseudoelectrodes, at least as long as the surfaces remain nonconducting. Discharges with external electrodes, therefore, tend to be more uniform in the lateral dimension than E discharges under otherwise similar conditions. The disadvantages, aside from the restriction to rapidly changing electric fields, are the limitations on thermal

† J. J. Thomson used the terms E discharge and H discharge, respectively, for the two types of electrodeless discharges.[46]

dissipation and electric breakdown of the insulating layers themselves.   As a result, continuous P discharges with pseudoelectrodes are limited to rather low power densities, i.e., of the order of a few watts per cubic centimeter, and hence to low degrees of ionization.

As pointed out before, when the frequency is so high that most of the current carriers are no longer swept to the plasma boundary by the electric field, the surface interaction becomes relatively unimportant for the regenerative processes in the discharge.   This means that all charged particles are produced by ionization in the volume, and even at low gas densities the plasma properties are not influenced by surface effects.   At the same time $\nabla \cdot \mathbf{j}$ is apt to have a substantial value throughout the plasma, or at least in a macroscopic boundary layer, instead of only on the insulator surface.   In other words, the plasma displays its intrinsic reactive properties even if it is not surrounded by an insulator.   This is the reason for our choice of the term polarization discharges.

When the applied frequency ω is larger than the electron collision frequency ν_c, the reactive properties dominate over the resistive ones. This leads to a severe fundamental limitation for pure high-frequency P discharges in low-density gases.   The response of free electrons to an oscillating electric field is basically inductive (negative permittivity) inasmuch as the current lags in phase behind the field.   In other words, plasma has a high-frequency dielectric constant which is less than unity and which becomes zero when the permittivity due to the free electrons just cancels that of free space.   As seen in Chap. 8, this occurs when $\omega^2 + \nu_c^2 = \omega_p^2$; that is, at this point the response of the medium is purely resistive [cf. Eq. (8.7)].   The effect is most pronounced when $\omega \gg \nu_c$, and the term *plasma resonance* is often used[48] (although in transverse waves the condition $\omega = \omega_p$ leads to the phenomenon of cutoff rather than to resonance). When $\omega_p > \omega$ and $\omega \gg \nu_c$, the electric field is strongly attenuated in the interior of the plasma and simple high-frequency P discharges are therefore not suitable for the production of plasmas with electron densities in excess of

$$n^e \approx n_{res} \equiv 1.22 \times 10^{-8} \left(\frac{\omega}{2\pi}\right)^2 \quad \text{cm}^{-3} \tag{10.22}$$

The above limitation can be removed with the help of various modifications.[49]   For example, the addition of a static magnetic field transverse to the applied high-frequency electric field has a drastic effect on the dielectric properties of the plasma.   It then becomes possible to find conditions where the electric field can penetrate into a plasma with electron densities well in excess of that given by Eq. (10.22).   Particularly attractive are those arrangements where use is made of the efficient energy transfer to the current carriers at their gyroresonance frequency.   Such methods are still the subject of research.[50,51]

10.7.4. Induction Discharges

The inherent disadvantages of internal electrodes are avoided altogether in configurations in which the driving electric field is everywhere parallel to the plasma surface.   This is the situation in the pure I discharges in which the electric field is caused by a changing magnetic field only.   The classical example is the so-called ring, or H, discharge described already by J. J. Thomson.   It is a low-pressure high-frequency glow that may be excited in a glass bulb surrounded by a coil carrying a large rf current.[46,52] To be truly electrodeless, the electrostatic field existing between the coil terminals must be shunted out, since otherwise certain aspects of a P discharge would be present simultaneously.   Of course, I discharges can take many different forms such as microwave glows in cylindrical cavities excited in a TE_{on} mode, toroidal-pinch discharges, stellarators, or theta pinches.   The plasma current must evidently always flow in closed loops.

In a way the I-discharge system can be considered as a current transformer in which the plasma represents the secondary winding consisting usually only of a single short-circuited turn with a distributed load.   It follows that it is difficult to produce high electric fields in this manner so that at high gas densities I discharges need special provisions for their ignition.   At very low gas densities and low frequencies, on the other hand, there is the possibility of electron runaway becoming important.[53] In fact, the system has been proposed for ultrahigh-current electron accelerators.[54]   It is indeed possible to induce transiently very high plasma currents in I discharges, in the 100- or even 1,000-kA range, when they are driven by a high voltage on a single-turn low-inductance primary.

In this configuration the attenuation of the electric field in the interior of the plasma may be interpreted as the skin-depth phenomenon; the penetration of the field decreases with the increasing plasma conductivity.   It would be erroneous, however, to conclude that plasmas produced by I discharges must necessarily be hollow.   The core of continuous rf glows induced in long cylinders fills in by diffusion and energy transport, for instance, and in transient high-current theta pinches the plasma near the axis is heated by compression and perhaps even by shock waves rather than by the discharge current directly.   As a result both at low and high power densities relatively uniform plasmas may be produced in this manner.   Intense I discharges have become very popular in high-temperature plasma research, because the gas is ionized and may be heated in a configuration that is inherently well suited for magnetic confinement and the contact with surrounding walls may be minimized if not avoided altogether.   For experiments with very high transient currents plasma energies of many hundreds of electron volts per particle have been reported.

The properties and behavior of plasmas produced in this manner are still the subject of vigorous research, predominantly in connection with controlled fusion.[55] We discuss them no further at this point. For details the reader is referred to the abundant literature in this field, particularly the proceedings of the international conferences on controlled fusion.[56]

10.8 SELF - IONIZED GASES

10.8.1. General Considerations

The plasmas described so far all require an external power source. It remains to discuss those important phenomena in which the entire energy may be regarded as already stored in the gas. All such plasmas must ultimately decay, because energy loss is in general unavoidable. Any plasma production may then be considered simply as internal conversion of kinetic or of available chemical energy into ionization energy. In terms of our basic conservation equation (10.3), not only do we find that in general $\nabla \cdot \mathbf{Q} \geq 0$ when integrated over the volume of interest, but we also have $\mathbf{E} \cdot \mathbf{j} \leq 0$. More properly, with reference to Eq. (10.3a), the surface integrals of neither $\mathbf{S}$ nor $\mathbf{Q}$ are negative in these cases. We include here all those cases in which mechanical work is done by a moving piston, because in a local frame of reference in which the piston is at rest the energy is already present in the form of mass motion. This generalization unifies the discussion of all plasmas powered by dissipation of mechanical energy.

When the conversion of mechanical energy to ionization involves electric currents, the processes are closely related to those in the electrical discharges discussed in the previous section. This situation arises, for instance, in the MHD generators described in Chap. 11. In such a generator the energy is derived from a mass flow transverse to a magnetic field, and the electric field appearing in the frame of the fluid is allowed to generate a current. In the frame of the magnetic field sources, however, $\mathbf{E} \cdot \mathbf{j}$ is zero or negative. The thermal nonequilibrium effects frequently observed can therefore be considered as gas-dynamic frictional heating (as in flow through a porous plug) just as well as electrical dissipation. These phenomena have recently had considerable attention in connection with MHD power generation.[57]

10.8.2. Shock Waves

The most important form of laboratory plasma production by mechanical means undoubtedly is the gas-dynamic shock wave. In such a wave the gas is irreversibly heated by an abrupt, strong compression. In the frame of reference in which the wave is instantaneously stationary the moving gas is decelerated by the steep pressure gradient of the wave, and a fraction of the energy of motion is thus converted into internal energy. The structure of the wave is of course determined by the relaxation processes involved in the conversion to the various degrees of freedom of the internal energy. Randomization of kinetic energy is accomplished in a few mean-free paths, but ionization requires energy transfer to electrons and, in turn, inelastic collisions between these electrons and the gas. Such rates can be quite slow,[58] and in general it is not very practical to use gas-dynamic shocks for the ionization of gases with an initial pressure much below 1 mm Hg. It is interesting to note in this connection that in gas-dynamic shocks the heavy particles tend to be hotter than the electrons, whereas in electric discharges the opposite is true.

The most common tool for the production of high-temperature gases by this method is the shock tube. In a shock tube the gas is initially at rest, i.e., in equilibrium, at a relatively low pressure. The energy is supplied by a rapidly moving piston capable of exerting a relatively high pressure. In conventional tubes the piston is formed by another body of gas which is initially at a very high pressure and which is suddenly allowed to expand into the low-pressure region. Under ideal conditions the resulting flow is one-dimensional, involving the propagation of a shock of constant strength through the low-pressure gas. The compressed and heated material travels along in a uniformly growing test region between the shock and the driver-gas interface.

If thermal equilibrium is approached in the test region, the conditions there are completely determined by the initial conditions. A detailed knowledge of the relaxation processes in the shock layer is not required, since the final state can be derived as a function of the shock strength or of the piston speed by direct integration of the equations of change (cf. Chap. 4) across the shock so that all derivatives vanish at the limits. The required thermal equilibrium implies the existence of an equation of state relating internal energy to pressure and temperature. The composition of the gas when chemical changes occur must be determined from the laws of mass action for all the possible reactions. In the case of ionization this reduces to our equation (10.1); i.e., we are dealing strictly with thermal plasmas here, and the properties of these need no special elaboration. Equilibrium conditions in hydrogen as functions of shock speed have been computed, for instance, by Turner.[59] The principles and techniques of shock tubes in general are well discussed in the pertinent literature.[60-62] We therefore restrict ourselves to a few remarks concerning the limitations of this method for plasma production purposes.

It is clear from Eq. (10.1) that very strong shocks are required to heat room-temperature gases to the point that appreciable degrees of ionization result. In the limit of very strong shocks the pressure and

internal energy of the unshocked gas can be neglected, and integration
of Eq. (10.3) across the shock yields, for $j = 0$ and on the assumption of
one-dimensional flow,

$$(\mathcal{E} + \tfrac{1}{2}\rho u^2)\,(u_s - u) = pu \tag{10.23}$$

where u_s is the shock speed in the frame of the undisturbed gas and all
other quantities apply to conditions in the test region.   In particular, the
velocity u of the gas is also the piston speed if this piston is impenetrable.
A similar integration of the equation of motion (4.70) results in

$$\rho u(u_s - u) = p \tag{10.24}$$

It follows that

$$\mathcal{E} = \tfrac{1}{2}\rho u^2 \tag{10.25}$$

which says that exactly half the work done by a fast piston goes into
internal energy and half is invested in kinetic energy of mass motion of
the compressed gas.   The latter can be converted into internal energy as
well if the shock is reflected from a stationary wall so that the gas is
brought to rest again.   Conditions behind reflected shocks in hydrogen
have also been computed by Turner.[63]

The limitations of conventional gas-dynamic shock tubes are now
obvious.   The effective piston speed u cannot exceed the mean thermal
speed of the driver by much.   Thus substantial degrees of ionization can
be achieved only in heavy, easily ionized gases, and the driver should have
as low a molecular weight as possible.   For example, 25 percent ionization
has been reached in argon using high-pressure helium as a driver.[58]   The
speed u can be raised to a certain extent by raising the temperature of the
driver gas, for instance, through the release of either chemical (combustion)[64] or electrical (discharge) energy.[65]   These means, however, are
usually not adequate to generate shocks so strong that they result in
completely ionized helium or hydrogen plasmas, where u would have to
exceed 5×10^6 cm/sec.   Such speeds are clearly exceedingly difficult to
achieve with any purely "mechanical-pressure" drive.   A possible exception is found in the electron-driven shocks discussed by Fowler et al.[66]
These, however, should be considered in a class by themselves, because
they do not allow thermal equilibrium in the region behind the shock.

In recent years considerable efforts have been made to develop
current-driven shock tubes in which the self-generated electric-motor
force of a large current acting on a body of ionized gas is used to replace
the high-pressure driver gas.   A résumé of this work has been given by
Kolb.[67]   Obviously these so-called magnetic pistons should be able to

move at much higher speeds than expanding gases.   The practical limitations of such systems are often set by insulation failure in some region behind the propagating discharge, because large electric fields must be sustained there.   Moreover, if a traveling high-current E discharge is used as a driver, surface ablation is likely to produce a substantial drag on the piston.

The most serious complications encountered in magnetically driven shock waves, however, stem from the fact that the electric discharges are likely to interfere somehow with the gas-dynamic shocks themselves. First of all, energetic radiation and fast electrons from the discharge region may penetrate to the shock front or even beyond into the unshocked gas, thus changing the conditions there in an uncontrolled manner.   Analysis is thus rendered practically impossible.   Second, the driving discharges are often violently unstable to the point that no steady shocks can be formed at all.   And finally, even when the first two difficulties are avoided, for instance, by operating at relatively high gas densities, the resulting shock is purely gas-dynamic only if none of the discharge current penetrates to the shock front.   This means the speed $u_s - u$ must be considerably larger than the speed of interdiffusion between plasma and magnetic field.   Very frequently there is insufficient separation between current-carrying region and shock front, and the resulting plasma must be considered as a compound product of an electric discharge and a shock-heated gas.   Very recently a comprehensive review of theoretical as well as experimental work on strong ionizing shocks has been given by Gross.[67a]

10.8.3. Flames

The last types of plasma production in the laboratory that need to be mentioned in this survey are probably by far the oldest known to man. Nevertheless, in some forms they may gain great technical significance in the near future.   These are the ionization phenomena frequently observed is gaseous combustion processes.[68]   Basically, we have to distinguish between two very different types of mechanisms here:

1.   Simple thermal ionization, near equilibrium, of any ionizable constituent is of course expected if the exothermal chemical reactions result in an adequate kinetic temperature of the entire gas.   The effect can be artificially enhanced by the addition of an easily ionizable admixture such as alkali vapors (seeding).   This method is of particular interest in MHD generator development as discussed in Chap. 11 and in the other pertinent literature.[69]

2.   The other mechanism involves a nonequilibrium process in which a particular chemical reaction liberates enough energy to cause ionization of one of the participants.   This process has been given the

name *chemiionization*,[70] and the most striking examples are found in the recombination of active nitrogen.[70,71]   However, some processes of this type may be involved whenever nonequilibrium ionization is found in chemically reacting gases.[72]   In some sense the so-called Penning effect, e.g.,

$$Ne^* + Ar \rightarrow Ne + Ar^+ + e$$

and the Hornbeck-Molnar effect,

$$He^* + He \rightarrow He_2^+ + e$$

occurring in some discharges may be classed as a form of chemiionization.[6]

In all these cases the energy is already present in the gas in a form that may be considered as available chemical energy.   It is clear, of course, that only very low degrees of ionization can result from such processes.   They are generally not of great interest as methods of ionization for plasma research per se, but they are important mechanisms in rendering gases electrically conducting.

10.8.4. Afterglows

Finally, in line with our system of classification according to the nature of the source terms in Eqs. (10.3) and (10.5), we must consider as a separate category all those plasmas in which no new energy whatever is available for transfer to ion production.   These plasmas must, of course, invariably decay from some initial state which must have been produced by any one or a combination of several of the mechanisms discussed before in this chapter.   We adopt the common term (*ionization*) *afterglow*, which should not be confused with the chemiluminescent afterglows such as the Lewis-Rayleigh (active nitrogen) glow mentioned in the previous section.   We include the afterglows among the self-ionized gases because the energy is already stored in the volume and, in fact, may be considered as a form of chemical energy in the broadest sense of the term.   It should be emphasized that a decaying discharge becomes a pure afterglow only after all dissipation of electromagnetic energy has ceased.   On the other hand, the hot thermalized plasma behind a gas-dynamic shock transition is always cooling at a finite rate and could very properly be included in this section.   Likewise, the assembled plasmas discussed in Sec. 10.5 are very closely related to the afterglows.

The term afterglow must thus not be limited to transient plasmas that decay locally in the same place in which they were created.   For instance, there are many experiments in which highly ionized gases are produced impulsively by transient high-power discharges in so-called plasma guns and from there are injected into specially provided receiving chambers for the study of plasma trapping and confinement.[73] In other recent investigations plasmas are produced by localized steady discharges and allowed to expand or stream into the main experimental region for a variety of purposes. Plasmajets and torches[74] as well as the plasma columns effusing from reflex discharges[38] and from so-called Duoplasma-trons[75] may be included here. In these cases the decaying plasmas are maintained locally in a steady state primarily by the convective term of Eq. (10.5). Inasmuch as these plasmas are still in contact with the generating discharges, however, additional energy flow from the source region, particularly in the electron component, can render the situation more complicated. In the latter event we are, strictly speaking, no longer dealing with genuine afterglows. Such plasmas may be called *extruded discharges*.

Except for completely thermalized gases with equilibrium ionization, afterglows are probably the best-understood plasmas in the laboratory. Hence they are well suited for a variety of fundamental investigations. Very weakly ionized decaying low-density plasmas have long been used for the study of recombination and diffusion processes in gases in which the temperature is controlled by the surrounding surfaces.[3-7] If collisional-radiative three-body recombination[8] is involved in the plasma decay, a fraction of the ionization energy is returned to the free electrons so that these tend to have a higher temperature than the gas or the ions. In those cases it is therefore necessary to determine the electron temperature separately, as was done in some recent spectroscopic work,[76] or to calculate it from known energy-transfer rates.[77]

If recombination occurs in a dense, highly ionized plasma, the decay is of course controlled primarily by the energy transport to the boundaries.[78] At high temperatures, on the other hand, e.g., above 30,000°K, recombination is usually negligible, and any noticeable plasma decay must be due to particle transport. Such afterglows are thus useful for direct measurements of confinement times. It should be emphasized that in afterglows particle residence times tend to be longer than in plasmas that are maintained by production mechanisms, because the same mechanisms often give rise to enhanced transport rates. In other words, afterglows obviously tend to be more quiescent and more nearly thermalized than the driven plasmas from which they were derived. It seems therefore advisable that, unless thermal plasmas are available, afterglows be used wherever possible if quantitative studies in basic plasma physics are attempted. Good examples are found in some of the experiments with hydromagnetic waves[79] as well as plasma waves[80] and in recent work on diffusion of plasma across magnetic field lines.[81]

REFERENCES

1. Hirschfelder, J. O., C. F. Curtiss, and R. B. Bird: "Molecular Theory of Gases and Liquids" pp. 496–501, Wiley, New York, 1954.
2. Hirschfelder, J. O., C. F. Curtiss, and R. B. Bird: "Molecular Theory of Gases and Liquids," pp. 697–698, Wiley, New York, 1954.
3. Massey, H. S. W., and E. H. S. Burhop: "Electronic and Ionic Impact Phenomena," Oxford University Press, New York, 1952.
4. Brown, S. C.: "Basic Data of Plasma Physics," M.I.T. and Wiley, Cambridge, Mass., New York, 1959.
5. Bates (ed.), D. R.: "Atomic and Molecular Processes," Academic, New York, 1962.
6. McDaniel, E. W.: "Collision Phenomena in Ionized Gases," Wiley, New York, 1964.
7. Hasted, J. B.: "Physics of Atomic Collisions," Butterworth, London, 1964.
8. Bates, D. R., A. E. Kingston, and R. W. P. McWhirter: *Proc. Roy. Soc. (London)*, **A267**, 297 (1962), **A270**, 155 (1962).
9. Langmuir, I., and K. H. Kingdon: *Proc. Roy. Soc. (London)*, **A107**, 61 (1925); J. B. Taylor: *Z. Physik*, **57**, 242 (1929).
10. Enriques, L., and F. Magistrelli: *Rev. Sci. Instr.*, **35**, 1708 (1964).
11. Sellen, Jr., J. M., W. Bernstein, and R. F. Kemp: *Rev. Sci. Instr.*, **36**, 316 (1965).
12. D'Angelo, N.: *Nuclear Fusion*, **3**, 147 (1963).
12a. Rynn, N., and N. D'Angelo: *Rev. Sci. Instr.*, **31**, 1326 (1960); N. Rynn: *Rev. Sci. Instr.*, **35**, 40 (1964).
13. Baker, D. A.: *Am. J. Phys.*, **32**, 347 (1964).
14. Weissler, G. L.: Photoionization in Gases and Photoelectric Emission from Solids, in "Handbuch der Physik," vol. 21, pp. 304–341, Springer, Berlin, 1956.
14a. Haught, A. F., and D. H. Polk: "High-temperature Plasmas Produced by Laser Beam Irradiation of Single Solid Particles," *Phys. Fluids* (1966), to be published.
14b. Leffert, C. B., D. B. Rees, and F. E. Jamerson: *J. Appl. Phys.*, **37**, 133 (1966).
15. Dunn, D. A., and S. A. Self: *J. Appl. Phys.*, **35**, 113 (1964).
16. Hedvall, P.: *J. Appl. Phys.*, **33**, 2426 (1962); W. D. Getty and L. D. Smullin: *J. Appl. Phys.*, **34**, 3421 (1963).
17. Merrill, H. G., and H. W. Webb: *Phys. Rev.*, **55**, 1191 (1939); A. Garscadden: *J. Electronics and Control*, **14**, 303 (1963).
18. Allis, W. P.: Motion of Ions and Electrons, in "Handbuch der Physik," pp. 383–444, Springer, Berlin, 1956.
19. Francis, G.: The Glow Discharge at Low Pressures, in "Handbuch der Physik," vol. 21, pp. 53–208, Springer, Berlin, 1956.
20. White, A. D.: *J. Appl. Phys.*, **30**, 711 (1959); D. G. Sturges and H. J. Oskam: *J. Appl. Phys.*, **35**, 2887 (1964).

21. Persson, K. B.: *J. Appl. Phys.*, **36**, 3086 (1965).
22. Robertson, H. S.: *Phys. Rev.*, **105**, 368 (1957); A. W. Cooper: *J. Appl. Phys.*, **35**, 2877 (1964); D. A. Lee et al.: *J. Appl. Phys.*, **37**, 377 (1966).
23. Somerville, J. M.: "The Electric Arc," Methuen, London, 1959; W. Elenbass: "The High Pressure Mercury Vapour Discharge," North Holland Publishing Company, Amsterdam, 1951.
24. Finkelnburg, W., and H. Maecker: Elektrische Bögen und Thermisches Plasma, in "Handbuch der Physik," vol. 22, pp. 254–444, Springer, Berlin, 1956.
25. Ecker, G.: Electrode Components of the Arc Discharge, in "Ergebnisse der exakten Naturwissenschaften," vol. 33, pp. 1–104, Springer, Berlin, 1961; I. G. Kesaev: "Cathode Processes in the Mercury Arc," Consultants Bureau, New York, 1964.
26. Ecker, G., W. Kröll, and O. Zöller: *Phys. Fluids*, **7**, 2001 (1964).
27. Finkelnburg, W.: Behavior of Matter at High Temperatures, in "High Temperature—a Tool for the Future," p. 39, Stanford Research Institute, Menlo Park, Calif., 1956.
28. Gibbons, R. A., and R. J. Mackin: In *Proc. Fifth Intern. Conf. on Ionization Phenomena in Gases*, Munich, 1961, vol. 2, p. 1769; C. Mahn, H. Ringler, R. Wienecke, S. Witkowski, and G. Zankl: *Z. Naturforsch.*, **19a**, 1202 (1964).
29. Luce, J. S.: In *Proc. Second UN Intern. Conf. on Peaceful Uses of Atomic Energy*, Geneva, 1958, vol. 31, p. 305.
30. Shkarofsky, I. P.: *Can. J. Phys.*, **39**, 1619 (1961).
31. Knauer, W.: *J. Appl. Phys.*, **33**, 2093 (1962).
32. Dow, D. G.: *J. Appl. Phys.*, **34**, 2395 (1963).
33. Buneman, O.: *Nature*, **165**, 474 (1950); G. G. Macfarlane and M. G. Hay: *Proc. Phys. Soc. (London)*, **63B**, 409 (1950); R. G. E. Hutter: "Beam and Wave Electronics in Microwave Tubes," chap. 12, Van Nostrand, Princeton, N.J., 1960; W. Knauer: *J. Appl. Phys.*, **37**, 602 (1966).
34. Knauer, W., A. Fafarman, and R. L. Poeschel: *Appl. Phys. Letters*, **3**, 111 (1963).
35. Leck, J. H.: "Pressure Measurement in Vacuum Systems," pp. 88–93, Institute of Physics, London, 1957.
36. Backus, J.: Theory and Operation of a Philips Ionization Gauge Type Discharge, in A. Guthrie and R. K. Wakerling (eds.), "Characteristics of Electrical Discharges in Magnetic Fields," chap. 11, McGraw-Hill, New York, 1949; G. Briffod, M. Gregoire, and S. Gruber: *J. Nuclear Energy*, **C6**, 329 (1964).
37. Bonnal, J. F.: In *Proc. Fifth Intern. Conf. on Ionization Phenomena in Gases*, Munich, 1961, vol. 2, p. 1787; R. Bingham, F. F. Chen, and W. L. Harries: Preliminary Studies of a Reflex Arc, *Princeton Univ. Plasma Phys. Lab. Rept.* MATT-63, 1962.
38. Hall, L. S., and A. L. Gardner: *Phys. Fluids*, **5**, 788 (1962).
39. Chen, F. F., and A. W. Cooper: *Phys. Rev. Letters*, **9**, 333 (1962).
40. Geller, R., and D. Pigache: *J. Nuclear Energy*, **C4**, 229 (1962).
41. Loeb, L. B.: "Electrical Coronas: Their Basic Physical Mechanisms," University of California Press, Berkeley, Calif., 1965.

42. Finkelnburg, W.: "Hochstromkohlebogen," Springer, Berlin, 1948.

43. Witkowski, S.: *Z. angew. Phys.*, **11**, 135 (1959); H. Maecker: In *Proc. Fifth Intern. Conf. on Ionization Phenomena in Gases*, Munich, 1961, vol. 2, p. 1793.

44. Alexeff, I., and R. V. Neidigh: *Phys. Rev.*, **129**, 516 (1963).

45. Paulikas, G. A., and R. V. Pyle: *Phys. Fluids*, **5**, 348 (1962).

46. Francis, G.: "Ionization Phenomena in Gases," Academic, New York, 1960.

47. Brown, S. C.: Breakdown in Gases: Alternating and High Frequency Fields, in "Handbuch der Physik," vol. 22, pp. 531–575, Springer, Berlin, 1956.

48. Cf. Ref. 4, p. 308; W. P. Allis, S. C. Brown, and E. Everhardt: *Phys. Rev.*, **84**, 519 (1951).

49. Consoli, T.: *Sixth Intern. Conf. on. Ionization Phenomena in Gases*, Paris, 1963, vol. 2, p. 455.

50. Hooke, W. M., and M. A. Rothman: *Nuclear Fusion*, **4**, 33 (1964).

51. Dandl, R. A., A. C. England, W. B. Ard, H. O. Eason, M. C. Becker, and G. M. Haas: *Nuclear Fusion*, **4**, 344 (1964).

52. Eckert, H. U.: *J. Appl. Phys.*, **33**, 2780 (1962).

53. Dreicer, H.: *Phys. Rev.*, **115**, 238 (1959), **117**, 329 (1960); E. R. Harrison: *J. Nuclear Energy*, **C1**, 105 (1960); G. Ecker and K. G. Müller: *Z. Naturforsch.*, **16a**, 246 (1961).

54. Budker, R. I.: *Soviet J. Atomic Energy*, **5**, 673 (1956); J. G. Linhardt: "Plasma Physics," p. 247, North Holland Publishing Company, Amsterdam, 1960; J. G. Linhardt: In *Proc. Fourth Intern. Conf. on Ionization Phenomena in Gases*, Uppsala, 1959, vol. 2, p. 981.

55. Rose, D. J., and M. Clark, Jr.: "Plasmas and Controlled Fusion," Wiley, New York, 1961; L. A. Artsimovich: "Controlled Thermonuclear Reactions," Gordon and Breach, Science Publishers, New York, 1964.

56. *Nuclear Fusion Suppl.*, pts. 1, 2, and 3, 1962; "Plasma Physics and Controlled Nuclear Fusion Research," vols. 1 and 2, Internat. Atomic Energy Agency, Vienna, 1966 (*Proc. Second Intern. Conf. on Controlled Fusion, Culham, England*, 1965).

57. Cf. Chap. 11, Refs. 45 to 45*f*.

58. Petschek, H., and S. Byron: *Ann. Phys.*, **1**, 270 (1957).

59. Turner, E. B.: Equilibrium Hydrodynamic Variables behind a Normal Shock Wave in Hydrogen, *Space Technol. Lab. Rept.* GM-TR-0165-00460, 1958.

60. Resler, E. L., S. C. Lin, and A. Kantrowitz: *J. Appl. Phys.*, **23**, 1390 (1952).

61. Gaydon, A. G., and I. R. Hurle: "The Shock Tube in High-temperature Chemical Physics," Reinhold, New York, 1963.

62. Ferri, A.: "Fundamental Data Obtained from Shock Tube Experiments," Pergamon, New York, 1961.

63. Turner, E. B.: Equilibrium Hydrodynamic Variables behind a Reflected Shock Wave in Hydrogen, *Space Technol. Lab. Rept.* TR-59-0000-0074, 1959.

64. Nagamatsu, H. T., R. E. Geiger, and R. E. Sheer: *Am. Rocket Soc. J.*, **29**, 332 (1959).

65. Camm, J. C., and P. H. Rose: *Phys. Fluids*, **6**, 663 (1963).
66. Fowler, R. G., G. W. Paxton, and H. G. Hughes: *Phys. Fluids*, **4**, 234 (1961); R. G. Fowler and B. D. Fried: *Phys. Fluids*, **4**, 767 (1961).
67. Kolb, A. C.: In *Proc. Fourth Intern. Conf. on Ionization Phenomena in Gases*, Uppsala, 1959, vol. 2, p. 1021.
67a. Gross, R. A.: *Revs. Modern Phys.*, **37**, 724 (1965).
68. Gaydon, A. G., and H. G. Wolfhard: "Flames," pp. 302–323, Chapman & Hall, London, 1960; K. E. Shuler: Ionization in High Temperature Gases, "Progress in Astronautics and Aeronautics," vol. 12, chap. 2, Academic, New York, 1963.
69. Shuler (ed.), K. E.: Ionization in High Temperature Gases, "Progress in Astronautics and Aeronautics," vol. 12, chap. 5, Academic, New York, 1963; G. J. Mullaney and N. R. Dibelius: In I. A. McGrath, R. G. Siddall, and M. W. Thring (eds.), "Advances in Magnetohydrodynamics," pp. 47–54, Macmillian, New York, 1963.
70. Gatz, C. F., F. T. Smith, and H. Wise: *J. Chem. Phys.*, **35**, 1500 (1961).
71. Kunkel, W. B., and A. L. Gardner: *J. Chem. Phys.*, **37**, 1785 (1962).
72. Von Engel, A., and J. R. Cozens: *Sixth Intern. Conf. on Ionization Phenomena in Gases*, Paris, 1963, vol. 1, p. 257; A. Von Engel and J. R. Cozens: *Proc. Phys. Soc. (London)*, **82**, 85 (1963).
73. Glasstone, S., and R. H. Lovberg: "Controlled Thermonuclear Reactions," pp. 145–152, Van Nostrand, Princeton, N.J., 1960; F. R. Scott and H. G. Voorhies: *Phys. Fluids*, **4**, 600 (1961); for a collection of papers, see also "Proceedings of an International Symposium on Plasma Guns," *Phys. Fluids*, **7**, Supplement pp. S1–S74 (1964).
74. John, R. R., and W. L. Bade: *Am. Rocket Soc. J.*, **31**, 4 (1961); T. Reed: *J. Appl. Phys.*, **32**, 821 (1961); P. R. Dennis et al.: "Plasma Jet Technology," Technology Survey NASA SP-5033, 1965.
75. Boeschoten, F., and F. Schwirzke: *Nuclear Fusion*, **2**, 54 (1962); J. H., Malmberg, N. W. Carlson, C. B. Wharton, and W. E. Drummond: *Sixth Intern. Conf. on Ionization Phenomena in Gases*, Paris, 1963, vol. 4, p. 229.
76. Hinnov, E., and J. G. Hirschberg: *Phys. Rev.*, **125**, 795 (1962); F. Robben, W. B. Kunkel, and L. Talbot: *Phys. Rev.*, **132**, 2363 (1963).
77. Bates, D. R., and A. E. Kingston: *Proc. Roy. Soc. (London)*, **279A**, 10, 32 (1964).
78. Cooper III, W. S., and W. B. Kunkel: *Phys. Rev.*, **138A**, 1022 (1965).
79. Wilcox, J. M., A. W. DeSilva, and W. S. Cooper III: *Phys. Fluids*, **4**, 1506 (1961).
80. Malmberg, J. H., and C. B. Wharton: *Phys. Rev. Letters*, **13**, 184 (1964).
81. Boeschoten, F.: *J. Nuclear Energy*, **C6**, 388 (1964).

11

POWER CONVERSION

ROSA

RICHARD J. ROSA, *Principal Research Scientist, Avco-Everett Research Laboratory, Everett, Massachusetts*

11

11.1 INTRODUCTION: THE PLASMA DIODE

This chapter will discuss both the plasma thermocouple and the magnetohydrodynamic generator as energy-conversion devices using a plasma. Greater emphasis is placed on the MHD generator, since this is the author's major field of interest.   However, an attempt has been made to provide an adequate bibliography for those who wish to go into either subject further.

The plasma diode consists of two electrodes, an emitter and a collector, in general placed as close together as it is practical to put them. The emitter is heated to boil off electrons, and the collector is cooled so that the electrons will condense on it.   The device is a heat engine, and hence the thermodynamic limitations summarized in the laws of Carnot apply to it.   It is convenient to discuss the operation of this device with the aid of an energy-level diagram such as shown in Fig. 11.1.   An approximate voltage scale, with negative voltage upward, is shown in this diagram to give the reader an idea of the energy involved.   The quantities ϕ_e and ϕ_c are the work functions of the emitter and collector. V is the voltage appearing at the terminals of the device.   In operation the emitter is heated until some of the electrons may surmount the potential barrier ϕ_e and escape from the emitter.   They must then find their way to the collector, where they fall through the potential ϕ_c.   Then

348

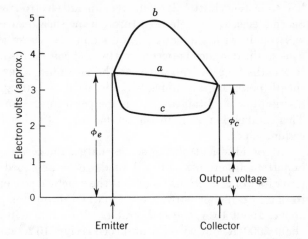

Fig. 11.1   Energy-level diagram for the thermionic diode.

the potential V is left to drive the electrons through the external load. If the device is operated in vacuum and only a small amount of current is drawn from the emitter, a flat potential distribution indicated by the line (*a*) would be obtained in the space between the electrodes. However, if one attempts to obtain a large current flow in the device, space-charge effects cause the potential to develop the humped appearance indicated by the line (*b*). This potential peak limits the amount of current that one can obtain in a vacuum diode. One would like very much to get as high a current density as possible in order to get reasonable power densities and efficiencies. Ten amperes per square centimeter is often quoted as a desirable figure.[1-3] In a vacuum diode one can reduce space-charge effects by placing the electrodes very close together, but one soon reaches a practical limit to this. An alternative approach is to introduce positive ions into the gap to neutralize the electron space charge. In principle, one could introduce just the right number of ions to reduce the potential curve between the electrodes back to the straight line (*a*), but this turns out to be an unstable situation for a plasma. The resulting potential distribution always turns out to be approximately like that shown at (*c*), where a plasma sheath has formed close to each electrode. It will be recalled that the sheath potential is basically the mechanism by which a plasma equalizes the rate at which it loses electrons and ions at its boundary and hence maintains approximate electrical neutrality.

At present the accepted method for introducing ions into the diode is simply to introduce a small amount of cesium vapor. This vapor becomes ionized in two ways. First, because the ionization potential of cesium is less than the work function of tungsten, upon colliding with the hot tungsten emitter it will give up an electron and become ionized. Second, electrons accelerated through the sheath surrounding the emitter will collide with the cesium atoms and also cause ionization. Since in general the distance between emitter and collector is many electron mean-free paths long, it appears that the electrons, after falling through the sheath potential, are quickly randomized and an approximate Maxwell distribution is obtained with a temperature of approximately 5000°K. This is much higher than the temperature of either the electrodes or the cesium vapor.

Besides neutralizing space charge, the cesium performs two other important functions. First it tends to be absorbed on the emitter and lower its work function. This has the practical advantage of allowing one to obtain a higher current emission at a lower temperature. As noted above, about 10 A/cm² is desirable. Pure tungsten must be operated at about 3000°K, but with the absorbed cesium 10 A/cm² can be obtained at roughly 2000°K. A second function performed by the cesium is that it reduces the work function of the collector. It can be seen from Fig. 11.1 that this will also improve the performance of the device. Of course, when the work function of the collector is reduced, care must be taken to avoid back emission of electrons by keeping its temperature low enough. In general a temperature of 900°K or lower is required.

Now a certain conflict of interest can be seen here. One wants to use a high pressure of cesium in order to get the absorbed layers on the electrodes and hence the reduced work functions. However, ideally one would like to have the electrons flow across the gap unimpeded, i.e., without suffering any collisions with the cesium atoms, and this, of course, implies low pressure. Practical experience to date seems to indicate that a relatively high pressure of about 1 mm Hg is the best compromise, with the result that the electron mean-free path is much less than the interelectrode spacing. Hence the electron motion in the gap is essentially a diffusion process under conditions not unlike those in the column of a medium-pressure glow or arc discharge.

In addition to the power-generating mechanism there are loss mechanisms by which heat is transferred from the emitter to the collector. The first of these is the energy that the electron gives up to the collector when it falls through the collector's work function. This we might call the *heat of condensation* of the electron. In addition, heat flows from emitter to collector by radiation, by conduction through the plasma, and by conduction through leads and supporting structure. It is in order to minimize the relative importance of these heat-conduction losses that a high power density is essential for high efficiency. One finds that high power density is very nearly synonymous with high current density. Current density is in part determined by the well-known Richardson equation, which predicts an exponential dependence of current density upon temperature.   Thus, even though heat losses tend also to increase with temperature, to the fourth power in the case of radiation, it appears that the higher the temperature the better the efficiency.   However, aside from the obvious engineering difficulties associated with high temperature, it becomes harder and harder to find materials to use with a suitably low work function as the temperature is raised.   For example, in the case of an electrode of tungsten in cesium vapor, as the tungsten temperature is raised, the cesium no longer is adsorbed on the surface, and hence the work function rises sharply.

So we see in general that there are a number of considerations that must be balanced, one against another, to obtain the optimum performance.   It appears from the various studies that the author has seen that predicted best performance from a plasma diode is a power density of about 30 W/cm² of electrode surface and an efficiency of about 30 percent. This is a very good figure, since the heat rejected by one of these devices is still at a quite high temperature and hence can be radiated with relative ease or can be used to do more useful work.   For example, if 30 percent efficient diodes were used to top a good 40 percent efficient steam plant, the net overall plant efficiency would work out to about 58 percent. However, the nuclear reactor that would be required for this would have to operate at relatively high temperatures and moreover would have a complex and fine-grained structure with each fuel element surrounded by plasma diodes.   Hence the economics of the situation is far from clear so far as the generation of bulk power is concerned.

There are two other devices that should be mentioned here.   The first is a plasma diode that uses an inert gas for space-charge neutralization instead of cesium vapor.[13-18]   The potential advantages of this are: (1) greater flexibility in the choice of electrode material as regards work function and operating temperature, resulting in particular in the ability to operate efficiently at a relatively low temperature like 1500°K; (2) less inhibition of current flow by the gas owing to the fact that noble gases, argon in particular, present a very small collision cross section to electrons.

Since noble gases have high ionization potentials, a major problem is how to ionize them.   Use of the radiation existing in a reactor has been suggested[13,17] and also use of an auxiliary discharge.[14,18]   It is not yet clear that sufficient ionization can be maintained with a low enough expenditure of energy.   However, theoretical studies indicate that it should be possible, especially if advantage is taken of the so-called Penning effect,[19] which increases the ionization efficiency in certain atomic gas mixtures.

The second device is derived from a phenomenon that has been called the *unipolar arc*.[20]   Apparently no applications have been seriously proposed for it, but it is nevertheless an interesting idea which differs at least conceptually from the plasma diode.   Again one has two electrodes immersed in a plasma, one of which, by virtue of high temperature or low work function, emits electrons, while the other does not. The emitting electrode will tend to float at a potential equal to that of the bulk plasma, but the other, in accordance with Langmuir probe theory, will tend to float negative by an amount of order kT/e. Then, if a load is connected between the two electrodes, and if heat is supplied to keep the plasma hot, a conversion of heat to electricity results.

The practical distinction between this device and a plasma diode is not altogether clear to the author except under conditions where kT/e is large compared with typical electrode work functions, i.e., of the order of 10 volts or more (a temperature far beyond the capabilities of any heat source we have today, although it could, in principle, be produced by a gaseous fission or fusion reactor). However, conceptually, the difference is that thermal energy is added to the plasma instead of to one of the electrodes, and performance characteristics such as output voltage are determined more by the temperature and other properties of the plasma than by those of the electrodes.

11.2 INTRODUCTION: MHD POWER GENERATION

A magnetohydrodynamic, or MHD, generator is not a direct conversion system and is not a static system in the sense that the plasma thermocouple is. It is, in fact, not basically so much different from the conventional power-generating systems employing a turbogenerator that are used for large-scale power generation today. As a source of electricity we shall see that the MHD generator is in some ways less than ideal, but it makes up for this by being potentially an extremely good turbine or "expansion engine." Hence it is perhaps unfortunate that it is not called an MHD turbine.

Figure 11.2 is an attempt to show the basic similarity and difference between the MHD generator and the turbogenerator. In both cases the object is to make a heated compressible fluid do work by pushing a current-carrying electrical conductor across a magnetic field.

When Faraday discovered electromagnetic induction, the most convenient electrical conductors were solid materials. This made it necessary to convert gas enthalpy first into mechanical energy, which could then be used to move the solid conductor. Faraday, it seems, understood perfectly well that the working fluid could, in principle, generate electric power directly by moving itself through the field of a magnet.[46] However, as in the case of many devices, it required more than one hundred years of subsequent technological advance and research into the properties of matter before there was a sufficient understanding of and facility for producing electrical conductivity in gases. There is in fact still a great

Turbogenerator   MHD generator

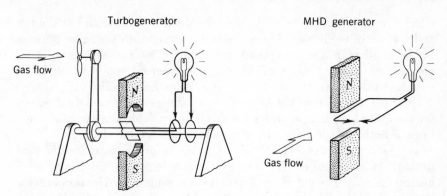

Fig. 11.2 Comparison between the turbogenerator and MHD generator.

deal that is poorly, if at all, understood, and there is still a need for better materials and other technical improvements.   However, we have reached the point where with existing technology it looks as though competitive generating systems and important applications on the ground, and perhaps in space, are within our grasp.

A simple form of MHD generator is sketched in Fig. 11.3.   It consists of a duct through which the gaseous working fluid flows, coils

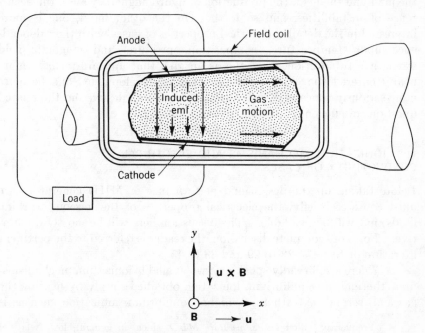

Fig. 11.3 Schematic of a dc MHD generator.

which produce a magnetic field across the duct, and electrodes at the top and bottom of the duct.   These electrodes serve much the same purpose as the brushes in a conventional generator.   The gas, by virtue of its motion through the magnetic field, has an emf generated in it which drives a current through it, the electrodes, and the external load.

Such a device would be used in a heat cycle, conventional except for the quite high temperatures involved, and would take over the functions of both the turbine and the generator in such a cycle.

As a branch of MHD, this work meets problems quite different from those encountered in astrophysics and in research on controlled nuclear fusion.   The difference is perhaps best summarized by the observation that in this work the magnetic Reynolds number R_m tends to be small compared with unity, whereas in fusion research it tends to be much larger than unity.   Here $R_m = \mu_0 \sigma u l$, where μ_0 is the permeability of free space, σ and u are the gas conductivity and velocity, and l is a characteristic length (cf. Chap. 7).   Practical mksa units are used in this chapter.

Since the whole field of MHD is quite new and since there has been relatively little experimental work in low-magnetic-Reynolds-number MHD[†] in particular, it is not possible to say just what will prove to be the most important phenomena.   In broad terms, low magnetic Reynolds number implies that one is less interested in complex interactions between the fluid and the field, the production of hydromagnetic waves, the occurrence of instabilities, and so forth.   On the other hand, one is more interested in the detailed electrical properties of gases, with their dependence upon temperature, composition, pressure, and magnetic field strength.   In addition, since this is an area that appears to have more nearly entered the category of engineering development, one becomes more seriously concerned with questions of overall design and the related questions of efficiency, reliability, and economy.

11.3 IONIZATION MECHANISMS AND ELECTRICAL CONDUCTIVITY OF GASES

Before taking up the design and performance of MHD generators, we must consider briefly, the electrical properties of the gaseous working fluids that will be available.   The discussion here will be mostly qualitative.   For more complete discussion, the reader is referred to the pertinent literature or to Refs. 22 to 29 and 38 to 45.

There are, broadly speaking, two available ionization mechanisms: first, thermal, or equilibrium, ionization obtained simply by heating the gas and, second, extrathermal, or nonequilibrium, ionization such as is

[†] Frequently called *low-temperature MHD*, since, in general, low (relative) temperature and low magnetic Reynolds number go together.

obtained in low-pressure gas discharges.   Extrathermal, or nonequilibrium, ionization is much more familiar than thermal ionization, since it occurs in such widely used devices as gas rectifiers and fluorescent lights. However, it is in fact a considerably more complicated phenomenon.   It is not at all clear that it could be used in a way that would yield a net power output.   We will return to this interesting question later after a discussion of the Hall effect.

In thermal ionization, which at present seems most practical, the ionization follows a mass-action law, as does molecular dissociation. Most common gases, such as air, CO, CO_2, or the noble gases have a relatively high ionization potential and hence do not ionize thermally until quite high temperatures are reached.   However, if a small amount (0.1 to 1 percent) of some easily ionizable material, such as an alkali-metal vapor, is added to the gas, a sufficient degree of ionization can be obtained at temperatures low enough to be withstood by some solid materials, to be produced in furnaces, and also conceivably to be produced in nuclear-fission reactors.

Figure 11.4 shows calculated values of conductivity for argon, plus 0.175 percent cesium vapor.   It can be seen that argon seeded with cesium vapor will have useful conductivity at temperatures greater than about 2000°K, whereas clean argon becomes essentially nonconducting at about 4000°K.   The term *useful conductivity* needs some defining. The value of gas conductivity determines the magnetic field strength and generator size necessary for the production of a given amount of power. If the conductivity is too low, the generator becomes excessively large and one then expects that the power required to maintain the field, heat transfer, and other losses, and also cost and weight, will become excessive. It is not possible to fix a definite value as being the lowest that is useful; however, it will become apparent in later discussions that gas conductivities, at least of the order of 1 mho/m and hence gas temperatures in the neighborhood of 2000°K, seem to be required.   For some purposes it will appear that one would like a considerably higher conductivity.   However, Fig. 11.4 shows that approximately 100 mhos/m is the highest that we are likely to get with existing thermal-energy sources.   Thus, as at least initial working assumptions, we may take it that the gaseous working fluid in our MHD generator will have a temperature of 2000 to 3000°K and an electrical conductivity of 1 to 100 mhos/m.

One question that frequently arises is: Why does one not get a higher electrical conductivity by increasing the amount of seed material or indeed by going to a working fluid that is 100 percent alkali-metal vapor? The reason this does not in general appear to be a good idea is that apparently all alkali-metal atoms have a very large cross section for collision with electrons in the energy range of interest for MHD generators.

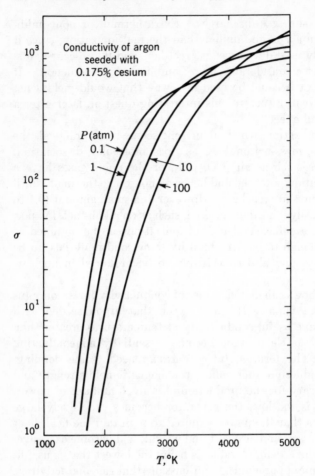

Fig. 11.4   Conductivity of argon plus 1 percent cesium.

Therefore, addition of seed material beyond a certain point causes the electron mobility to go down faster than the electron concentration goes up.   The net result is then a decrease in electrical conductivity.   Maximum conductivity is obtained when the fractional seed concentration is equal to the ratio between the collision cross section of the parent gas and the collision cross section of the seed atom.   For most gases this ratio appears to be 0.01 or less.

11.4 PRINCIPLES OF OPERATION OF MHD GENERATORS

The equations governing the essentially one-dimensional flow in a magnetohydrodynamic generator as sketched in Fig. 11.3 are

Energy: $$\rho u \frac{d}{dx}\left(\frac{u^2}{2} + C_p T\right) = \mathbf{j} \cdot \mathbf{E} \qquad (11.1a)$$

Momentum: $$\mathbf{x}_1 \rho u \frac{du}{dx} + \text{grad}\, P = \mathbf{j} \times \mathbf{B} \qquad (11.2a)$$

Continuity: $$\rho u A = \text{const} \qquad (11.3)$$

Ohm's law: $$\mathbf{j} = \sigma(\mathbf{u} \times \mathbf{B} + \mathbf{E}) - \frac{\omega\tau}{B} \mathbf{j} \times \mathbf{B} \qquad (11.4)$$

where P, T, ρ, u = gas pressure, temperature, density, and velocity
   C_p = constant-pressure heat capacity
   $\mathbf{j}$ = current density
   $\mathbf{E}, \mathbf{B}$ = electric and magnetic fields
   A = channel cross section
   ω, τ = cyclotron frequency and mean-free time of an electron in the gas

The symbol $\mathbf{x}_1$ denotes a unit vector in the direction of flow.

These equations can be solved in closed form for some special conditions and for more general conditions may readily be integrated numerically. A simplifying feature of magnetohydrodynamic theory under the conditions of interest here is that the magnetic field strength B can usually, to good approximation, be taken as a given quantity fixed by external conditions, and not influenced appreciably by the current patterns that arise in the gas. On the other hand, it is frequently necessary to allow for both the temperature and pressure dependence of the electrical conductivity σ.

As in a turbine, the flow velocity through a magnetohydrodynamic generator is or can be more or less a constant, the primary effect being a drop in pressure as the gas forces itself through the successive turbine stages of the former or the magnetic field of the latter. The assumption of strictly constant velocity is useful in that it greatly simplifies the equations of motion and makes clear the processes occurring in the generator. The equations then are

Energy: $$\rho u \frac{dh}{dx} = \mathbf{j} \cdot \mathbf{E} \qquad (11.1b)$$

Momentum: $$\frac{dP}{dx} = j_y B \qquad (11.2b)$$

Continuity: $$\rho A = \text{const} \qquad (11.3b)$$

where h = gas enthalpy per unit mass
The continuity equation now gives the area increase which must be provided to ensure constant velocity.

The rate at which work is done by the gas in pushing itself through the magnetic field is obtained by multiplying Eq. (11.2b) by u,

$$u \frac{dP}{dx} = j_y u B$$

The rate at which gas enthalpy is extracted as electrical energy is given by Eq. (11.1b). Dividing the latter by the former (on the assumption that $j_x = j_z = 0$) gives

$$\eta_e = \frac{\rho u(dh/dx)}{u(dp/dx)} = \rho \frac{dh}{dp} = \frac{E}{uB} \tag{11.5}$$

where η_e is usually called the *loading parameter* or *electrical efficiency* of the generator, i.e., the fraction of the electric power generated that is actually delivered to the load, the difference being dissipated in the internal resistance of the generator itself.   It is to be noted that in a conventional generator the power lost in the internal impedance is low-grade heat, which must be disposed of, and this not without difficulty.   On the other hand, in a magnetohydrodynamic generator this heat remains within the working fluid.   However, it does result in an increased pressure drop or a departure from isentropy or reversibility.   Equation (11.5) shows in fact that η_e is approximately equal to the "small-stage" efficiency of a turbine.   It becomes exactly equal in the limit as $u \to 0$, if there are no other losses.   Of course, in an actual machine, there will be other sources of loss.   Some of these will be discussed later.

11.4.1. Hall-current Generators

The assumption that $j_x = j_z = 0$ made in deriving Eq. (11.5) implied in effect that the term involving $\omega\tau$ in Eq. (11.4) could be neglected or that j_x and j_z could by some means be prevented from flowing.[49]   In other words, the generator was assumed to be what one might call a *normal-current*, or *Faraday*, *generator*.   Now we shall examine another possible type of MHD generator.

    The term involving $\omega\tau$ expresses the fact that ionized gases like solid conductors exhibit Hall effects that are proportional to the magnetic field strength.   Gases at field strengths above a certain density-dependent value exhibit, or should exhibit, the effects much more strongly than is usual in solids.   For a further discussion of the origin of the Hall effect, the reader is referred to other chapters in this volume or to Refs. 22 to 29.

    One possibility introduced by the Hall effect is to make a so-called Hall-current generator.   In this type of generator the normal current, i.e., the current that flows in the $\mathbf{u} \times \mathbf{B}$ direction, is allowed to flow freely. This might be done, for example, by short circuiting across the load shown in the sketch in Fig. 11.3, or by arranging a suitable channel configuration such as a coaxial geometry sketched in Fig. 11.5.   Then two new electrodes are inserted; one upstream and one downstream, with the load connected between them.   Thus power is extracted from the generator via the Hall or axial current.   Now the work done by the gas pushing itself

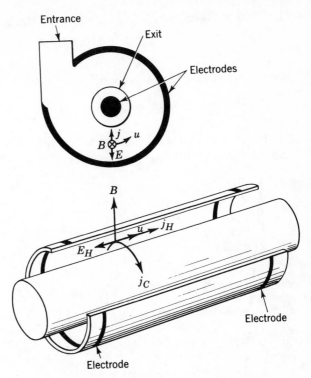

Fig. 11.5 The normal-current vortex geometry and the Hall-current coaxial geometry.

through the magnetic field is still $|j_y uB|$, but the power output is now $-j_x E_x$. To analyze these further, we need to write the Ohm's law equation (11.4) in component form as follows:

$$j_x = \sigma E_x - \omega_e \tau_e j_y \tag{11.6}$$
$$j_y = \sigma uB + \omega_e \tau_e j_x \tag{11.7}$$

With these, we can write the expression for the electrical efficiency in terms of the output current j_x as follows:

$$\eta_e = \omega_e \tau_e \left| \frac{j_x}{\sigma uB} \right| \left(\frac{1 - \dfrac{1 + \omega_e^2 \tau_e^2}{\omega_e \tau_e} \cdot \left| \dfrac{j_x}{\sigma uB} \right|}{1 - \omega_e \tau_e |j_x/\sigma uB|} \right) \tag{11.8}$$

We can also express output voltage E_x as a function of output current as follows:

$$E_x = \omega_e \tau_e uB + \frac{1 + \omega_e^2 \tau_e^2}{\sigma} j_x \tag{11.9}$$

Also, output power, which is $j_x E_x$, will have the following form:

$$-j_x E_x = -\omega_e \tau_e u B j_x - \frac{1 + \omega_e^2 \tau_e^2}{\sigma} j_x^2 \tag{11.10}$$

And in terms of j_x, the pressure drop has the following form:

$$\frac{dP}{dx} = j_y B = (\sigma u B + \omega_e \tau_e j_x) B \tag{11.11}$$

Figures 11.6 to 11.8 show plots of these equations under the assumption that $\omega \tau$ is about 10. Also shown are the way these quantities would vary

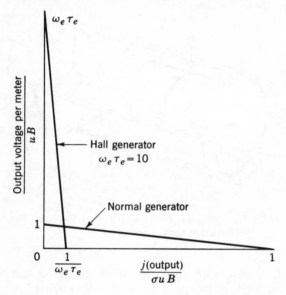

Fig. 11.6 Comparison of the normal- and Hall-current generators. Output voltage vs. output current per unit cube.

with output current for a normal-current generator which had the same σ, u, and B. One can notice several quite startling differences. First the Hall generator tends to be a high-voltage low-current device, as compared with a normal-current generator. Second, the efficiency tends to increase with output current rather than decrease as is the case in the normal-current generator, except at high currents, where the efficiency of the Hall generator drops abruptly to zero. Finally, the pressure drop, instead of increasing as more current is drawn, tends to decrease. Thus we observe that, as the load on the generator is varied, a normal-current generator behaves at least qualitatively the same as a conventional turbogenerator would, but the Hall generator in many ways behaves in an

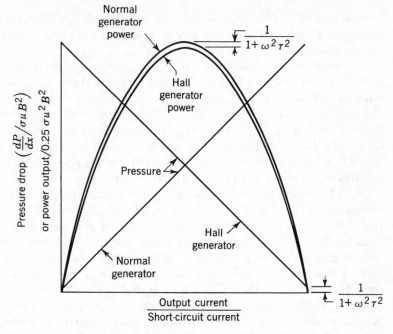

Fig. 11.7   Comparison of the normal- and Hall-current generators.   Pressure drop and power output per unit cube.

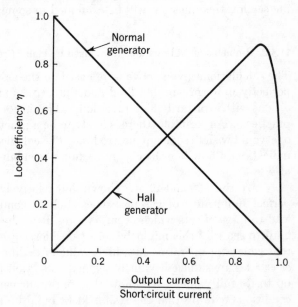

Fig. 11.8   Comparison of the normal- and Hall-current
generators.   Local efficiency.

inverse fashion.   This is not necessarily a drawback, but it would clearly present some interesting control problems for a power plant incorporating such a machine.

Another characteristic of a Hall generator that might be unpleasant is the fact that as shown by Figs. 11.6 and 11.8 operation at high efficiency requires operation at an output voltage which is only a small fraction of the no-load or open-circuit voltage.   This, then, would imply that the device would have very poor voltage regulation.   In fact, one might say that a Hall-current generator tends to be a constant-current source, whereas a normal-current MHD generator as well as a conventional rotating machine tends to be a constant-voltage source.

The advantages of the Hall generator are that it delivers its power through only two sets of leads as compared with the multiplicity of leads required by a normal generator when operated at any value of $\omega\tau$ greater than about $\frac{1}{2}$.   In addition it may prove possible for other reasons to operate a Hall-current generator at higher values of $\omega\tau$ than the normal-current machine, and this, in turn, implies operation at higher power densities.   However, there is still a great deal of work that must be done before a decision can be made as to which type of generator is most advantageous.   In particular, it is still an open question as to whether or not one will ever be able to operate any type of machine at a sufficiently high $\omega\tau$ to ensure good operation in the Hall-current mode.   In the following section this question will be examined in somewhat more detail.

11.4.2. Problems and Limitations Introduced by Hall Effects and Ion Slip

The calculations given above were based on the assumption that we had a perfectly uniform working fluid and that we could apply boundary conditions with arbitrarily high precision.   This is not likely to be the case in practice.   For example, there is likely to be a practical limit to how finely one would wish to segment electrodes.   The calculations in Ref. 61 give an indication of how fine the electrodes must be subdivided for a given value of $\omega\tau$.

In Refs. 29 and 49, it is shown that relatively small departures from perfect uniformity of gas properties such as conductivity and $\omega\tau$ could lead to large effects which occur as a result of the Hall effect.   It is not hard to see why this might be so.   Consider, for example, a normal-current generator operating at a high value of $\omega\tau$ like 10.   Assume that the electrodes are sufficiently finely segmented so that the Hall field can build up to its full value.   We know that optimum performance under these conditions requires that the Hall field E_x build up to a value that reduces the Hall current j_x to zero.   Equation (11.6) shows that, to do this, E_x must build up to a value $\omega\tau$ times $uB - E_y$.   In other words, E_x will be a quantity approximately one order of magnitude larger than $uB - E_y$. Then we see from Eq. (11.6) that this large quantity multiplied by the conductivity σ must exactly balance another large quantity, the product of normal current j_y and $\omega\tau$. Thus, the performance of the generator depends on the difference between two large quantities being precisely zero. If the gas properties and $\omega\tau$ are not perfectly uniform throughout the generator, it is probably impossible to have the two large quantities balance out completely everywhere, so some internal circulation of Hall current is to be expected.

Let us consider one special form of nonuniformity, namely, that the conductivity σ varies in the y direction but in no other direction. This is not a very general form of nonuniformity, and, as is shown in Ref. 29, it is the form which leads to the worst possible results. Hence the results are perhaps more pessimistic than need be; however, they will give us at least a qualitative picture of what might happen. For this analysis it is worthwhile to introduce a more complete conductivity equation than Eq. (11.4). This equation, which includes the fact that the ions as well as the electrons have some mobility, is

$$ \mathbf{j} = \sigma\mathbf{E} - \frac{\omega_e\tau_e}{B}\mathbf{j} \times \mathbf{B} + \frac{\omega_e\tau_e\omega_i\tau_i}{B^2}(\mathbf{j} \times \mathbf{B}) \times \mathbf{B} \qquad (11.12) $$

Again, the reader should consult Refs. 22, 23, or 25 for a more complete discussion of the origin of this equation. Here all the symbols have the same meaning as before except for the new symbol $\omega_i\tau_i$, which is the $\omega\tau$ for the positive ions. Equation (11.12) can be split into component forms analogous to what was done for Eq. (11.4). The result is

$$ j_y = \Sigma E_y + \Omega j_x \qquad (11.13) $$
$$ j_x = \Sigma E_x - \Omega j_y \qquad (11.14) $$

Here, two new symbols have been introduced,

$$ \Sigma = \frac{\sigma}{1 + \omega_i\tau_i\omega_e\tau_e} \qquad (11.15) $$

$$ \Omega = \frac{\omega_e\tau_e}{1 + \omega_i\tau_i\omega_e\tau_e} \qquad (11.16) $$

It can be seen that these two new symbols are equal to the gas conductivity and to $\omega_e\tau_e$ up to the point where $\omega_i\tau_i \times \omega_e\tau_e$ approaches and exceeds the value 1. This does not in general happen until $\omega_e\tau_e$ has exceeded a value of 10 or more, since

$$ \frac{\omega_e\tau_e}{\omega_i\tau_i} \approx \sqrt{\frac{m_i}{m_e}} \approx 100 \text{ to } 1{,}000 $$

where m_e, m_i = electron and ion masses

Now, although the above equations hold at any given point in a nonuniform gas, they may not hold for the gas as a whole. To obtain the appropriate equations, we must form an average over y. When this is done, one gets the following result,[29]

$$\langle j_x \rangle_{\text{av}} = \langle \Sigma \rangle_{\text{av}} E_x - \langle \Omega \rangle_{\text{av}} j_y \tag{11.17}$$

$$\langle j_y \rangle_{\text{av}} = \frac{\langle \Sigma \rangle_{\text{av}}}{G} \langle E_y \rangle_{\text{av}} + \frac{\langle \Omega \rangle_{\text{av}}}{G} \langle j_x \rangle_{\text{av}} \tag{11.18}$$

where $\quad G = \langle \Sigma \rangle_{\text{av}} \left\langle \dfrac{1 + \Omega^2}{\Sigma} \right\rangle_{\text{av}} - \langle \Omega \rangle_{\text{av}}^2 \tag{11.19}$

One sees that with the substitution of appropriate average quantities these equations are formally the same as the previous equations except that the equation for the current component in the direction of property variation acts as if both Σ and Ω had been reduced by the factor $1/G$. The function G contains in it all the effects introduced by the inhomogeneity. To calculate its value, one needs to know the profiles of conductivity and $\omega\tau$ across or along the gas stream. The important point is that G is in general a function of $\omega_e\tau_e$ and introduces into Ohm's law for the nonuniform gas a dependence on $\omega_e\tau_e$ that does not exist in a uniform gas. Physically, what has happened is that a current flowing in a direction perpendicular to the planes of nonuniformity sets up circulating currents in the parallel direction within the planes. This additional mechanism for energy dissipation causes an apparent decrease in the scalar conductivity and also apparently in the effective value of Ω.

Now let us consider the operation of a normal-current generator, i.e., a generator with electrodes sufficiently well segmented so that the average value of the Hall current is equal to zero. Using the above equations, one finds that the power output per unit volume $\mathbf{j} \cdot \mathbf{E}$ is given by the following expression:

$$\mathbf{j} \cdot \mathbf{E} = \eta(1 - \eta) \frac{\langle \Sigma \rangle_{\text{av}}}{G} u^2 B^2 \tag{11.20}$$

Let us compare this with the power output W_1 that would be delivered by an ideal generator with a perfectly uniform gas when the magnetic field strength was sufficient to make $\omega\tau$ equal to 1. One finds that the ratio of actual power output to the ideal power output for $\omega\tau$ equals 1 is given by the following expression:

$$\frac{W}{W_i} = \frac{\langle \Sigma \rangle_{\text{av}}}{\langle \sigma \rangle_{\text{av}}} \frac{\omega_e^2 \tau_e^2}{G} \tag{11.21}$$

Here we have used the fact that B/B_1 is equal to $\omega_e\tau_e$. A plot of this equation is shown in Fig. 11.9. The assumptions used in making this plot were that the ratio between $\omega_e\tau_e$ and $\omega_i\tau_i$ was equal to 1,000 and that the nonuniformity consisted of gas layers whose conductivity alternated between a value σ_1 and σ_2.   The ratio between σ_1 and σ_2 is given by the parameter K.   From this figure one sees that for a uniform gas, i.e., for K equal to 1, the power increases as B^2 until ion slip causes it to saturate at a value equal to $\omega_e\tau_e/\omega_i\tau_i$ times W_1.   For a nonuniform gas the power out-

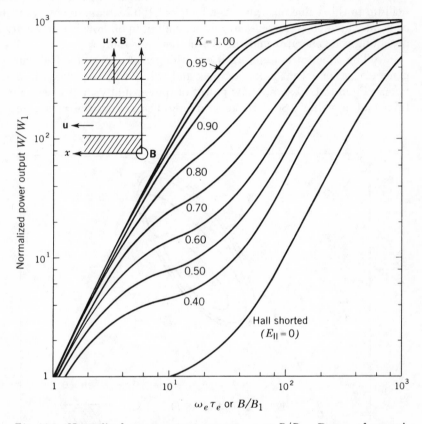

Fig. 11.9   Normalized power output versus $\omega_e\tau_e$ or B/B_1.   Degree of gas uniformity is $K \equiv \sigma_1/\sigma_2$.

put begins to saturate at a lower value.   However, when the positive ions begin to slip, the effect of nonuniformity decreases, causing the slope of the power-output curve to steepen again and finally saturate at the same level as did the uniform gas.   This is at first sight reassuring, but the logarithmic scale of the abscissa is to be noted.   It is not often that one will be willing to "buy" the order-of-magnitude increase in field strength required to make a 50 percent nonuniform gas, for example, produce as much power as a uniform one.

As a numerical example of typical magnitudes of the quantities involved, in a kerosene oxygen flame at 1 atm pressure the magnetic field B_1 that makes $\omega_e \tau_e$ equal to 1 will be about 10,000 G, and the power output at this field strength will be about 10,000 kW/m³.   Now it is possible that fields of up to 100,000 G will eventually become practical in an engineering sense, but we see from Fig. 11.9 that increasing the magnetic field up to this value may give us a factor of 100 increase in power output or it may give us no increase at all, depending upon how clever we are in designing the generator and the thermal-energy source.

The effect of nonuniformity on the performance of a Hall-current generator is somewhat more complex and will not be considered in detail here.   Figure 11.10 shows how power output would vary if η were maintained at a value 0.80.   In general it appears that the power-output

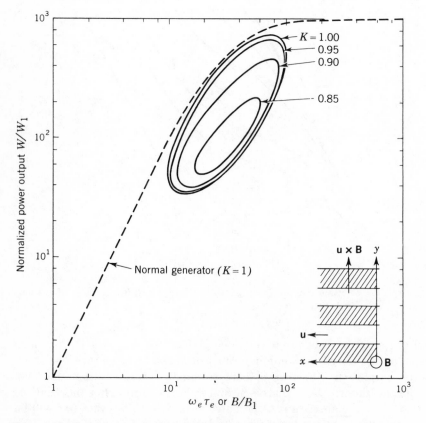

Fig. 11.10 Hall-current generator having 80 percent local efficiency.   Normalized power output versus $\omega_e \tau_e$ or B/B_1.   Degree of gas uniformity is $K = \sigma_1/\sigma_2$.

density is affected in the same way as it is in the normal-current generator. However, one also observes that for a given efficiency the Hall-current generator can operate only over a restricted range of $\omega_e\tau_e$, and one observes that the usable range of $\omega_e\tau_e$ is further restricted by the presence of non-uniformity.   Figure 11.11 shows attainable values of efficiency as a func-

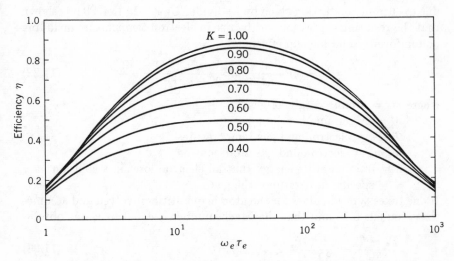

Fig. 11.11   Hall-current generator.   Maximum attainable local efficiency versus $\omega_e\tau_e$ for various degrees of gas uniformity K.

tion of $\omega_e\tau_e$ for the same assumptions that were used in drawing Figs. 11.9 and 11.10.   However, this figure holds for variations of conductivity in either the cross-channel direction or the x (that is, axial) direction.

It should be repeated that a quite idealized model of nonuniformity was used which gives the largest possible effect.   However, these calculations at least show qualitatively the kind of problems that are to be expected as $\omega_e\tau_e$ increases, if there is any departure from perfect insulation of the Hall field.

11.4.3. The Production of Extrathermal Ionization

Since even in seeded gas the temperatures required for good conductivity are uncomfortably high by present engineering standards, it would be desirable to be able to obtain good conductivity at lower temperatures. Since in low-pressure gas discharges it is commonly observed that the electron temperature can be very much higher than the temperature of the gas, the question arises as to whether or not one can do this in an MHD device to increase the gas conductivity over the value it would have in

thermodynamic equilibrium.   It turns out that this is closely related to
the question of how the MHD device behaves at high values of $\omega_e \tau_e$.

To calculate the increase in electron temperature, we must balance
the rate at which electrons lose energy by collisions with the other particles
in the gas and the rate at which they gain energy from the applied electric
fields and motional emfs.   The latter quantity is just the Joule dissipation
due to current conduction given by j_e^2 divided by σ.   In Ref. 29 it is shown
that the resulting expression for the rise in electron temperature including
nonuniformity and ion slip effects is

$$\frac{T_e - T_0}{T_0} = (1 - \eta)^2 \frac{2m_e}{m_0 \delta} \frac{\gamma}{3} M^2 \frac{\omega_e^2 \tau_e^2}{1 + \omega_e^2 \tau_e^2} \frac{1 + \langle \Omega \rangle_{\mathrm{av}}^2}{G} \tag{11.22}$$

where m_0 = atomic weight of gas
  M = Mach number
  T_e, T_0 = electron and gas temperatures
  m_e, m_0 = electron and gas atom masses
  δ = fraction of its energy that an electron loses in a collision
  γ = specific heat ratio of the gas
If one takes typical values for the atomic quantities involved and assumes
that η equals 0.75 and that the Mach number is equal to 1, one obtains

$$\frac{T_e - T_0}{T_0} \approx \frac{1}{30} \frac{\langle \Omega \rangle_{\mathrm{av}}^2}{G} \lesssim \tfrac{1}{30} \omega_e^2 \tau_e^2 \tag{11.23}$$

It can be seen from this that under the best of conditions large increases in
electron temperature may require operation at $\omega_e \tau_e$ of 5 or more.   In
general, Eq. (11.22) shows that the fractional increase in electron temperature will be proportional to

$$\frac{\Delta T_e}{\Delta T_{e1}} = \frac{\omega_e^2 \tau_e^2}{1 + \omega_e^2 \tau_e^2} \frac{1 + \langle \Omega \rangle^2}{1 \quad G} \tag{11.24}$$

where ΔT_{e1} may be defined as the temperature rise one would get in a
uniform gas at $\omega_e \tau_e = 1$.   It can be seen from Fig. 11.12 that, if this
quantity must indeed approach 30 to obtain significant electron heating,
the gas must be no less than 75 percent uniform.   Moreover, it must
remain this uniform during the nonequilibrium ionization process.   In
addition, the curves indicate that there is not a very great margin for
error in our assumptions about the electron-atom processes involved.
One quite definite indication is that, since the energy-loss parameter δ is
likely to be one or two orders of magnitude greater in molecular gases than
in atomic gases, significant electron heating in molecular gases is not likely
to be practical.

There is another way one might use extrathermal ionization.   This
is to ionize the gas initially by fields or simply by having it initially at a

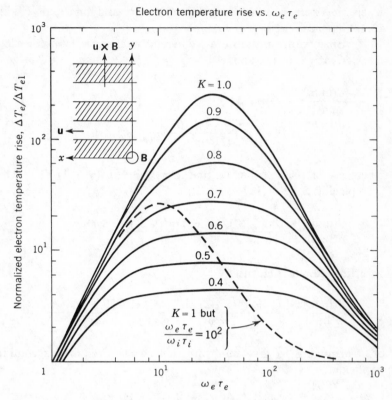

Fig. 11.12   Normalized electron temperature rise as a function of $\omega_e \tau_e$ for various degrees of gas uniformity K.

high temperature and then try to achieve power extraction from the gas before the electrons and ions have a chance to recombine. We shall see that again the value of $\omega_e \tau_e$ plays an important role.

The equation for the relaxation of nonequilibrium ionization is usually written

$$\frac{dn_e}{dt} = -\alpha n_e^2 \tag{11.25}$$

where α = recombination coefficient, cm³/sec. It may or may not be a function of n_e, pressure and temperature depending upon the atomic processes involved.

The rate at which an MHD generator extracts enthalpy from the gas is

$$\rho u \frac{dh}{dx} = \rho \frac{dh}{dt} = -\mathbf{j} \cdot \mathbf{E} = -\eta(1 - \eta) \frac{\Sigma}{G} u^2 B^2 \tag{11.26}$$

where the expression including nonuniformity and ion slip derived in Sec. 11.4.2 has been used.

Since we are most interested in relative changes, let us form logarithmic derivatives of the above.

$$\frac{d(\ln n_e)}{dt} = -\alpha n_e \tag{11.27}$$

$$\frac{d(\ln h)}{dt} = -\frac{\eta(1-\eta)\Sigma B^2}{\rho[\frac{1}{2} + 1/(\gamma-1)M^2]G} \tag{11.28}$$

where use is made of the fact that $h = u^2[\frac{1}{2} + 1/(\gamma-1)M^2]$. The ratio is [replacing Σ by $\sigma/(1 + \omega_e \tau_e \omega_i \tau_i)$]

$$\frac{d(\ln n_e)}{d(\ln h)} = \frac{\alpha n_e[\frac{1}{2} + 1/(\gamma-1)M^2]\rho G(1 + \omega_e \tau_e \omega_i \tau_i)}{\eta(1-\eta)\sigma B^2} \tag{11.29}$$

Inserting atomic quantities,

$$\frac{d(\ln n_e)}{d(\ln h)} = \frac{(\frac{1}{2} + 1/(\gamma-1)M^2}{\eta(1-\eta)} \frac{m}{m_e} \frac{\alpha}{Qc_e} \frac{G(1 + \omega_e \tau_e \omega_i \tau_i)}{\omega_e^2 \tau_e^2} \tag{11.30}$$

or, rearranging to give an expression for the required recombination coefficient,

$$\frac{\alpha}{Qc_e} = \frac{\eta(1-\eta)}{\frac{1}{2} + 1/(\gamma-1)M^2} \frac{m_e}{m} \frac{d(\ln n_e)}{d(\ln h)} \frac{W}{W_1} \tag{11.31}$$

where W/W_1 is the power-density ratio including the effects of nonuniformity and ion slip.

Now $\eta(1-\eta)$ will always nearly equal 0.25, the factor $\frac{1}{2} + 1/(\gamma-1)M^2$ will never differ very much from 1, and if for $h_f/h_i = \frac{3}{4}$ we allow $n_{e_f}/n_{e_i} = \frac{1}{10}$, then $d(\ln n_e)/d(\ln h) \approx 7.5$. Hence

$$\frac{\alpha}{Qc_e} \lesssim \frac{1 \times 10^{-3}}{A} \frac{W}{W_1} \tag{11.32}$$

where A = average atomic weight of the gas
or

$$\alpha \lesssim 3 \times 10^{-12} \frac{Q'}{A} \frac{W}{W_1} \tag{11.33}$$

where Q' is the electron-collision cross section in units of square angstroms (10^{-16} cm^2).

For helium, $Q' = 6$, and $A = 4$; so

$$\alpha(\text{cm}^3/\text{sec}) \gtrless 4.5 \times 10^{-12} \frac{W}{W_1}$$

For argon, $Q' = 0.6$,[32,33] and $A = 40$; so

$$\alpha(\text{cm}^3/\text{sec}) \gtrless 4.5 \times 10^{-14} \frac{W}{W_1} \qquad (11.34)$$

For cesium, $Q' = 200$, and $A = 133$; so

$$\alpha(\text{cm}^3/\text{sec}) \gtrless 4.5 \times 10^{-12} \frac{W}{W_1}$$

just as for helium.   The fact that the numerical coefficient for argon is much smaller than the others must be weighed against the fact that in argon the ratio $\omega_e \tau_e / \omega_i \tau_i$ is in general very large.   Hence, in principle, very large values of W/W_1 can be achieved.

Electron-ion recombination is a subject that has been studied for many years.   Unfortunately the conditions under which it has been studied have usually differed in important respects from the conditions which are apt to exist in an MHD generator.

Recently a theory of electron third-body recombination has been developed.[36b]   Several experiments at various places have exhibited rates in helium in apparent agreement with this theory; hence one has some confidence in using it to extrapolate to typical MHD generator conditions. If this is done, it appears that α might be as low as 10^{-10} cm^3/sec.   Referring to the expressions given above, one then finds that (in helium) this will be adequate if W/W_1 is greater than 20.   This value of W/W_1 has been achieved in our small-scale Hall-generator experiments (Ref. 62a). Hence, prospects for getting at least some modest (but significant) gains due to ionization frozen at the stagnation level seem reasonably good.

11.5 OVERALL DESIGN CONSIDERATIONS

One can draw some general conclusions about the size, weight, performance, and operating conditions for an MHD generator from an examination of the problems of heat transfer in the generator and dissipation in the field coil that supplies the magnetic field.   The processes of heat transfer and coil dissipation are in competition with the process of power generation.   It will be seen that the generation process can win this competition but that the conditions required for victory play a very important part in determining what the performance of MHD generators can be and what applications they are best suited for.   First of all, notice that the amount of power generated is proportional to the volume of the device, whereas the amount of heat transferred to the walls and also the amount of power dissipated in the field coils are, to at least a first approximation, proportional to the surface area of the device. This immediately leads one to expect that performance will improve as physical size or as rated power output is increased. Now let us try to make this statement more quantitative. Ideally, the power output of an MHD generator is

$$W = \eta(1 - \eta)\sigma u^2 B^2 V \tag{11.35}$$

where V = total volume of generator

Assuming that the flow is turbulent the heat-transfer loss to the walls will be given by

$$Q \approx 0.003 \rho u C_p \, \Delta T \, S \tag{11.36}$$

where ΔT = difference between gas temperature and wall temperature

S = total surface area of the device

Now let us make the optimistic assumption that the surface to volume ratio is equal to that for a sphere. Let us also eliminate the magnetic field strength B from these equations by use of the formula

$$\omega_e \tau_e = \frac{eB}{m_e} \frac{\lambda}{c_c} \equiv \theta \frac{B}{\rho/\rho_{\text{STP}}} \tag{11.37}$$

The justification for introducing the quantity θ is that under the conditions most likely to be encountered in an MHD generator it is a relatively slowly varying function of temperature, pressure, etc. Hence, for this rough analysis we can consider it to be a constant.

Now, if we combine the above two equations so as to eliminate S and V, we obtain the following equation for rated power output:

$$W = \left(\frac{0.014 \rho_{\text{STP}} C_p \, \Delta T}{Q/W}\right)^3 \frac{\theta^4}{\eta^2(1 - \eta)^2 \sigma^2 u \rho/\rho_{\text{STP}} \omega_e^4 \tau_e^4} \tag{11.38}$$

Representative numerical values are $\sigma = 50$ mho/m, $u = 2,500$ m/sec, $\Delta T = 1000°C$, $\theta = \frac{3}{10}$. Then, if we say that we want the wall heat-transfer loss to be no more than 10 percent of the power generated, we obtain

$$W \approx \frac{10^8}{(\rho/\rho_{\text{STP}})\omega_e^4 \tau_e^4} \quad \text{W} \tag{11.39}$$

If, now, in addition we specify that the density is 1 atm and that $\omega_e \tau_e = 1$, we conclude that the design power output must be 100 MW. One can see the possibility of a substantial reduction of this figure by, for example, increasing the density or increasing $\omega\tau$, but the conclusion is inescapable that the MHD generator is primarily a device for generating power on a large scale. Nevertheless it is interesting to note how strongly one's estimate of what a reasonable value of $\omega\tau$ is will affect one's estimate of what is a minimum useful value of design power output.   One should note also that the gas density is probably not something that one can choose arbitrarily.   For example, if the density is too high, then the heat transfer per unit area may become uncomfortably large, or if the density is too low, then such associated equipment as heat exchangers and compressors may become uncomfortably large.   Choice of the density also has an effect upon the geometric shape of the generator, which in a more detailed analysis would become an important consideration.

One should also observe that pressure losses due to wall friction will become prohibitive at just about the same time that heat loss does.   Therefore, one would arrive at roughly the same conclusion about generator size even if heat loss were eliminated by making the walls as hot as the gas.

11.5.1. The Magnetic Field

Let us assume that the generator channel is somehow fitted into a spherical volume of radius R.   This is not a very reasonable geometry for a generator, but for present purposes it will give order-of-magnitude results close enough to what is likely to be achieved in practice.   If a field coil is wound around this spherical volume with the current distributed uniformly in the axial direction and if the winding thickness is also uniform with respect to axial position, a crescent winding cross section results. One can then show that the magnetic field in the cavity will be uniform and given by the following formula,

$$B = \tfrac{2}{3}\mu_0 J \tag{11.40}$$

where J = current density per unit axial distance
The power dissipated in this coil will be

$$W_d = \frac{\pi}{4}\frac{J^2}{\sigma_s t} 4\pi R^2 \tag{11.41}$$

where σ_s is the electrical conductivity of the winding material and t is the thickness of the winding.   As before, the power generated will be

$$W = \eta(1 - \eta)\sigma u^2 B^2 V \tag{11.42}$$

where in this case the volume V is equal to $\tfrac{4}{3}\pi R^3$.

Now, to arrive at a minimum size for a generator, we have to make some assumption about the winding thickness t.   The assumption we shall make is that this winding thickness is, at most, equal to the radius of the interior cavity.   This can be justified on the basis of more detailed calculations, which show that, after the winding thickness has been made this large, a further increase in thickness, even to infinity, does not appreciably reduce the power required to produce a given magnetic field. Now, if we eliminate the dimension R between the above two equations, we arrive at the following equation for generator rated power output:

$$W = 1.55 \frac{B^2}{\sqrt{\eta(1-\eta)\sigma}\, u} \left(\frac{1}{\sigma_s\mu_0{}^2 W_d/W}\right)^{3/2} \qquad \text{W} \tag{11.43}$$

If we substitute into this formula the same numbers as we used in the estimate of the heat transfer, we arrive at the conclusion that the minimum rated power output for a generator in which the field-coil dissipation is no more than 10 percent of the power output is approximately 100 kW. Evidently field-coil considerations do not place as high a value on minimum rated power as heat transfer.   This is especially true in that there are a number of ways in which one may be able to reduce field-coil dissipation still further.   The first possibility involves cooling the conductors to some relatively low temperature.   For example, at the temperature of liquid oxygen, the electrical conductivity of copper would be approximately a factor of 10 higher than at room temperature.   The second and more exciting possibility is that one may be able to use superconductors to supply the magnetic field.   Recent progress in the development of superconducting material makes this possibility seem a very good one.

Another quantity of interest is the specific power output or power output per unit weight of an MHD generator.   Since an MHD generator consists essentially of nothing but a duct and a field coil, it seems not unreasonable to expect that the major portion of the weight of the generator will just be the weight of this coil.   For the spherical coil that we have been considering, the mass will be

$$M \approx \pi^2 R^2 t \rho_s \tag{11.44}$$

where ρ_s = density of conductor

Now let us use this equation to eliminate t from the previous equations instead of restricting t to be equal to R.   Then if we combine these equations to eliminate R we obtain

$$\frac{W}{M} \approx 3 \times 10^{-2}\mu_0{}^2 \frac{\sigma_s}{\rho_s} W^{2/3} B^{2/3} [\eta(1-\eta)\sigma u^2]^{4/3} \frac{W_d}{W} \tag{11.45}$$

Assuming the more or less typical values $\eta = 0.75$, $\sigma = 100$ mhos/m, and $u = 3{,}000$ m/sec and in addition $\sigma_s = 3.5 \times 10^7$ mhos/m, $\rho_s = 2.7 \times 10^3$ kg/m^3 (aluminum at 293°K), $B = 10$ kG, and a total coil dissipation not exceeding 1 percent of rated output (that is, $W_d/W = 0.01$), one obtains

$$\frac{W}{M} \approx 0.7 W^{2/3} \qquad \text{W/kg}$$

The latest modern utility plants put out about 10^9 W. Using this figure for W yields

$$\frac{W}{M} \approx 0.7 \times 10^6 \text{ W/kg or 700 kW/kg}$$

a figure which is of the order of the specific mechanical power output of a good chemical rocket engine! This leads one to speculate on the possible application of MHD generators in space, in spite of the fact that their tendency to large size seems to make applications in space seem unlikely. We shall return to this interesting question later.

If one makes a more careful calculation, assuming that the generator, instead of being round, has the more realistic length to diameter ratio of 10, and if one takes into account the geometric effects that occur when the coil winding thickness becomes comparable with the channel dimensions, a curve of specific power vs. rated power as shown in Fig. 11.13 results. Here, curve (1) assumes a coil at 293°K, curve (2) assumes a coil chilled sufficiently to give a factor of 10 increase in conductivity, and curve (3) assumes a coil of superconductor that carries 10^5 A/m^2, a value that seems not at all unreasonable in view of recent developments in the field. It is interesting to observe that the advantage of the superconductor tends to diminish as size increases, but it is, nevertheless, very great at all the more interesting power levels.

11.5.2. The Production of AC Power

In contrast to conventional rotating machinery, an MHD generator as sketched in Fig. 11.3 is essentially a homopolar device which does not naturally produce alternating current. Numerous ways of forcing it to do so can be conceived of. However, at the present time it seems as though the most simple and economical procedure is to convert from direct to alternating current by conventional techniques external to the generator. For example, the most obvious direct method would be to alternate the magnetic field. However, this is shown to be impractical by the following argument:

The power generated in unit volume of the duct is

$$W_g \approx \tfrac{1}{4}\sigma u^2 B^2 \tag{11.46}$$

The reactive power required to alternate unit volume of the field at frequency f is

$$W_B \approx 2\pi f \frac{B^2}{2\mu_0} \tag{11.47}$$

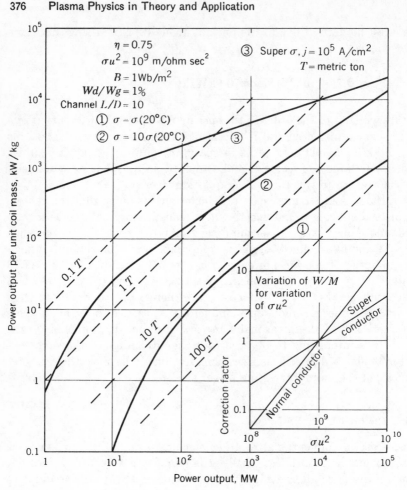

Fig. 11.13 Power output per unit coil mass vs. size or design output.

So for $f = 60$ sec^{-1}

$$\frac{W_B}{W_g} \approx \frac{750}{\mu_0 \sigma u^2} \tag{11.48}$$

For a coal-burning power plant the most likely values for σ and u are about 50 mhos/m and 1,000 m/sec; hence

$$\frac{W_B}{W_g} \approx 10 \tag{11.49}$$

In practice, because of fringing fields and fields within the winding itself, this ratio may actually approach 100. Now, although W_B is only "imaginary" power, the equipment required to handle it is very real.   It is technically possible to build the necessary high-Q coil and associated capacitive storage system but it appears to be much more difficult and expensive than external inversion.   These remarks apply to any MHD generator that employs an alternating field, including the various types of induction generator that have been proposed from time to time.   If a way could be found to increase the conductivity of gases by one or preferably two orders of magnitude, the situation would be more promising.

11.5.3. Losses in an MHD Generator and Development Problems

We have discussed two major sources of loss in an MHD generator: heat-transfer loss to the walls and Joule heat dissipation in the field coil.   We have also mentioned the small-stage efficiency which is to some extent at one's disposal, i.e., which can be varied by varying the value of load resistance presented to the generator, and we have seen how the Hall effect and effects arising from it influence these various parameters. There are several other possible loss mechanisms which we shall consider briefly.   The first of these is eddy-current loss.

In an MHD device the moving conductor is a three-dimensional continuum rather than a bundle of essentially one-dimensional wires. Therefore, eddy currents may occur whenever there is a rapid space varia-tion in electric or magnetic field strength or in gas velocity.   For the type of generator shown in Fig. 11.3 such variations can be largely avoided except at the entrance and exit.   These entering and leaving losses have been analyzed and calculated in Refs. 50 and 51.   Under typical condi-tions it appears that they will be roughly 10 percent of the entrance stagnation pressure.   This is not too serious in a typical power cycle. Moreover, in some cases it should be practical to put vanes in the entrance nozzle and diffuser to block the eddy-current circulation.   The implica-tion, however, is that, if one goes to channels with a very short L/D, eddy-current losses may become very serious.

One also expects aerodynamic losses such as occur in a conventional turbine to arise owing to viscous drag or the production of eddies.   Since an MHD generator has the possibility of being aerodynamically much cleaner than a turbine, such losses may be relatively small.   However, as noted above, Reynolds' analogy leads one to expect that a consideration of viscous drag would tend to set about the same limit on minimum genera-tor size as is set by heat losses.   In addition, if supersonic velocity seems desirable under some conditions, one will have to consider diffuser losses.

Another potential source of loss is in the space-charge sheath around electrodes, in particular the cathode.   However, one expects the electrode drop to be of the order of 10 V, and since typically the generator output voltage will be of the order of 1,000 V, the electrode drop represents a loss of only approximately 1 percent. Another source of loss may be associated with the tendency for the current at the cathode to narrow down to a small spot and thereby cause an appreciable resistive voltage drop in the gas in the neighborhood of the electrode. This again is not likely to be a major source of loss, but it may be something that one should keep in mind when trying for the ultimate in efficiency.

As in the development of most pieces of machinery, it is likely that, before an overall efficiency in the desirable range of 70 to 80 percent is attained, MHD generator development will go through a period of refinement and engineering compromise to minimize or eliminate the losses discussed above, plus others which will doubtless arise in practice. At the present time, experimental devices have been built that perform well enough to prove that the basic assumptions underlying their design are valid and that a functioning generating system could be built with the materials and techniques now at hand. Nevertheless, there is room for improvements of an engineering nature, and there are some basic unanswered questions. For instance, it is important to know how a generator using a monatomic gas will behave at high $\omega\tau$. Will the nonequilibrium effects that occur help or hinder the generator process?[45e,62a]

11.6   MHD GENERATORS FOR COMMERCIAL POWER GENERATION

For commercial power generation,[72] the MHD generator offers several conceptual advantages. First, the ability to withstand high gas temperatures implies the possibility for high thermodynamic efficiency. Second, basic simplicity and a lack of highly stressed, hot moving parts and close tolerances may eventually result in a plant that is cheaper, easier to maintain, and generally less complex than existing plants, although it must be admitted that the kind of MHD plants that we can see how to build today are not notably less complex than their conventional predecessors. Finally, the MHD generator appears to be uniquely well suited to the production of power on a large scale. Hence, with the growth of our electric power industry and the increase in the size of power plants being built, the advantages of the MHD generator over other mechanisms for power generation should become more and more marked.

The modern central-station steam electric plant has been developed to an extreme degree of engineering sophistication. With economic consideration always dominant, the efficiency of such plants has been pushed to 40 percent. While capital cost may somewhat alter the performance demanded of any new method, we may presume that efficiency considerably in excess of 40 percent will be necessary. The MHD concept offers such promise, with indicated cycle efficiencies for generators operating on combustion products as high as 56 percent.   Figure 11.14 shows an MHD power cycle developed by Avco in cooperation with the American Electric Power Company.   In this cycle the combustion products of fuel and

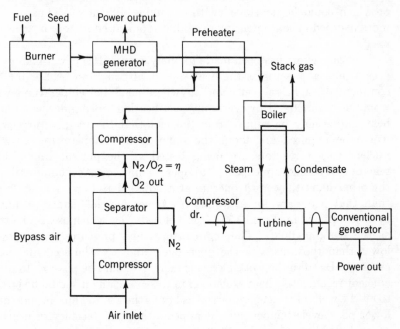

Fig. 11.14   A coal-burning commercial MHD power cycle using oxygen-enriched air.

oxygen-enriched air are passed first through an MHD generator and then into a conventional boiler.   Here part of the heat remaining after passage through the MHD generator is used to make steam that runs a more or less conventional steam plant.   The rest of the heat is used to preheat the incoming air and oxygen to the highest temperature that is practical with the materials at hand.   At present this temperature appears to be about 1100°K.   With this preheat temperature it is then necessary to enrich the incoming air so that the ratio between nitrogen and oxygen is approximately 2:1 in order to get a sufficiently high combustion temperature. If one could devise a preheater that could work up to about 2000°K, the use of oxygen enrichment would not be necessary (although it might still prove advantageous).   However, if one is to use the commercially important chemical fuels with their high concentration of ash and slag, preheating to greater than 1100 or possibly 1400°K does not seem possible at the present time.   In any case, cycle studies have shown that an MHD power plant with an attached oxygen-separating plant would have a capital cost competitive with existing power plants and a thermal efficiency of approximately 52 percent.   Perhaps the chief advantage of this oxygen cycle is that, except for the MHD generator itself, all the components of the plant are more or less standard items which are available at present or which could be developed without undue difficulty.   Thus one feels more justified in concentrating most development efforts on the generator itself.

Some effort has been made to analyze an MHD power plant which uses a nuclear reactor as the heat source, although no reactor built for commercial use has as yet come anywhere near the temperatures which would be required.   However, some justification for considering a nuclear heat source may be gained from the fact that the temperatures which would be required are about the same as the temperatures at which nuclear rocket reactors are being tested at the present time.   These reactors, of course, are expected to operate only for a few minutes, but on the other hand they must operate at a power density very much higher than that necessary for a ground-based commercial plant.   A nuclear cycle is outlined in Fig. 11.15.   It is a closed cycle which uses an inert gas as the working fluid.   The temperatures and pressures are somewhat lower than those used in the open cycle above.   The seed material is cesium.   Because the working gas is inert, it should be possible to put an efficient regenerative heat exchanger in this cycle, and it should be possible to let the walls of the generator run as hot as the gas.   Thus, in spite of the lower peak cycle temperature, it appears than an efficiency of nearly 60 percent could be attained.   Moreover, except for the nuclear reactor itself the materials problems in this cycle appear much less severe than those in the combustion cycle.   Thus it seems certain that, if a reactor suitable for commercial use at 2500°K could be built, a very attractive MHD power plant using it could be built.   A recent study of nuclear MHD plants is contained in Ref. 76.

11.6.1. Space-flight Applications[67-75]

It is conceivable that in addition to becoming the "prime mover" for the electric power industry, the MHD generator might become the prime mover for hypersonic and space flight.   This may seem, offhand, like a rather grandiose supposition, but as a matter of fact today's electric power plants and today's supersonic aircraft both use the same prime mover.   Hence the supposition has at least the moral support of history.

More concrete support derives from the possibilities, previously mentioned, that an MHD generator may be able to handle very high temperatures and produce very high power outputs per pound of weight. These are properties that seem to recommend it highly for space applica-

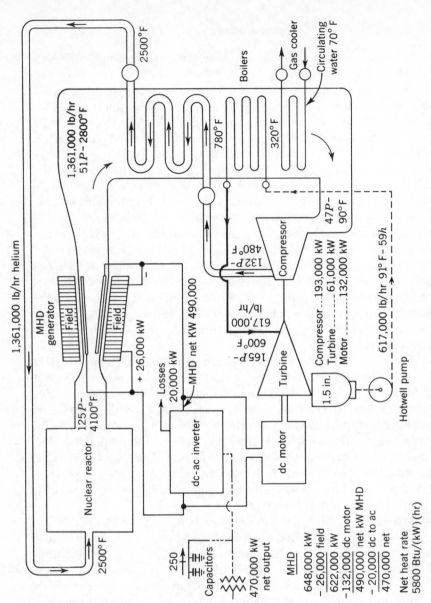

1,361,000 lb/hr helium

MHD generator

Field | Field

Nuclear reactor

125 P -
4100°F

2500° F

2500° F

1,361,000 lb/hr
51 P - 2800° F

Boilers

Gas cooler

Circulating
water 70° F

780° F

320° F

47 P -
90° F

Compressor

132 P -
480° F

165 P -
600° F

617,000
lb/hr

Turbine

1.5 in.

Hotwell pump

Compressor...193,000 kW
Turbine-------61,000 kW
Motor --------132,000 kW

617,000 lb/hr 91° F - 59h

dc motor

+ 26,000 kW

Losses
20,000 kW

MHD net KW 490,000

dc-ac inverter

250

Capacitors

470,000 kW
net output

MHD

648,000 kW
- 26,000 field
622,000 kW
-132,000 dc motor
490,000 net kW MHD
- 20,000 dc to ac
470,000 net

Net heat rate
5800 Btu/(kW)(hr)

Fig. 11.15 Conceptual design of a nuclear-MHD closed-cycle power plant. The working fluid is helium seeded with cesium.

tions, although the accompanying necessity for large size appears to be a drawback, at least at the present time.

The most obvious possibility for space application would be to use the generator in a nuclear closed cycle, much as was outlined above for commercial ground application, and to employ the output to drive a high specific-impulse device such as an ion rocket or an MHD accelerator. This is the MHD counterpart of more conventional systems that have been proposed for some time by Stulinger and others, for use where a very low thrust-to-weight ratio is tolerable. One finds that the design of such a system is largely controlled by the necessity for radiating the heat rejected by the cycle. With an MHD generator in the system, it should be possible to substantially increase the temperature of the radiator and hence gain a substantial reduction in the weight of the power system.

The foregoing represents a use of the MHD generator essentially as a source of electric power. However, as was mentioned earlier, this device is actually a better turbine than it is a generator. It seems to the author that equally interesting possibilities lie in using it in a manner more nearly analogous to the way the turbine is now used or has been proposed for use in flight.

Shown in Fig. 11.16 is a propulsion scheme that is a nuclear-MHD analog of the turborocket or to a lesser extent the turbo-fan-jet engine.[68,69] In addition to having in principle a much higher propulsive efficiency than a pure rocket when in an atmosphere, this scheme might employ advanced types of high-temperature nuclear reactor which cannot easily be used in any other way. These advanced concepts, such as the rotating fluidized bed and the gas-core reactor, suffer from the fact that they would in all probability lose fissionable material at a prohibitively high rate. However, if they were used in a scheme such as shown here, one would have an opportunity to attempt recovery of what escaped.

Operation of this type of power plant is conceived to be as follows: The propellant is introduced to the reactor under pressure and heated to a high temperature of perhaps 3 to 4000°K. It is then expanded and does work in an MHD generator, which reduces its temperature to about 2000°K. At this point the vapor pressure of uranium is low enough so that there is a reasonable chance that it can be recovered from the propellant stream. The propellant is then exhausted through a nozzle, and the electric power from the generator is used to drive an air-breathing accelerator of some sort.

This air-breathing accelerator might take the form of a simple ramjet. The ramjet designer would in this case have at his disposal a "fuel" (electric energy) that from a thermodynamic viewpoint is far superior to chemical fuels. Thus efficient ramjet performance at hypersonic velocity might be readily attainable.

For the cycle temperatures and pressures indicated in Fig. 11.16 this type of power plant might achieve truly spectacular performance as a satellite booster, expending no more propellant in relation to payload and vehicle weight than does a modern jet airliner on a transcontinental flight.

Energy Exchange by MHD from a Gas-core Reactor to Propellant

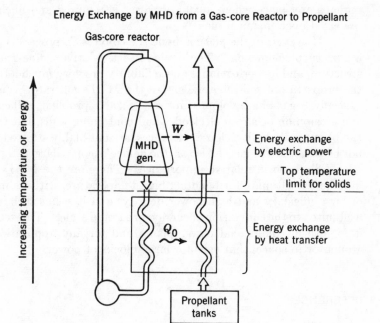

Fig. 11.16   A nuclear propulsion scheme which uses MHD to transfer energy and momentum between two gas streams when the temperature is too high for the use of solids.

11.7 CONCLUSION

The following is an attempt at an unbiased summary of the principal characteristics and fields of application for the plasma diode and MHD generator.   As the uneven development of these two devices proceeds, it is likely that, for some applications, first one and then the other will look best.   However, there are certain fundamental and rather large differences between the two which should eventually determine what each is primarily used for.

First an MHD scheme is not really a direct conversion system. It is likely to be large, to have a number of components, and to be neither particularly light in weight nor particularly efficient when operated at low power levels.   The plasma diode is completely static, its efficiency is pretty much independent of its size, and it shows promise of substantially higher specific power output than conventional methods of power generation. Also, it can reject heat at relatively high temperature, while at the same time, in at least some of its forms, it does not require as high a source temperature as does an MHD generator. Thus, at least at low power levels, it seems ideally suited for use in space, remote locations, or perhaps any place where auxiliary power is required and nuclear fuel is preferred over chemical fuel.

However, in the plasma diode the conversion process is essentially a surface phenomenon. The device has a relatively fine-grained solid structure, and its performance and reliability are very much a function of the properties of solid materials. In the MHD generator, conversion is basically a gas-phase volume process. Sizable problems of heat transfer and erosion do occur on electrodes and containing walls, but these parts can be ruggedly built to loose tolerances, are not highly stressed, and need not be as hot as the gas. Also, as the device becomes larger, these surface effects diminish in relative importance. Therefore the MHD generator may be the ultimate for handling high temperature, with its implication of high efficiency and high power density, and it seems to be especially well suited to the conversion of energy on a grand scale. This recommends it for use in commercial power plants and for any application, on the ground or in space, that calls for large blocks of power.

REFERENCES

Plasma Diode

1. Nottingham, W. E., G. N. Hatsopoulos, and E. N. Carabateas: *Aerospace Eng.*, **14** (July, 1961). Introductory discussion.
2. Ranken, W. A., G. M. Grover, and E. W. Salmi: *J. Appl. Phys.*, **31**, 2140 (December, 1960).
3. Kaye, J., and J. A. Welsh: "Direct Conversion of Heat to Electricity," Wiley, New York, 1960. A collection of articles by a number of workers in the field. Includes both theory and experimental results.
4. Wilson, V. C.: *J. Appl. Phys.*, **30**, 475 (1959).
5. Grattidge, W.: 1960 *IRE Conv. Record*, pt. 9. General discussion and work at General Electric.
6. Beller, W.: *Missiles and Rockets*, **8** (23), 22 (June 5, 1961).
7. Beller, W.: *Missiles and Rockets*, **9** (5), 24 (July 31, 1961). Describes recent low-temperature diode.
8. Stout, V. L.: In P. H. Egli (ed.), "Thermoelectricity," chap. 13, p. 207, Wiley, New York, 1960. Discusses emission properties of electrodes.
9. Rasor, N. S.: *IAS Paper* 61-72. Survey of experimental work on cesium diode.
10. Dobretsov, L. N.: Soviet Physics, *Tech. Phys.*, **5** (4), 343 (October, 1960).

After a brief statement—"we did it first"—he gives a good comprehensive review of largely American work in this field.

11. Morgulis, N. D.: Soviet Phys. *Uspekhi*, **3** (2), 251 (September–October, 1960). Review.

12. Ranken, W. A., and T. G. Frank: *Aerospace Eng.*, May, 1960, p. 58. Use in nuclear reactors for space applications.

12a. Bussard, R. W.: Some Considerations of Dynamic Behavior in the Plasma Thermocouple, *J. Appl. Phys.*, **33**, 606 (February, 1962).

Noble-gas Diodes

13. Jamerson, F. E.: *Rept. Twenty-first Ann. Conf. on Phys. Electronics*, M.I.T., Mar. 29–31, 1961, p. 121; see also Chap. 10, Ref. 14*b*.

14. Bernstein, W.: *Rept. Twenty-first Ann. Conf. on Phys. Electronics*, M.I.T., Mar. 29–31, 1961, p. 124.

15. Schultz, R. D.: *Rept. Twenty-first Ann. Conf. on Phys. Electronics*, M.I.T., Mar. 29–31, 1961, pp. 146, and 156.

16. Gabor, D.: *Nature*, **189**, 868 (1961).

17. Jablonski, F. E., C. B. Leffert, R. Silver, R. F. Hill, and D. H. Loughridge: *J. Appl. Phys.*, **30**, 2017 (1959).

18. Moss, H.: *Brit. J. Electronics*, **2**, 305 (1957).

19. Druyvesteyn, M. J., and F. M. Penning: *Revs. Modern Phys.*, **12**, 88 (1940).

Unipolar Arc

20. Robson, A. E., and P. C. Thonemann: *Proc. Phys. Soc.*, **73**, 508 (1959).

21. Waymouth, J. F.: Paper given at the Fourteenth Annual Gaseous Electronics Conference, Schenectady, N.Y., October, 1961. Sylvania Corp., Salem, Mass.

Electrical Conductivity—Hall Effects

22. Cowling, T. G.: "Magnetohydrodynamics," chap. 6, Interscience, New York, 1957.

23. Schlüter, A.: *Naturforsch. Z.*, **5A**, 72 (1950), **6A**, 73 (1951).

24. Spitzer, Jr., L.: "Physics of Fully Ionized Gases," Interscience, New York, 2d ed., 1962.

25. Kemp, N. H., and H. E. Petschek: *J. Fluid Mech.*, **4**, 553 (1958).

26. Ching-Sheng Wu: *Calif. Inst. Techn. Jet Propulsion Lab. Technical Rept.* 32-23, May, 1960. Generalized Ohm's law of plasmas.

27. Edwards, B. N.: *Univ. Calif.* AFCRC-TN-59-783, September, 1958. Study of Hall effect in glow discharge.

28. Harris, L. P., and J. D. Cobine: *ASME Preprint* 60-WA-329, November, 1960. Hall effect in different generator geometries.

29. Rosa, R. J.: The Hall and Ion Slip Effects in a Non-uniform Gas, *Avco-Everett Research Lab. Research Rept.* 121, December, 1961; *Phys. Fluids*, **5**, 1081 (September, 1962).

29a.   Sutton, G. W.: Hall Effect in a Lorentz Gas, *Phys. Fluids*, **4**, 1273 (October, 1961).

29b.   Shkarofsky, I. P.: Values of the Transport Coefficients in a Plasma for Any Degree of Ionization Based on a Maxwellian Distribution, *Can. J. Phys.*, **39**, 1619 (1961).

Data from Gas Discharge

30.   Bekefi, G., and S. C. Brown: *Phys. Rev.*, **112**, 159 (1958).   Collision cross section of H_2, obtained with microwave techniques..

31.   Phelps, A. V., J. L. Pack, and L. S. Frost: *Phys. Rev.*, **117**, 470 (1960).   Drift of electrons in He.

32.   Bowe, J. C.: *Phys. Rev.*, **117**, 1411, 1416 (1960).   Drift of electrons in N_2, He, Ne, A, K, and xenon.

33.   Pack, J. L., and A. V. Phelps: *Phys. Rev.*, **121**, 798 (1961).   Drift of electrons in He, Ne, A, H_2, and N_2.

34.   Salmona, A., and M. J. Seaton: *Proc. Phys. Soc.*, **77**, 617 (1961).   Electron collision with Na—theory.

35.   Moiseiwitsch, B. L.: *Proc. Phys. Soc.*, **77**, 721 (1961).   Electron collision with He—theory.

36.   Brown, S. C.: "Basic Data of Plasma Physics," Technology Press and Wiley, Cambridge, Mass., and New York, 1959; H. S. W. Massey and E. H. S. Burhop: "Electronic and Ionic Impact Phenomena," Oxford, 1962.

36a.   Bates (ed.), D. R.: "Atomic and Molecular Processes," Academic, New York, 1962.

36b.   McDaniel, E.: "Collision Phenomena in Ionized Gases," Wiley, New York, 1964.

37.   Mohler, F. L.: *Bur. Standards J. Research*, **21**, 873 (1938). σ measurements in low-pressure, high current-density cesium plasmas.

Thermal and Extrathermal Ionization.   Seeding

38.   Lin, S. C., E. L. Resler, and A. Kantrowitz: *J. Appl. Phys.*, **26**, 96 (1955).   Shock-tube experiments on argon.

39.   Basu, E.: *Phys. Fluids*, **3**, 456 (1959).   Seeded detonation waves.

40.   Rosa, R. J.: Ph.D. thesis, Cornell University, Ithaca, N.Y., 1956.   Seeding —theory, nonequilibrium ionization.

41.   Meyer, R. X.: *Space Technology Lab. Rept.* GM-TR-0127-0042, June, 1958.   Calculated, air to 24,000°K.

42.   Dibelius, N. R., E. A. Luebke, and G. Mullaney: *Proc. Second Symposium on Eng. Magnetohydrodynamics*, March, 1961.

43.   Gaydon, A. G., and H. G. Wolfhard: "Flames," chap. 13, Chapman & Hall, London, 1953.   Review of early studies of ionization in fuel—air flames.

44.   Eschenroeder, A. G., and J. W. Daiber: *ARS Preprint* 1458-60, December, 1960.   Ionization nonequilibrium in expanding gas flow.

45.   Kerrebrock, J. L.: *Proc. Second Symposium on Engineering Magneto-hydrodynamics*, March, 1961. Experimental study of nonequilibrium in seeded gas flow.

45*a*.   Robben, F.: *Phys. Fluids*, **5**, 1308 (1962).

45*b*.   Ben Daniel, D., and S. Tamor: *Phys. Fluids*, **5**, 500 (1962); *Gen. Elec. Rept.* 62 RL-2922 E, 1962.

45*c*.   Ben Daniel, D., and C. Bishop: *Phys. Fluids*, **6**, 300 (1963).

45*d*.   Rosa, R. J.: Nonequilibrium Ionization in MHD Generators, *Avco-Everett Research Lab. Rept.* AMP 108, May, 1963; *Proc. IEEE*, **51** (5), 744–784 (May, 1963).

45*e*.   Kerrebrock, J. L.: Segmented Electrode Losses in MHD Generators with Nonequilibrium Ionization, *Avco-Everett Research Lab. Research Rept.* 201, January, 1965.

45*f*.   Cool, T. A., and E. E. Zukoski: *Phys. Fluids*, **9**, 780 (1966).

MHD Power Generation—MHD Channel Flow

46.   Faraday, M.: Experimental Researches in Electricity, reprinted from *Phil. Trans.*, 2d ed., **1**, 38, 55, 56 (1831–1838), R. and J. E. Taylor, London, 1849.

47.   Rosa, R. J.: *Avco-Everett Research Note* 132, June, 1959.

48.   Rosa, R. J.: *J. Appl. Phys.*, **31**, 735 (1960). Generator experiment.

49.   Rosa, R. J.: *Phys. Fluids*, **4**, 182 (1960). Generator design consideration.

50.   Fishman, F.: End Effects in MHD Channel Flow, *Avco-Everett Research Lab. Rept.* 78, June, 1959.

51.   Sutton, G. W.: *Gen. Elec. Rept.* TISR595D431. July, 1959. End effects in MHD channels.

52.   Way, S.: *Proc. Second Symposium on Eng. Aspects of Magnetohydro-dynamics.* To be published. Small-combustion-generator experiment.

53.   *Proc. Symposium on Eng. Aspects of Magnetohydrodynamics.* T. Brogan, A. Kantrowitz, R. Rosa, and Z. J. Stekly: *Second Symposium*, March, 1961; T. R. Brogan, J. F. Louis, R. J. Rosa, and Z. J. Stekly: *Third Symposium*, March, 1962; J. F. Louis, J. Lothrop, and T. R. Brogan: *Fourth Symposium*, April, 1964; J. F. Louis, G. Gal., and P. R. Blackburn: *Fifth Symposium*, April, 1964; J. Klepeis and R. J. Rosa: *Fifth Symposium*, April, 1964. This series of papers describes the yearly progress in the joint Avco-Utilities project to develop MHD for commercial power.

54.   Coerdt, J., W. C. Davis, R. L. Graig, and J. E. McCune: *ARS Preprint* 1319-60, September, 1960. The vortex geometry.

55.   Karlowitz, B., et al.: U.S. patent 2,210,918, Aug. 13, 1940.

56.   Hains, F. D.: *Boeing Sci. Rept.* DI-82-0047, May, 1960. Viscosity and heat conduction in MHD channel.

57.   Fay, J. A.: Hall Effects in a Boundary Layer, *Avco-Everett Research Lab. Rept.* 81, 1959.

58.   Podolsky, B., and A. Sherman: *Appl. Sci. Rev.*, **B9**, 77 (1961). MHD channel flow.

59.   Elsasser, W. M.: *Phys. Rev.*, **95**, 1, 1954. Magnetic Reynolds number.

60.   Shercliff, J. A.: *J. Fluid Mech.*, **9**, 481 (1960). Flow in oblique fields.

61. Hurwitz, H., R. W. Kilb, and G. W. Sutton: *J. Appl. Phys.*, **32**, 205 (February, 1961). Effects of tensor conductivity.

62. Pain, H. J., and P. R. Smy: *J. Fluid Mech.*, **10**, 51 (February, 1961). Power generation in argon shock tube.

62a. Klepeis, J., and R. J. Rosa: Experimental Studies of Strong Hall Effects and $V \times B$ Induced Ionization, *AIAA J.*, **3**, 1659 (September, 1965). Describes experiments at high $\omega\tau \approx 10$ in segmented and disk Hall generators.

63. Deissler, R. G.: *NASA* TND-680, March, 1961. Generator-channel flow analysis.

64. Foshag, F. C., and A. E. Were: *Gen. Elec. Rept.* R59SD447. Generator experiment with arc tunnel.

65. Patrick, R. M., and T. R. Brogan: *J. Fluid Mech.*, **5**, 289 (1959). Argon shock tube—quantitative studies.

66. Brogan, T.: Master's thesis, Cornell University, Ithaca, N.Y., 1956; see also Refs. 23–25, 38. Power generation in argon shock tube.

Applications for MHD Generators and Related Devices

67. Rosa, R. J.: Advanced Propulsion Systems, in M. Alperin and G. P. Sutton (eds.), *Proc. Symposium*, Los Angeles, Calif., December, 1957, p. 175.

68. Rosa, R. J.: *ARS J.*, **31**, 884 (July, 1961).

69. Rosa, R. J.: Application of MHD Generators to Rocket Propulsion, *Avco-Everett Research Lab. Rept.* 111, August, 1961.

70. Lewellen, W. S., and W. R. Grabowsky: *ARS Preprint* 1738-61, May, 1961. Closed-cycle MHD space power plants—vortex generator.

71. Kunkel, W. B.: *Space Technology Lab. Rept.* TR 59-0000-00608, February, 1959. MHD rocket-nozzle vectoring.

72. Kantrowitz, A., and P. Sporn: *Power*, **103**, 62 (November, 1959). Application of MHD to commercial power generation.

73. Steg, L., and G. W. Sutton: *Astronautics*, **5**, 22 (1960). Application of MHD to commercial power generation.

74. Huth, J. H.: *ARS Preprint* 1147-60, May, 1960. MHD power from rocket engines.

75. Rosa, R. J.: Nuclear Fueled MHD Power Plants, *Avco-Everett Research Lab. Rept.* 87; 1960 *IRE Intern. Conv. Record*, vol. 8, pt. 9, p. 72.

76. Cooper, R. S., and L. A. Booth: Nuclear Reactor Magnetohydrodynamic Power Generator, *Los Alamos Scientific Laboratory Report* LA-3368, Jan. 20, 1966.

12

IONIC AND PLASMA PROPULSION
FOR SPACE VEHICLES

KNECHTLI, BREWER, and CURRIE

RONALD C. KNECHTLI, *Head, Plasma Physics Department, Hughes Research Laboratories, Malibu, California*

GEORGE R. BREWER, *Manager, Ion Propulsion Department, Hughes Research Laboratories, Malibu, California*

MALCOLM R. CURRIE, *Vice-president and Director, Electron Dynamics Laboratory, Hughes Research Laboratories, Malibu, California*

12

12.1 INTRODUCTION

The subject of electrical propulsion represents an exciting new area of applied science and technology. Electrically propelled spacecrafts provide the only feasible means for exploring the further reaches of our solar system, for economically transporting large manned expeditions and heavy payloads to our neighboring planets, and for probing the solar system away from the ecliptic plane.

It is inevitable that vast new concepts such as this build on fundamental knowledge and experience in a variety of fields and, in turn, provide strong motivation for vigorous new work and progress in these areas. This is certainly the case here. From an overall viewpoint a nuclear-electric propulsion system consists of three major components—a lightweight nuclear reactor, an efficient method for converting thermal energy to electrical energy, and the thrust unit itself. The first two of these are giving strong impetus to fundamental and applied work in the areas of new types of high-temperature reactors and methods of direct energy conversion such as thermionic and MHD devices.

The third major component, the electrical thrust unit, is the subject

---

This chapter is based on a review paper published under the same title in *Proc. IRE*, **49**, 1789 (1961), and revised and brought up to date by the authors.

of this chapter. As will be shown, it builds directly on the broad field of electronics concerned with plasma physics and devices, beam and particle dynamics, electron optics, and electron tube technology. It is the purpose of this paper to discuss electrical propulsion devices from this point of view. Rather than present a detailed picture of the increasing number of specific engineering developments in this field, it is intended that the discussion relate the general directions of current efforts to basic problems in plasma physics, ion-beam generation and control, and particle dynamics. We take the point of view that the rapid development and future success of electrical propulsion will depend upon continued fundamental research in these areas and that, conversely, progress in electrical propulsion will have a significant effect on these basic areas in terms of new understandings and new tools for attacking problems of fundamental importance.

The concept of the ion rocket engine has existed for many years. In 1906, R. H. Goddard mentioned it in his notebook,[1] and in 1929, H. Oberth devoted a chapter of his book "Wege zur Raumschiffahrt,"[2] to the concept. However, active research and development effort has taken place only during the last few years. Although a large and growing body of literature on the subject exists, no attempt at completeness of reference is made here.

The importance of electrical propulsion can be seen from the following basic relations. The equation of motion in free space for a vehicle whose mass M and velocity v are functions of time is

$$M \, dv = -v_{ex} \, dM \tag{12.1}$$

For an initial mass M_i starting from rest, Eq. (12.1) gives

$$v = v_{ex} \ln \frac{M_i}{M} \tag{12.2}$$

Thus, the effectiveness of mass expenditure to gain velocity is a function of exhaust velocity; i.e., from a mass point of view, a given velocity increment is achieved more efficiently with a system having high exhaust velocity.

With low-exhaust-velocity chemical rockets, the mass loss is very high. These systems, of course, have high thrust and are needed for overcoming gravitational forces and injecting payloads into orbit. However, for many projected space missions starting from an earth orbit, the tremendous mass of a chemical or even a nuclear rocket, which is necessary to deliver a heavy payload, is completely unfeasible. The exhaust velocity of a chemical rocket is limited to values below about 4,000 m/sec; the primary limitation is imposed by the maximum reaction temperature and minimum molecular weight of the fuel. By choosing hydrogen for the propellant, the direct nuclear heat-transfer rocket is expected to increase the exhaust velocity to perhaps 8,000 m/sec; here, the temperature limitations of the material establish an upper limit. Because of their high thrust, chemical and nuclear rockets will be important for near-earth missions. However, many interplanetary and deep-space scientific missions requiring heavy payloads are impossible to accomplish with foreseeable chemical or nuclear vehicles. Low-thrust propulsion systems operating over long periods of time and having very high exhaust velocity are mandatory in such cases.

It is clear that the acceleration of ions to any desired velocity by electrostatic or electromagnetic fields is unrelated to thermal heating; i.e., in the former process, the exhaust velocity can be adjusted to any desired value. The power required to accelerate the particles is proportional to the product of thrust and velocity. The weight of the nuclear-electric power supply increases roughly in proportion to power. Therefore, depending upon the mission, there exists an optimum exhaust velocity or specific impulse which results in minimum overall weight (i.e., compromise between power-system weight and propellant weight). This optimum exhaust velocity is larger for longer missions and heavier payloads but is invariably above several tens of thousands of meters per second, varying from roughly 50,000 m/sec for a 5,000-lb payload in a Mars orbit using the Atlas-Centaur booster and a 60-kW power supply to more than 100,000 m/sec for a 15-ton terminal mass in a Jupiter orbit using Saturn and a 1-MW power supply. Thus, nuclear-electric propulsion systems constitute an ideal solution for the types of missions discussed above.

Many detailed calculations have shown that, for a given overall initial weight and, for example, for a round-trip mission to and from a Mars parking orbit, payloads of up to four or more times those possible with direct nuclear rockets or, conversely, much shorter mission durations with the same payload weight will result. For more ambitious space experiments (e.g., to Jupiter) the figures are even more impressive. Often a factor of 10 increase in payload or a considerable reduction in initial vehicle weight can be attained. It should be noted that these calculations include the nuclear-electric power supply as part of the propulsion system. However, the use of the power supply for wideband data transmission and for primary power for other purposes at the destination makes the effective payload-weight ratios between electric and nuclear systems even higher.

There are several varying approaches to the electrical generation of thrust. These are usually classified as *electrothermal*, *electrostatic*, and *electromagnetic*. All have in common the requirements for very long life, high power efficiency, and high propellant-utilization efficiency.

Electrothermal devices include resisto-jets, arc jets, and magnetic-nozzle devices. Resisto-jets and arc jets are not true electric thrust units in the sense of overcoming material temperature limitations and thus permitting very high exhaust velocities.   In these the propellant gas (hydrogen) is heated to high temperatures either directly by a high current arc (arc jet) or indirectly by a resistive heating element (resisto-jet).   In either case the gas is then expanded through a nozzle where the thermal energy is converted to directed kinetic energy.   It is limited to exhaust velocities in the neighborhood of 10,000 m/sec, the life decreasing very rapidly beyond that point because of excessive heating of the nozzle material.   This latter limitation, however, is avoided in magnetic-nozzle devices.   Magnetic-nozzle devices use relatively low pressure discharges (typically, less than 1 torr) in which the plasma is heated electrically and partially isolated from the anode or nozzle by a magnetic field.   The thermal energy of the plasma is converted into directed kinetic energy by expansion through the properly shaped magnetic field, which thus acts as a magnetic nozzle.   The plasma heating can be produced either by direct current in a low-pressure dc crossed-field discharge, or by radio frequency through electron cyclotron resonance heating.   While magnetic-nozzle devices constitute a promising class of electrothermal devices and have recently received increased attention, their discussion will be omitted from this chapter.   Only electrostatic and electromagnetic devices will be considered here.

The electrostatic ion engine appears to offer the possibility of efficiently achieving higher exhaust velocities than the other types, i.e., in the range of several tens of thousands of meters per second and higher. Electrostatic ion propulsion is very simple in concept, since it involves the imparting of energy to ions by electrostatic acceleration.   To serve as a reference, Fig. 12.1 shows exhaust velocity as a function of net accelerating voltage for some of the common propellants.   However, in transforming this essentially simple idea to practical fruition, there are a number of problems involving fundamental considerations in plasma physics and electronics.   First, an efficient ion source is required.   A number of such sources are possible: contact ionization, Penning discharge, magnetically stabilized arcs, and various combinations of these.   All are the subject of intensive research efforts, the results of which will add significantly to the technology of this and related fields in physics and electronics. Second, sophisticated high-perveance ion-optical acceleration and focusing systems are mandatory, with a perfection exceeding even that required by electron guns and beams for use in high-power microwave tubes.   Third, space-charge-neutralization mechanisms are required which introduce intriguing new concepts such as synthetic nonthermal plasmas and the basic questions relating to them.

The other major class of electrical propulsion engines involves the direct acceleration of plasmas by magnetic fields.   Again, a number of

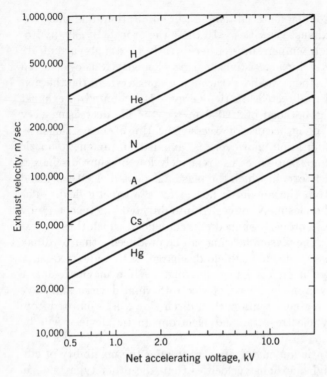

Fig. 12.1   Curves showing exhaust velocity as a function of
net accelerating voltage for various propellant
gases (singly ionized) in electrostatic ion engine.

mechanisms for acceleration are possible, including the use of dc crossed
electric and magnetic fields (MHD types), traveling-wave systems, and
quasistationary rf systems.   These involve problems which are fundamental to other fields (e.g., energy conversion and high-temperature
plasma generation and containment) as well as to the attainment of high-
efficiency thrust devices.

12.2 BASIC PROBLEMS

In the process of developing electric propulsion systems, there are several
technical objectives and problems which are common to the various types
of systems outlined above.   Obviously, the primary objective involves the
devising of field configurations (electric and/or magnetic) capable of
accelerating charged particles in a controlled manner to the desired exhaust
velocity.

One of the primary considerations in the invention of propulsion
systems is efficiency—efficiency in the use both of electrical power and of
propellant.

The second major consideration is neutralization of the charges associated with the particles of the exhaust beam. The class of electric engines in which the acceleration of the charged particles results from magnetic fields will presumably exhaust a neutral beam; i.e., the accelerated and exhausted medium will be very close to a true plasma. In the case of electrostatic rockets, on the other hand, the acceleration of the charged particles is accomplished by means of electric fields; therefore only one species can be accelerated in a given region. In this case, the charges must be separated. In order to prevent buildup of space-charge forces at the engine exhaust, the accelerated charges must be neutralized by the injection of particles of the opposite sign. (We do mean, not that the positive and negative particles must recombine, but merely that in any given region of the exhaust beam there must be an equal number of positive and negative charges.)

Since the electrostatic rocket imparts acceleration to the positive and negative charges separately, some mechanism must be provided to ionize the neutral propellant particles. As we shall see, the ionization mechanisms used in the engines to be considered are contact ionization of cesium on tungsten and ionization by electron impact.

Another fundamental limitation is the erosion of electrodes involved in the acceleration of the charge particles, due to impingement of the charged particles on these electrodes. The electric propulsion engines that we are describing here will be used for space vehicles in interplanetary travel, and the mission times involved are measured in years. Therefore, a serious constraint on the design of the engines is that the total erosion of the electrodes must be small enough so that it does not interfere with engine operation over lifetimes of tens of thousands of hours. This imposes, for example, the condition of very careful control over the ion optical characteristics of the accelerator structure so that the impingement of the accelerated ions on any of the surrounding electrodes may be avoided. In the electrostatic rocket, a maximum of about one ion can be allowed to strike the accelerating electrode for each 1,000 ions emitted from the ionizer, i.e., an interception ratio of about 10^{-3}.

In the rest of this chapter, we shall discuss the several types of electric propulsion engines from the point of view of these basic objectives and attempt to show how these fundamental problems are being approached.

12.3 ELECTROSTATIC ACCELERATION

Electrostatic acceleration is characterized by the separate acceleration by electrostatic fields of the positive and negative charges; i.e., ions are created by some ionization process prior to acceleration. In the cesium-type engine this ionization occurs at a hot tungsten surface; in the electron bombardment engine the ionization mechanism is electron impact.   This class of engines also has the common requirement for neutralization of the exhaust ion beam by the injection of charged particles of the opposite polarity.   First we shall describe this common problem of neutralization and then discuss several of the electrostatic engines being investigated today.

12.3.1. Neutralization

As mentioned earlier, in the electrostatic rocket the ionized propellant particles are accelerated separately from the negative particles, after which electrons are injected to neutralize the positively charged beam. While the real test of the requirement for and the effectiveness of neutralization must await a space flight, some demonstrations of neutralization in practical engines have been made in laboratory tests.   We shall present here a brief picture of the neutralization problem; it will be clear, however, that many important questions remain unanswered.   In addition to its obvious application to electrical propulsion, a neutralized and directed beam of ions represents a new and novel form of plasma with many interesting properties and with potential uses so far almost completely unexplored.

There are two types of neutralization which must be clearly differentiated:

1.   Current neutralization.   A positive ion beam which is ejected from an electrically isolated space vehicle will produce on this vehicle a net negative charge.   This charge will build up very quickly to such a magnitude that all the ions will be pulled back.   It is necessary, therefore, to eject an equal current of negative charges along with the positive ions. This step is essential in order to maintain the entire space vehicle electrically neutral.   In principle, we could accomplish this current neutralization merely by ejecting negative particles (e.g., electrons) from any part of the vehicle, and they would ultimately be attracted toward the cloud of positive ions.

2.   Space-charge neutralization.   The question of the need for detailed neutralization of the space charge in the ejected ion beam still is a controversial problem.   The following discussion will be limited to the need for space-charge neutralization, since this is the aspect of the neutralization problem requiring more discussion and study at present; it is tacitly assumed here that the requirement of current neutralization is met exactly.   Since this neutralized ion beam can be thought of as possessing many of the properties of a very energetic nonthermalized synthetic plasma, with interesting possible applications other than propulsion, we

shall treat this subject in a fair amount of detail and include some of the experiments which have been performed.

In order to obtain charge neutrality of the ion beam, electrons must be injected either into or close to the ion beam.

The fundamental purpose of neutralizing an ion beam, of course, is to eliminate, or to reduce to the point of ineffectiveness, any electric fields in the vicinity of this beam which can diminish the thrust obtainable from the ion engine. Since the thrust is proportional to the product of the beam current and the ejected velocity of the propellant, any fields which act to reduce either the beam current or the ejected velocity will adversely affect the thrust; mere transverse expansion of the beam after leaving the engine will not produce such a counterthrust.

At this point we can distinguish two broad classes of ion beams, giving rise to different neutralization problems:

1. The broad beam. This class of beam corresponds to an aspect ratio† considerably greater than unity; i.e., the area of the beam is considerably greater than the square of the spacing between the ionizer and the accelerating electrode. This beam in idealized form can be thought of as originating in a large accelerator structure with a grid over the accelerator electrode aperture. Because of the shape of this beam, the electric-flux lines will be predominantly axial, i.e., from the positive charges in the beam to negative charges on the outer surface of the accelerator electrode (see Fig. 12.2a). These axial-field lines will act to reduce the velocity of the ejected particles and therefore to diminish the thrust obtainable. In fact, if the proper boundary conditions are forced along the outer edge of the beam (to ensure that all the flux lines will be truly axial in direction), the ions will be stopped and will form a "virtual cathode" at a distance from the accelerator electrode equal to the accelerator ionizer spacing.[3] A close analogy can be drawn between this "one-dimensional" flow of ions and the flow of charged particles between planar grid surfaces, as treated by Fay, Samuel, and Shockley.[4]

This "broad" type of ion beam obviously exhibits attractive advantages, in that it provides relatively high values of thrust density (thrust per unit of ion source area); however, if unneutralized, the strong axial fields can reduce the exhaust velocity of the propellant and/or, in the extreme case, produce "turnaround"[5] so that particles are returned to the accelerator, thus reducing the exhaust current.

† It is well known that the perveance of an electron- or ion-gun structure is dependent only on a dimensionless factor relating to the geometrical form of this structure. Specifically, it is the area of the emitter A_i divided by the square of the emitter-anode spacing d (for a plane-parallel system); the aspect ratio R is defined as $R = \sqrt{4A_i/d^2}$.

2. The "thin" beam. A "thin" beam is one for which the transverse dimension at a point near the engine is small compared with the acceleration distance (that is, $R \ll 1$). The electric-flux lines associated with the charges of this beam will be predominantly transverse to the beam boundary (see Fig. 12.2b) so that the space charge will result in a transverse beam spread rather than in a decrease in axial velocity of the particles, as in the broad beam. A thin beam can travel long distances from the engine unneutralized. This beam has the severe disadvantage, however, of very low thrust density and is therefore impractical for space propulsion.

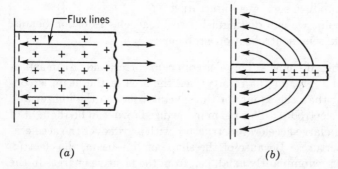

(a)   (b)

Fig. 12.2 Figures showing the effect of the shape factor of the ion beam on the electric field. (a) Schematic configuration: broad beam; (b) shape of the flux lines in the case of a thin-strip, or pencil, beam.

Most practical electrostatic ion engines will in fact consist neither of thin nor of thick beams. They will consist of clusters of beams whose aspect ratio may be of the order of unity. The behavior of an individual beam with an aspect ratio of the order of unity is intermediate between that of a thin and that of a thick beam (i.e., both beam spreading and turnaround can occur to some extent). A cluster of such beams, however, is expected to behave as a thick beam, with the same space-charge neutralization problem as the thick beam.

The broad or the clustered types of beams, in which neutralization appears to be a definite requirement, will be discussed below. But first let us try to provide a more quantitative justification for the need for charge neutralization in an ejected ion beam by means of the following examples:

1. One serious consequence of the lack of neutralization will be the transverse spreading of the beam due to space-charge forces. A typical

cesium ion beam of current density 10 mA/cm² at an exhaust potential of 10 kV and with an initial diameter of 9 cm will double in diameter in a distance of less than 2 cm.   This spreading is so rapid as to invalidate the simple beam-spread theory used, but the example does serve to illustrate the seriousness of the spreading problem.   While such spreading may not reduce the engine thrust, it must be considered in the engine design in order to avoid any interception, etc.

2.   If the same beam is considered as in one-dimensional flow, with entirely axial electric fields, the velocity of the ions will be reduced to zero owing to the space-charge fields, in a distance of only 0.7 cm!   This turnaround will return particles to the engine and will obviously reduce or even eliminate the thrust.

We shall assume now that neutralizing electrons are injected either into or in close proximity to the ion beam.   The next question is: How intimately should the electrons and ions be mixed?   Since the currents in the electron and ion beams must be equal and opposite, we can write

$$I_i = q_i v_i = -I_e = -q_e \hat{v}_e$$

where q = charge per unit length of electrons or ions
$\quad v$ = velocity
$\quad \hat{v}$ = average velocity
If the electrons are mixed intimately so that the current density j in the ion and electron beams is equal, then

$$j_i = \rho_i v_i = -j_e = -\rho_e \hat{v}_e$$

where ρ = charge per unit volume
We really do not know as yet whether the neutralization must be effected on a microscopic scale, with equal values of ion and electron charge per unit volume, or merely with equal charge per unit length.   The choice will depend on the seriousness of possible instabilities which may be excited by less than perfect neutralization.

In any event, these equalities require in turn that the ion velocity be equal to the electron velocity.   The next question to be resolved, then, is whether equality of velocity must be true on an absolute basis, with essentially all the electrons moving with the velocity of, and in the same direction as, the ions, or whether it is sufficient to require that the average velocity of the electrons be equal to, and in the direction of, the ion motion. The latter possibility would allow the existence of double-streaming electrons, i.e., electrons moving both away from and toward the engine with a net average drift velocity away from the engine.   One can understand how this situation might arise by thinking of the neutralized ion beam as a "plasma bottle"[6] (see Fig. 12.3), with the electrons moving around inside and being reflected at the edges and at the end by an electron sheath, thus giving rise to both transverse and axial components of electron velocity. If the electrons cannot be returned from the end of the plasma column, i.e., if they are forever lost from the system, then the neutralization requirement means that the electron and ion velocities must be equal on an instantaneous basis.   For most reasonable values of specific impulse, the ejection

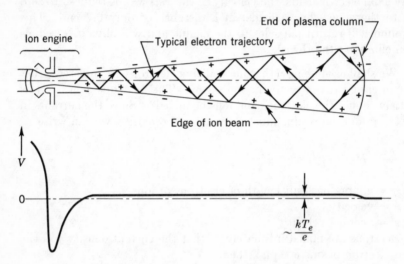

Fig. 12.3   Schematic representation of the plasma-bottle concept of a
neutralized ion beam.   The plasma sheath at the end and edges
and the potential hill at the engine end will confine the electrons
to the interior of this bottle.   These reflections can give rise to
a counterstreaming effect as shown.   The potential in the
plasma column will be uniform at a value of a few times kT_e/e
above the space potential.

velocity of the ions is considerably less than the mean thermal velocity of the electrons emitted from the thermionic emitter.   Therefore, if the electron and ion velocities must be equal on an instantaneous basis, some kind of interaction mechanism must be used to reduce the mean velocity of the electrons to that of the ions.   The term *mean velocity* denotes here the average over the velocity distribution of the electrons, but the electrons all have the same direction: away from the engine.

In the counterstreaming case, where the electrons are contained within a plasma bottle, we see that there are two possibilities.   First, this mixture could result in a truly neutral plasma with the net volume charge density equal to zero throughout the volume of the bottle.   This condition obviously satisfies all the requirements of neutralization and, if stable, forms the ideal situation.   The second possibility is that the beam is neutral on the average only; i.e., there can be small volumes throughout the bottle in which the net charge is not zero, though by integrating over the entire bottle the total net charge will be equal to zero.   The local fields resulting from the localized regions of nonzero charge density can give rise to instabilities and may cause the electrons either to return to the engine or to scatter transversely.   In general, it is believed that this type of plasma will probably be satisfactory if the instabilities are of such a nature that they do not grow indefinitely in amplitude with either time or distance.

Let us discuss the plasma-bottle concept in a little more detail, with the object of showing that in this model charge neutralization can occur in either a laboratory or a space situation and under conditions where the injected axial electron velocity exceeds the ion velocity.   Consider that the ejected "plasma" consists of three particle streams: (1) the outgoing ion stream of current density j_i, velocity v_i, and an electron gas with a velocity distribution which may be considered to be made up of two counterstreaming components; (2) the outgoing electron stream of current density j_e' and average stream velocity v_e'; and (3) a returning electron stream of current density j_e'' and average stream velocity v_e''. The outgoing electrons are reflected by the sheath at the end of the plasma bottle (considered, for the moment, to be of finite extent) so that the reflected electron velocity is given (under the assumption of ion mass much larger than electron mass) by

$$v_e'' = -v_e' + 2v_i \tag{12.3}$$

which can be seen easily by transforming to a coordinate system moving with the end of the plasma sheath, i.e., at velocity v_i.

We have assumed that the total electron and ion currents are equal. There is no loss of generality in assuming that the current densities are also equal; which must certainly be true on the average over the beam cross section.   Then

$$j_i = \rho_i v_i = -(j_e' + j_e'') = -(\rho_e' v_e' + \rho_e'' v_e'') \tag{12.4}$$

We can define an average electron velocity $\hat{v}_e$ as

$$j_e' + j_e'' \equiv \hat{j}_{et} = \hat{v}_e(\rho_e' + \rho_e'') \tag{12.5}$$

where $\hat{j}_{et}$ is the total net electron current density, which has been assumed equal to the ion current density.   By combining (12.4) and (12.5), we find that

$$\hat{v}_e = -\frac{\rho_i v_i}{\rho_e' + \rho_e''} \tag{12.6}$$

The condition for the ion beam to be charge-neutralized is that

$$-\rho_i = \rho_e' + \rho_e'' \tag{12.7}$$

i.e., that the net charge density be zero.   Therefore, under the neutralized condition we see from (12.6) and (12.7) that

$$\hat{v}_e = v_i \tag{12.8}$$

and this condition applies independent of the velocity of the incident electrons, v_e'.   The importance of this conclusion lies in the fact that in general the ion velocity is less than the mean thermal velocity of emission of an electron from a hot cathode.   Therefore, if the above conclusion were not valid, some mechanism for slowing down the injected electrons below their mean thermal velocity would be required in order to ensure that $v_e = v_i$.

In the above analysis it was not necessary to use (12.3); therefore, the conclusion applies not only to the steady-state space situation, where the end of the ion beam is moving away from the engine, and to the transient situation in a ground-based laboratory test during the time of transit of the ions from the engine to the beam collector electrode, but also to the steady-state laboratory test, where the collector electrode is electrically "floating" in potential so that an electron-reflecting sheath exists at the collector surface.   From (12.4), and (12.7), we can show by using (12.3) that $\rho_e' = \rho_e''$, that is, that the densities of incident and reflected streams are equal in the space situation.   In the laboratory floating-collector case, (12.3) becomes $v_e' = -v_e''$, and $\rho_e' = -\frac{1}{2}\rho_i(1 + v_i/v_e')$, $\rho_e'' = -\frac{1}{2}\rho_i(1 - v_i/v_e')$ (that is, the densities of incident and reflected electron streams are not equal); the reflected density is lower because electrons are collected by the collector electrode.

As mentioned earlier, a second possible model of the neutralized ion beam (in addition to the plasma bottle with counterstreaming electrons discussed above) involves the neutralization by equality of the ion velocity and the average electron velocity with the electrons moving randomly away from the engine.   This condition requires, however, that the electrons move very slowly.   (In fact, if the ions and electrons have the same average directed velocity, this would correspond to the average directed velocity of electrons emitted from a cathode at a temperature of the order of 100°K, for a cesium-ion beam at 2,500 V!)   Even in this case, however, processes like those described for the plasma-bottle model appear inescapable at the beam front and edges.

In conclusion, it may be stated that, regardless of thermalizing interactions between the injected neutralizing electrons and the ion or plasma beam, neutralization can take place.   In the absence of such interactions the plasma-bottle model suffices to account for neutralization; in the presence of such interaction, the plasma-bottle effect may not be required.   Whether such interactions do take place and what their nature is still remain subjects of active investigation.[7-9]   From the standpoint of ion-beam neutralization this appears, however, to be a rather academic question, as neutralization should be able to take place without as well as with such thermalizing interactions.

We shall now discuss briefly some of the experimental techniques which have been used to inject electrons into an ion beam and to measure

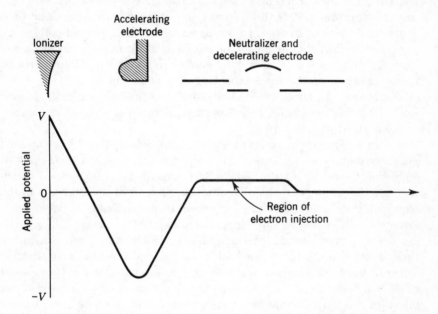

Fig. 12.4   Potential distribution along the axis of a cesium-ion engine; the neutralizing electrodes shown produce the potential trap into which electrons are injected.

the degree of neutrality of that beam.   Perhaps the most sophisticated system which has been used involves the injection of electrons from both sides of an ion beam into a region constituting a potential trap for electrons.[10]   This trap can be created by an arrangement of electrodes similar to that shown in Fig. 12.6, which creates a potential distribution as shown in Fig. 12.4.   It is seen that this potential distribution represents the accelerator-decelerator mode of operation, in which the neutralizing system is maintained at a potential higher than that of the accelerator electrode to provide a potential "hill" that prevents electron flow into the accelerator region.   In addition, the potentials on the several neutralizing electrodes are adjusted to provide a region free of axial fields, into which the electrons are injected, followed by a small potential hill that the electrons must pass over in order to leave the trap.

The effectiveness of the system shown in Figs. 12.4 and 12.6 in neutralizing both solid cylindrical beams and large, thin, hollow ion beams has been demonstrated by measuring the transverse spread of the beam due to space-charge forces.[11]   When electrons are injected under the correct potential conditions, this transverse spread can be made zero, indicating essentially no space-charge forces in the ion beam.   Under these neutralized conditions, it has also been found possible to isolate the ion-beam collector from the rest of the system electrically and still to maintain the neutralized state of the beam (i.e., to have the beam remain collapsed).

Somewhat similar means of injecting electrons into an ion beam have also been used, e.g., wrapping an electron-emitting filament around the outer edge of the beam, or arranging such a filament through the center of the beam.   In these experiments the electron-emitting cathode was located outside the engine, i.e., downstream from the point of emergence of the beam from the engine.

In a very careful series of experiments, Sellen et al.[12] have applied pulse techniques to studies of the neutralization problem for beams of high aspect ratio ($R \approx 10$ to 30).   By means of the pulse method which they use they have been able to study the behavior of the ion beam in the transient state before and after injection of neutralizing electrons from symmetrically arranged impregnated cathodes close to the edge of the ion beam.   They found, for example, that with careful adjustment of these neutralizing emitters, in both potential and position, it is possible to neutralize a broad béam in such a way as to eliminate the turnaround effect which occurs without the electron injection (i.e., without electron injection the space-charge forces cause this broad beam to turn around and return to the engine after moving only a short distance, while with neutralizing electrons the beam will move all the way to the collector). It is possible that these experiments also involved the creation of a potential trap in the ion beam by the potential and placement of the electron emitters.

In addition to the pulsed-beam neutralization experiments described above, Ward and Hubach[13] have measured the degree of neutralization of a hollow-ring cesium-ion beam by pulsing the beam and measuring the current induced in a surrounding loop by the pulse of charge passing through it.   The oscilloscope traces in Fig. 12.5 show these induced currents.   This figure shows the collector, grid, and loop (top to bottom) currents for the unneutralized beam (i.e., without injection of neutralizing electrons in the engine); it is seen that current is induced in the loop as the pulse of charge approaches and departs from the loop.   The grid and collector currents are very low in this unneutralized case because the

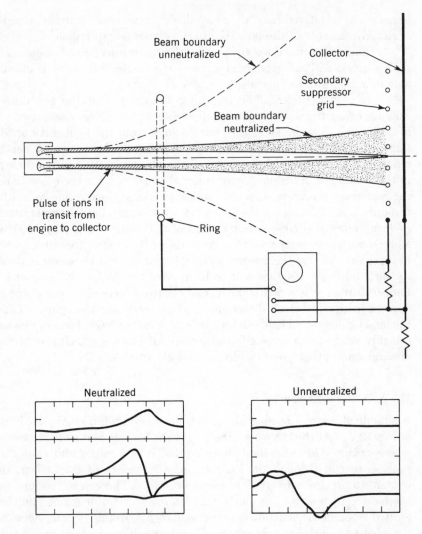

Fig. 12.5 Schematic diagram of the pulsed-beam experiment. The currents induced in the collector, grid, and ring electrode (top to bottom) are reproduced from oscilloscope traces.

space-charge spread of the beam between the engine and collector is so large that only a small part of the beam actually strikes the collector. When the beam is neutralized by injection of electrons (in a trap system of the type shown in Fig. 12.6), the loop current vanishes, as shown in Fig. 12.5 (no net charge in the beam), but the grid and collector exhibit displacement and conduction currents, respectively, as expected. The beam could be altered from the neutralized to the unneutralized state by slight adjustments of the potential in the electron-trap region.

These experiments show that, with proper injection of electrons in both a broad and an intermediate beam, charge neutralization is obtainable in a laboratory experiment.

It is evident from the foregoing discussion that the neutralized exhaust beam from an ion engine possesses some of the characteristics of a plasma. The greatest similarity results from the neutrality condition, i.e., equal electron and ion densities. Also, the formation of sheaths reflecting the electrons at the edges of the plasma bottle has an analogy in the ambipolar diffusion of conventional plasmas. On the other hand, the ion beam possesses directed motion of high velocity (ion velocities typically 7×10^6 cm/sec); the ions will have a longitudinal "temperature" (velocity spread) corresponding to that of the emitter (typically 1400°K) and a transverse temperature several times this value (because of the beam compression in the accelerator). Measurement of the electron velocity distribution[13] has shown it to be approximately Maxwellian, with a temperature in excess of 10,000°K, and electrons have also been detected moving in the "reverse" direction, i.e., from collector to engine. Thus, the ion-engine system has provided us with a rather novel form of plasma, a highly energetic or directed nonthermalized plasma, which should find application in other areas of physics and electronics.

12.3.2. Cesium-ion Engines

A simplified schematic diagram of a typical cesium-ion engine is shown in Fig. 12.6. At the left we see the propellant heating and storage system, where cesium, the propellant, is converted into a vapor and caused to flow up to and through the ionizer. At the surface of the ionizer, the cesium atoms are converted into cesium ions. These positive ions are accelerated by means of an electric field in the accelerator region and then ejected through the neutralizer region as shown. In this system, electrons are injected from a cathode surrounding the ion beam and are mixed with the ions in the electron trap created by the electrodes shown.

The electrode system marked "neutralizer region" in Fig. 12.6 serves two functions: (1) The shapes and potential differences of the electrodes have been designed so as to achieve most efficiently the injection of electrons for neutralization. (2) The electrodes are held at an average potential well above the accelerator in order to decelerate the ion beam. Essentially all modern cesium-ion engines operate in this so-called accel-decel mode. In this mode, the final exhaust of the propellant takes place at or near the potential V_f corresponding to the final, or decelerator,

electrode, but between this electrode and the ionizer the accelerator elec-
trode is arranged at potential V_a.   The fraction V_a/V_f is called the *accel-decel ratio*; typical values are approximately 2 to 10.   This system will
produce a potential distribution along the beam similar to that illustrated
in Fig. 12.4.   This potential distribution is required because of the limited
perveance of practical ion guns, and in order to prevent the injected neutralizing electrons from being attracted into the accelerator region and

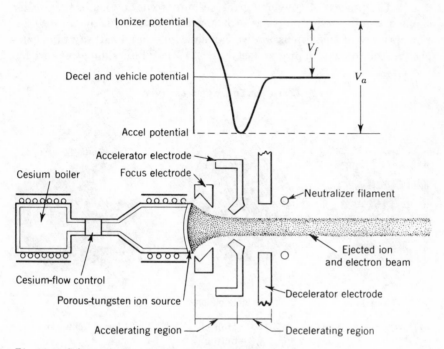

Fig. 12.6 Schematic diagram showing the various components of a cesium-ion engine.

thereby bombarding the ionizer, which would result in subsequent power
loss and possible overheating.

    There are several geometrical forms in which the ionizer, and there-
fore the entire engine, can be arranged to produce the required area: (1)
a planar structure with grids across the accelerator aperture, (2) an array
of small solid-beam accelerators with an open accelerator aperture (Fig.
12.7), (3) an accelerator to produce a long, narrow strip beam (Fig. 12.8),
and (4) a structure producing a thin hollow-ring beam (Fig. 12.9).

    The most promising configuration seems to be the strip-beam con-
figuration.   This configuration is easily amenable to clustering in a number of parallel strip beams.   It has the basic advantage of using a one-

dimensionally convergent optical system, which permits operation at higher perveance or with less critical mechanical tolerances than with solid beams of circular cross section and two-dimensional convergence. A further and major advantage of the strip-beam configuration is that it is amenable to clustering in configurations which have inherently longer life (as far as sputter erosion is concerned) than clusters of beams with circular cross section.

The process of converting the cesium atoms to ions in the ionizer is one of the interesting physics problems associated with the ion engine. For purposes of this discussion we can show one practical way of accomplishing this ionization process (see Fig. 12.10). The cesium atoms are fed

Contact Ion Engine Configurations

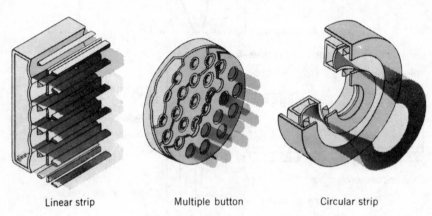

Linear strip   Multiple button   Circular strip

Fig. 12.7 Linear strip.   *Fig. 12.8* Multiple   *Fig. 12.9* Circular strip.
                              button.

from behind a porous tungsten pellet; they migrate along the walls of the very thin capillary channels between the grains of tungsten and emerge on the top surface, from which they are emitted predominantly as ions. This process of contact ionization of alkali metals on hot tungsten has been studied in great detail by Taylor and Langmuir;[14] the valence electrons of the low-ionization-potential alkali metals remain with the higher-work-function tungsten surface as the particle is evaporated from this surface. Thus, the current-voltage characteristics appear[15] as in Fig. 12.11, where they are seen to be similar to the saturation characteristics of an electron emitter. Below the level of maximum space-charge-limited emission, the cesium supply is excessive, and many particles evaporate as neutrals; above this level, all the ions which the surface is capable of supplying can be drawn off so that the neutral efflux is low.

Cesium atoms ionized by contact
with hot tungsten surface

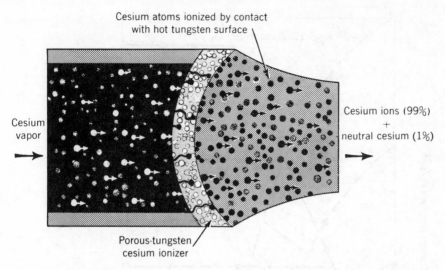

Cesium
vapor

Cesium ions (99%)
+
neutral cesium (1%)

Porous-tungsten
cesium ionizer

Fig. 12.10   A simplified diagram of a porous-tungsten contact ionizer for cesium. The cesium atoms diffuse through the hot (1150°C) tungsten and migrate onto the surface.   Under proper conditions, the evaporation from this surface consists predominantly of ions, with a few neutral cesium atoms.

Although cesium has many undesirable properties, it is usually chosen as the propellant because of its low ionization potential: low ionization potential permits efficient contact ionization.

The very accurate control of the trajectories of the accelerated ions, so that they flow in a uniform and "laminar" manner through the accelerator and neutralizer regions of the "gun," is of essential importance to keep to the tolerable level of less than 10^{-3} the fraction of accelerated ions intercepted by the accelerating electrode.   The design of such ion guns is somewhat analogous to the problem of designing optical lenses that will produce images free of aberrations.   Such close control of the ion-optical design of the accelerator structure makes use of techniques which were found essential in the design of high-current-density electron guns for beam-type microwave tubes.

Perhaps the most significant advance in the design of electron guns, including the effects of space charge, was made by Pierce.[16]   Although the Pierce design technique has proved extremely useful and has formed the basis for the design of many guns over the past ten years or so, it suffers from a fundamental limitation when applied to the design of those high-perveance guns which do not have a grid across the accelerator electrode aperture.   That is, the ideal Pierce gun can be divided into two regions: (1) the region of rectilinear electron flow, in which the potential is given by the Langmuir relations for space-charge-limited flow for the

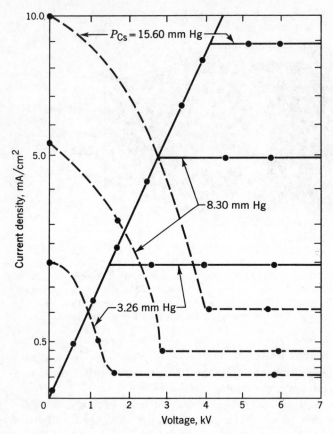

Fig. 12.11 Porous-tungsten cesium-ion emitter characteristics as a function of accelerating voltage. The solid lines show the ion current for several values of cesium flow rates (or cesium vapor pressure), and the dashed lines show the evaporation rate of neutral cesium particles. (*Data courtesy of Dr. Otto Husmann.*)

geometry used and a unipotential electrode is designed to effect the proper electrical boundary conditions at the beam edge, and (2) a thin, electric aperture lens in the vicinity of the anode.   When this method of analysis and design is used, it is assumed that trajectories in the two regions can be joined smoothly and that no appreciable distortion of the ideally shaped electric fields exists away from the immediate vicinity of the anode aperture.   Since higher-perveance guns do exhibit a considerable distortion of electric field due to the aperture in the accelerating electrode, this simple design criterion must be modified somewhat.   Modified techniques

of design of high-perveance electron guns,[17] including the use of an auto-
matic ion trajectory tracer with space-charge simulation, have been found
to be similarly successful in the design of ion accelerators.   For example,
by the use of these proved design techniques, ion guns have been built
which exhibit laminar beam flow and direct electrode interception values
of 10^{-4} of the emitted current, values which are lower by orders of magnitude than those observed in engines designed by less careful techniques,
and which are good enough now for a long mission.   These excellent
optical characteristics must of course be obtained consistent with uniform
emitter current density.

In addition to imperfect ion optics, another cause of ion interception on the engine electrodes can arise from ions which deviate from their
laminar or design trajectories as a result of emission from the ionizer with
a transverse component of velocity.   This transverse thermal spreading
is a rapidly decreasing function of accelerator potential; with correct
design of the accelerator electrode aperture, it can be made negligible at
usual operating voltages.

A third cause of ion interception results from ions created in the
flow region of the accelerator and decelerator by charge-exchange reaction
between the cesium atoms in this region and the moving ions.   Since
the accelerator structure is usually designed so as to control the trajectories of ions originating at the ionizer, those charge-exchange ions
created elsewhere in the flow region can be accelerated so that they impinge
on the engine structure, thus producing erosion.   There are at least two
ways of controlling this process: (1) reducing the atom density in the
flow region by decreasing the neutral efflux from the ionizer, and (2)
performing additional design studies (e.g., on a trajectory tracer) on
the engine so as to deflect the charge-exchange ions from the critical
electrodes.

The process of cesium ionization by contact with hot tungsten provides one of the principal advantages of the cesium-ion type of electrostatic propulsion engine, in that the ionization is effected at a well-defined
and carefully shaped equipotential surface and that the neutral efflux and
subsequent charge-exchange interception can be kept relatively low.   This
emitter surface can form the basis for accurate design of the ion-optical
characteristics of the accelerator in much the same way as a space-charge-
limited electron cathode forms the basis for design of Pierce-type electron
guns.

The cesium-ion engine is seen to be amenable to accurate design,
based on both analytical and carefully controlled experimental techniques,
which yields very predictable engine performance.   These engines are
capable of supplying ion beams of attractive densities (particle densities
of the order of 10^{11} ions/cm^3 and current densities of 20 to 30 mA/cm^2

in the exhaust beam).   These ion beams have been successfully neutralized (at least as far as laboratory tests can determine).

As shown in Fig. 12.12, the cesium-ion engine can be made to be quite efficient.   The principal energy loss is in the form of power required to heat the ionizer to the necessary emitting temperature.   Some additional power is required to heat the neutralizer filament and the cesium boiler, but, for reasons mentioned before, no power dissipation due to current interception on the electrodes is tolerable.   Figure 12.12 shows the

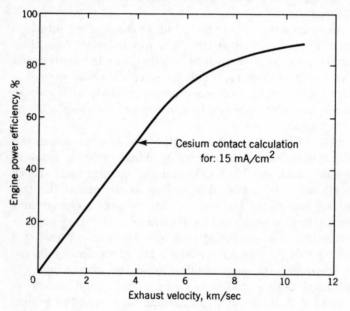

Fig. 12.12   Power efficiency of cesium-ion engine for an ionizer current density of 15 mA/cm².

efficiency (power into the beam to produce thrust, divided by the total input power) as a function of exhaust velocity of the exhaust beam.   This curve is derived from calculations based on experimental data on operating engines and other sources.   It is seen that efficiency values in excess of 80 percent are possible for exhaust velocities of interest for space missions.

12.3.3. Penning-discharge Ion Engines

A second important class of electrostatic ion engines employs a fundamentally different type of ion source in which the propellant gas is ionized by electron impact.   Since for any reasonable geometries and particle densities the mean-free path for ionization (of the order of meters) is much larger than the dimensions of the ionization chamber, it is evident that the efficiency of such a source depends upon increasing the mean lifetime of the primary electrons. Such an increase can be accomplished by the use of Penning-type discharges in which magnetic and electric fields are arranged so as to trap the electrons and thus increase the ionization probability. Such sources have been found to have good power efficiency (in terms of energy expended per ion produced) and relatively high propellant-utilization efficiency, together with the other obvious advantages of mechanical simplicity and ability to use a number of different types of propellant gases. An important problem concerns the design of high-quality ion-optical systems for low electrode interception, and hence for long life.

The most promising engine of this type is that invented by Kaufman;[18] it is shown schematically in Fig. 12.13. The ionization chamber is

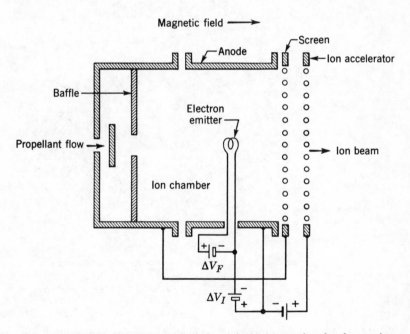

Fig. 12.13 Schematic illustration of original electron-bombardment ion engine invented by H. Kaufman.

composed of a cylinder, with the neutral propellant gas introduced at one end and the ions extracted at the other. The center portion of the cylinder (i.e., the anode for the discharge) is maintained at a positive potential with respect to the ends of the cylinder, thus providing electron trapping in the longitudinal direction. Electrons are emitted from a filament located on the axis.   The longitudinal magnetic field prevents the electrons from reaching the anode directly; in the crossed electric and magnetic fields, they tend to spiral around the axis, thus increasing their mean lifetime and the probability of making an ionizing collision.   A plasma is formed which fills the discharge chamber out to the screen; ions are extracted from this plasma boundary by an array of accelerating electrodes.

The energy range of interest for the primary electrons is dependent upon the ionization potentials of the propellant gas.   To date, both mercury and cesium vapor have been utilized extensively in this type of ion engine.

Figure 12.14 shows a mercury Kaufman engine with a liquid mercury cathode as an electron emitter.[19]   In this cathode, the mercury

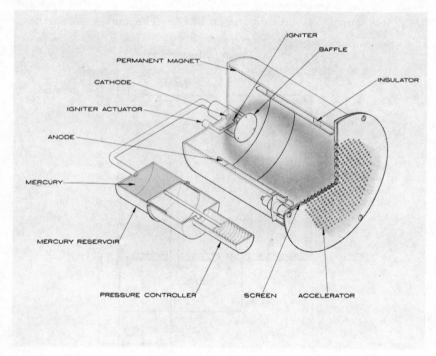

Fig. 12.14   Kaufman electron-bombardment mercury-ion engine with liquid mercury cathode.

surface is maintained in position independently of gravity, in a small capillary orifice.   The same mercury surface provides the electrons for this discharge through a mercury pool arc, and the mercury vapor required for the expellant.   With this cathode, there is no life problem, as has been established by life tests in excess of 3,000 hours.[19]   Typical performance obtained in such life tests with an engine having a 20 cm i.d. anode is as follows: a 600 mA ion-beam current at a net accelerating voltage of 6,000 V and an expellant utilization efficiency of 80 percent was obtained with an electrical efficiency of 89 percent.   The magnetic field was about 50 G at the ion extractor; the discharge current was 13.5 A, and the discharge voltage was 32 V.   The corresponding thrust was about 0.1 N $(20 \times 10^{-3}$ lb).   Still substantially better performance has been obtained since this life test.

Penning-discharge ion engines with other expellants than mercury are also being investigated.   Particular attention has been given to cesium as an expellant, because of its low ionization potential and low work function.   The low ionization potential should lead to a higher ionization efficiency; the low work function permits the use of a cesium-covered surface as the thermionic cathode for the discharge.   In order to maintain at the cathode surface the cesium-vapor pressure required for thermionic emission, one approach developed at Electro Optical Systems laboratories is to separate the cathode from the main discharge chamber by a constriction. A sketch of such a cesium Penning-discharge ion engine under development at Electro Optical Systems is shown in Fig. 12.15.   Performance of

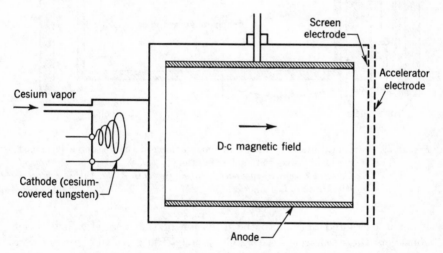

Fig. 12.15   Principle of electron-bombardment ion engine with cesium expellant and cesiated cathode.

such cesium-ion engines[20] has been found comparable with that of the best mercury engines, insofar as the product of electrical efficiency and expellant-utilization efficiency has been found to be at least as high at the same exhaust velocity.   Cesium engines are attractive because, like liquid mercury cathode engines, they have demonstrated cathode life with good engine performance.[20]   A disadvantage of cesium is a lower atomic mass than mercury, resulting in higher current density and ion interception for a given thrust density. Cesium-covered cathodes also are sensitive to impurities.[20] Finally, precautions are needed to prevent electron back-emission from cesium-covered accelerator electrodes.[20]

Another kind of ion source which utilizes a P.I.G.-type discharge is the so-called grid-type ion source used at the Oak Ridge National Laboratory in separating stable isotopes and in controlled fusion research. This type of source is shown schematically in Fig. 12.16. Here the ion

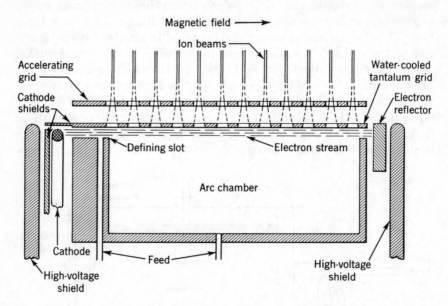

Fig. 12.16 Schematic drawing of the grid-type Penning-discharge source developed at the Oak Ridge National Laboratory. (*J. S. Luce, Ion Erosion of Accelerating Electrodes in Space Vehicles, presented at the American Rocket Society Meeting, Los Angeles, Calif., May 9, 1960, Paper 1159-60.*)

beam is extracted normal to the direction of the magnetic field which constricts the discharge. Although high ($\sim$100 mA/cm²) current densities are obtained at typical acceleration potentials of the order of 25 kV, this configuration has not been applied to the range of parameters of interest for propulsion purposes. Luce,[21] however, has conducted some electrode erosion studies using this source and has discussed its application to ion engines.

An essential problem in the development of the Penning-discharge ion engine for long life is that of devising an adequate ion-optical system, for extraction of the ions from the plasma with a minimum of interception of accelerated ions. To understand the process of ion extraction from a plasma, it is useful to consider the system of two apertured electrodes shown in Fig. 12.17. It is customary to call the electrode adjacent to the plasma-generating region the *screen electrode* and the one on the other (downstream) side the *accelerator* (*accel*) electrode. The apertures are both large compared with the Debye length. First, consider both electrodes to be immersed in the plasma and slightly negative (e.g., floating) with respect to the plasma. The ion sheaths on both electrodes are relatively thin, and the plasma diffuses through both electrodes as shown in

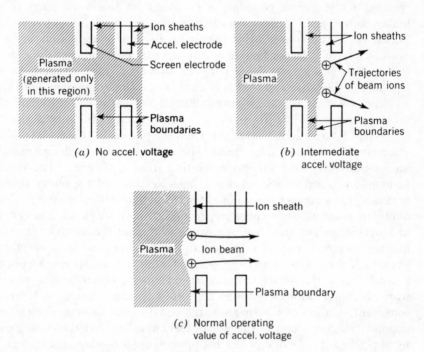

(a) No accel. voltage

(b) Intermediate accel. voltage

(c) Normal operating value of accel. voltage

Fig. 12.17 Plasma boundary positions in an ion-extraction system for various values of the accel voltage.

Fig. 12.17*a*. Suppose now that an accel voltage (i.e., a potential difference between the screen and accel electrodes) is being applied. As the accel electrode is made successively more negative, the sheath thickness at its surface increases; it then becomes comparable with the aperture diameter, and at a certain potential difference between the electrodes the sheaths surrounding opposite edges of the accel aperture meet, leading to the situation of Fig. 12.17*b*. Here the plasma is no longer penetrating the accel aperture, and a (divergent) ion beam begins to emerge from it. As the accel electrode is made still more negative, the plasma boundary recedes farther, the accel- and screen-electrode sheaths merge, and at last the situation shown in Fig. 12.17c is reached where the plasma region is limited to the upstream side of the screen electrode. This makes it possible to extract (from the part of the plasma boundary extending across the screen aperture) an ion beam which passes through the accel aperture with relatively little interception by the accel electrode. It is this situation which is of interest for applications requiring long electrode life.

It is apparent from this model that a knowledge of the shape and position of the plasma boundary is necessary for the exact design of an ion-extraction and -acceleration system leading to minimum interception. Fortunately, the conditions determining to a first approximation the position of this plasma boundary surface can be expressed rigorously: this surface is uniquely defined by the two conditions,[22] that it be an equipotential surface and that it be field-free. One consequence of these conditions is that the space-charge-limited ion current density (on the accelerator side of the surface) equals at each point of the surface the ion saturation current density available from the plasma at this point. With these conditions taken into account, it is now possible to design exactly an ion-optical system for ion extraction from a plasma. The design technique is based on self-consistent field-plotting and trajectory-tracing techniques analogous to those used for the design of electron guns. The only important difference between the design of an ion-extraction system and an electron gun (besides appropriate scaling for the relative masses of ions and electrons, respectively) is that, for the ion-extraction system, the position of the emitter surface (plasma boundary) is not specified a priori; it results from the determination of the field-free equipotential surface, which is found through iterative trajectory tracings leading to the self-consistent solution of the space-charge equations in the ion-accelerating region. The usefulness of this design technique has been demonstrated[22] by applying it to the design of a high-perveance ion-extraction system in which an accel interception of 10^{-3} (0.1 percent) was obtained with an ion current density of 45 mA/cm² at the screen aperture, at an accel voltage of 3.2 kV, with Xe⁺ ions.

In addition to direct interception, which can be minimized by the design techniques outlined above, interception of charge-exchange ions is a serious problem. Charge-exchange ions are generated in the accelerated and extracted ion beam by collision of the accelerated ions with neutral molecules or atoms escaping from the discharge chamber through the ion-extraction orifice. The trajectories of charge-exchange ions can also be determined; their detrimental effect is minimized[22] by deflection through an appropriately designed "diverter" electrode and by keeping the decel voltage close to the accel voltage (i.e., avoiding high accel-decel voltage ratios). While ion interception in Penning-discharge ion engines built and tested to date is typically of the order of 0.5 to 1 percent at useful ion current densities, the results obtained to date by means of the more sophisticated ion-optical design procedures outlined above make it reasonable to hope for sufficient improvements in the reduction of this ion interception to make engine life of the order of 10,000 h or more feasible.

12.3.4. The Duoplasmatron-type Ion Engine

A third major class of sources which is being extensively investigated and developed for the electrostatic acceleration of dense ion beams is based on the so-called Duoplasmatron concept of Von Ardenne.[23] As in the Penning-type source, ionization is achieved by electron bombardment, and the ion beam is extracted from a plasma. Of all such bombardment sources, the Duoplasmatron is capable of producing the highest ion current densities. Densities two or three orders of magnitude greater than those attained by means of contact ionization, i.e., of the order of amperes or even tens of amperes per square centimeter, are possible (although they have not yet been achieved in well-focused beams). This characteristic, together with an inherently high ionization efficiency and the prospects for eventually reaching high power efficiencies, makes this class of sources of considerable interest for development as part of an ion thrust system for propulsion. Although major problems exist in reducing electrode erosion, increasing emission area, and controlling and focusing the output beam, it is believed that these may not constitute fundamental limitations and that, with intensive and imaginative effort, the inherent promise of the Duoplasmatron-type ion engine may eventually be realized.

Figure 12.18 illustrates the basic Duoplasmatron configuration. A low-pressure arc is produced between a filament (or thermionic cathode) and the anode, which typically operates in a potential range from somewhat above the ionization potential of the working gas to about 100 V. An "intermediate electrode," conical in shape and either held at an intermediate potential or sometimes floating, serves to constrict the discharge and increase the plasma density. This is the form originally suggested by Von Ardenne in 1946 for use as an ion source for experiments in isotope separation. The addition of a high (essentially axial) magnetic field between the tip of the baffle and the anode (Von Ardenne[23]) acts to constrain the discharge further. These dual constraints—mechanical and magnetic—produce a very dense plasma in the extraction orifice. Since neutral particles have to pass through this cloud of hot, dense plasma to escape from the source, the propellant-utilization efficiency can be high. A very dense ion beam is obtained from the plasma by means of large negative potential (typically 10 to 60 kV) applied to the extractor electrode.

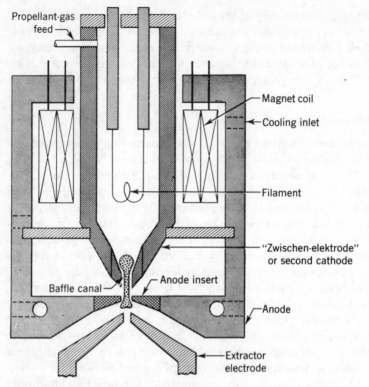

Propellant-gas feed

Magnet coil

Cooling inlet

Filament

"Zwischen-elektrode" or second cathode

Baffle canal

Anode insert

Anode

Extractor electrode

Fig. 12.18   Basic Duoplasmatron configuration.

The complex geometry of the Duoplasmatron makes exact analysis of its operation impossible.   However, the basic mechanisms which determine its characteristics can be discussed qualitatively and can serve as a guide in its development.   The arc inside the source has the typical characteristics of a low-pressure discharge in which the electron mean-free path is of the same order as the discharge length.   A positive sheath close to the cathode supports a potential difference approximately equal to the ionization potential of the working gas.   Following this cathode drop there exists a region of small potential gradient which contains an essentially neutral plasma; here, ionization as well as randomizing elastic collisions occur.   The pressure in the arc region is typically of the order of ten to several hundred microns.

In such a discharge per se the plasma density would not normally be sufficiently great to permit extraction of large ion currents.   As mentioned previously, the latter is accomplished through the use of magnetic and mechanical constraints.   The magnetic field has two functions: it gives rise to axial electron reflection, and it exerts a magnetic confining effect on the arc which serves to constrict it and to reduce radial diffusion losses.   Both effects increase the plasma density.

Figure 12.19 indicates schematically a typical magnetic field configuration in the neighborhood of the baffle canal and the anode.[24]   The

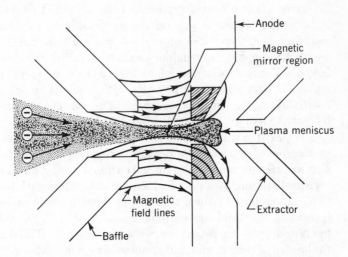

Fig. 12.19   Magnetic-field configuration in the neighborhood of the baffle canal and anode of a Duoplasmatron ion source.   The converging magnetic field constitutes a magnetic "mirror" for reflection of electrons.   It also constricts the discharge and reduces radial diffusion losses.

magnetic field, usually of the order of several kilogauss, is highly convergent in this region.   In fact, it can be regarded as a magnetic-mirror configuration in which electrons entering with an initial transverse-velocity component can be reflected.   Total reflection can occur for a large part of the incident electron current.

The electrons that are trapped by the magnetic field and reflected along their original path can be reflected again by the original electric field existing in the arc discharge, thus giving rise to electron trapping similar to that present in the Penning discharge.   Such trapping may last until collisions or field inhomogeneities allow the electrons to reach the anode.   The probability of ionization by the primary electrons is increased by increasing their effective lifetime, which, combined with a decreased rate of radial electron diffusion, leads to higher ionization density and power efficiency.   High-energy axial electrons which penetrate the mirror region can be reflected back into the plasma by the negative extractor potential.

It is evident that the magnetic field can also exert a confining effect on the plasma which it guides through the baffle and extraction orifice and reduces the effective plasma-boundary surface area and diffusion losses. The current density in the baffle canal is greater than that on either side of the baffle; most of the current passing through the baffle consists of electrons. The potential gradient which supports this current is sustained by a positive space charge on the anode side of the canal and a corresponding negative space charge on the cathode side.[25] As the canal length is increased, the tendency toward neutrality in this region becomes larger. Various estimates place the electron temperature in the baffle region of the order of 10^5 °K, with somewhat lower temperatures ($\sim 10^4$ °K) in the extraction region. The ion temperature was estimated by Von Ardenne[23] at several times 10^3 °K.

Only the axially symmetric pinhole-type source has been considered up to now. This source suffers from one major defect for application to electrostatic ion propulsion: the ion current density which it provides is too high for handling by a practical ion-optical system with tolerable interception. In order to reduce the current density while maintaining or even increasing the total ion current, much effort has been devoted to configurations with increased extraction-hole area. This has naturally led to the idea of rectangular and annular-slit configurations.[26] The resulting discharge configurations with which acceptably low current densities have been obtained (below 100 mA/cm²) depart substantially from the original Duoplasmatron and resemble the Penning-discharge ion engine. The main differences between these modified Duoplasmatron engines and the Penning-discharge engines are that the modified Duoplasmatrons operate at higher plasma density, higher ion current density, and higher magnetic field and possess only a single extraction slit. The best results have been obtained to date with engines having an extraction slit of the order of 1″ long and 2 mm wide.[26] These best results are an arc power expenditure of 360 eV/ion with an expellant- (mercury-) utilization efficiency of 90 percent and a magnetic field of 2 to 3 kG. The ion current density at the extraction slit was 70 to 100 mA/cm².

The problems still to be solved to make such an ion engine practical are cathode life, ion interception in the extraction system, and improved area utilization (while at the extraction slit the ion current density is 70 to 100 mA/cm², the total ion current divided by the total frontal area of the engine results in only a few milliamperes per square centimeter because of the relatively large magnetic circuit required). It may be noted that the ion-optical problem to be solved is basically the same as that of the Penning-discharge ion engine discussed previously. The systematic design techniques developed for the extraction of ions from a plasma should also be applicable to this type of engine, and the handling of current densities even as high as 70 mA/cm² may not be hopeless, in the light of the excellent results already obtained[22] up to 45 mA/cm².

12.3.5. Oscillating-electron Engines

BASIC PRINCIPLE   All the engines of the class which we shall call *oscillating-electron engines* have in common the presence of high-energy axially oscillating electrons.   A number of them have in common other features which in many respects may be regarded as dual to those of the ion engines described above.   It seems appropriate to begin the discussion of this class of engine with an idealized model, in order to emphasize their common features.   After analysis of this model, a detailed description of actual engines from a more unified view will be possible.

The principle of the oscillating-electron engine in its idealized form is shown in Fig. 12.20.   Electrons emitted from a cathode are acclerated

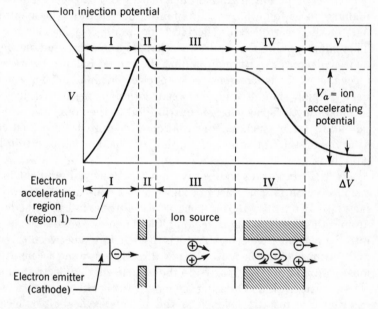

Fig. 12.20   Principle of oscillating-electron engine.

electrostatically and focused into a beam in region I.   They are slightly decelerated in region II and are made to pass through region III in an axial stream.   Ions which neutralize the electron space charge are then injected into region III.   The potential maximum to the left of region III prevents the flow of ions from region III to the cathode.   In region IV a potential gradient is caused in the plasma by the high-energy electrons trying to escape the plasma. The potential gradient reflects those electrons which do not have sufficient energy to pass the potential barrier and accelerates the ions. Most of the reflected electrons will then be reflected again by the cathode and oscillate back and forth between the cathode and region IV until they are ultimately collected in region II or III by one of the positive electrodes. The accelerated ions are ejected axially to the right of region IV with an energy approximately corresponding to the potential difference between the cathode and region III. Also, those electrons which are energetic enough to pass the potential ΔV will escape, together with the ions; in fact, the potential ΔV is adjusted to just such a value that the number of electrons escaping from region IV will be equal to the number required for neutralization of the ejected ion beam.

The possible existence of a potential gradient of finite axial extent in a plasma beam was first pointed out by Rose and by Saltz et al.[27] Qualitatively, if an electron velocity distribution of finite width exists, the existence of a potential gradient in a neutral or near-neutral plasma can be understood as follows: As the potential decreases in the axial direction, part of the electrons are reflected; this diminishes the electron density. This same potential decrease accelerates the ions, and hence the ion density decreases. By maintaining the proper balance between the two mechanisms, it is possible to keep electron and ion density equal or nearly equal; the difference between both densities is then just sufficient to make up the space-charge contribution to the gradient of electrostatic potential, when such a contribution is required for a self-consistent solution of the equations of motion of the particles and of Poisson's equation.

It is important to observe that the potential gradient in a region such as region IV can exist in a plasma as a consequence of two separate factors. In one case, the presence of an electrode close to cathode potential around the plasma beam in region IV and penetration of the vacuum field into the plasma are essential. Electrostatic forces between this electrode and the ions cannot be ignored; they directly cause the acceleration of the ions. Momentum is imparted to the vehicle via this electrode and via the electrostatic fields accelerating the ions. In the other case, penetration of external electrostatic fields into the plasma is negligible; electrostatic forces between plasma and electrodes in region IV are insignificant, and most of the potential gradient in region IV results from space-charge fields. Momentum is then imparted to the ions in region IV through these space-charge fields. In other words, it is transferred in region IV from the energetic electrons entering region IV to the accelerated ions leaving region IV via the space-charge field existing in region IV. The momentum of the high-energy electrons entering region IV from region III has been imparted to them by electrostatic forces in region I. The momentum imparted to the vehicle therefore corresponds in this idealized case to the vehicle's reaction to these forces in region I.

A quantitative analysis of the conditions for momentum transfer by space-charge forces in region IV (momentum transfer from energetic electrons to the ions) is given in the Appendix (Sec. 12.7). This analysis is based upon the observation that, in the absence of electrostatic forces between plasma and electrodes in region IV, all the momentum gained by the accelerated ions leaving region IV comes from the energetic electrons entering region IV from region III. This leads, as shown in the Appendix, to the following interesting relation between the density of oscillating-electron current j'_{e3} flowing toward region IV (from the left) and the ion current density j_i accelerated through region IV,

$$-j'_{e3} \simeq \frac{1}{2} \sqrt{\frac{m_i}{m_e}} \sqrt[4]{\frac{W_{i4}}{W_{i3}}} \cdot j_i \qquad (12.9)$$

where m_i = ion mass

m_e = electron mass

W_{i3} = average energy of ions entering region IV from region III

W_{i4} = average energy of ions ejected to right after acceleration through region IV

Equation (12.9) shows that the electron current density j'_{e3}, most of which is reflected by the potential gradient in region IV, is much larger than the ion current density accelerated through this region. This is in agreement with what one intuitively expects when considering the transfer of momentum between light electrons and heavy ions as described above. The net electron current density emerging to the right of region III, together with the ion current density j_i, is given by

$$|j_e| = |j'_{e3}| - |j''_{e3}| = j_i \qquad (12.10)$$

This relation results from the condition that the total current density $j_i + j_e$ emerging from the plasma accelerator be zero, a condition enforced by the processes outlined in the earlier discussion of ion-beam neutralization.

It may be of some interest to observe that, for desirable ion current densities j_i, the electron current density j'_{e3} becomes quite large; e.g., for $j_i = 1$ A/cm^2, $\sqrt{m_i/m_e} = 500$, and $W_{i4}/W_{i3} = 10^4$, $j'_{e3} \simeq 2{,}500$ A/cm^2. This result, however, is of no particular concern, as j'_{e3} consists of currents due to multiple reflections between the cathode and region IV. The net current density which must be emitted by the cathode in this example equals only $|j_e| = |j'_e| - |j''_e| = j_i = 1$ A/cm^2, plus the contribution corresponding to the current ultimately collected on the positive electrode.

Although some emphasis has been given above to the direct transfer of momentum from electrons to ions in region IV by space-charge forces, it is not yet clear which conditions prevail in actual oscillating-electron engines.   It may be plausible, in fact, that part of the momentum transfer to the ions occurs directly through electrostatic forces between electrodes and ions, another part of the momentum being transferred from the energetic electrons through space-charge forces.   In either case, however, oscillating electrons exist in region III and are ultimately collected by a positive electrode.   It is clear that this collected (or "intercepted") electron current must, in any event, be kept as small as possible for reasons of efficiency; it represents a power loss which is basically unnecessary. For this reason, good focusing of the beam of oscillating electrons in regions II and III is essential.   In the engines considered so far, this is done by means of a dc magnetic field, as in a Penning discharge.   If the cathode is itself immersed in the focusing dc magnetic field, it is of interest to ask how the plasma beam can escape from the diverging magnetic field at the end of the engine without being itself caused to diverge intolerably. Although a theoretical answer to this question does not seem to exist yet, Meyerand's experiments[27,28] (described below) on oscillating-electron ion engines seem to indicate that, in fact, such a plasma-beam divergence can be avoided.

A further quantity to be evaluated is the plasma density $n_i = n_e$ required in region III for a given ion current density j_i in the plasma beam escaping to the right of region IV.   If the rate of ion generation in region IV can be neglected as compared with the rate of ion injection or generation in region III, it is seen that

$$n_i = \frac{j_i}{e\bar{v}_{i3}} \tag{12.11}$$

where $\bar{v}_{i3}$ is the average axial velocity of ions leaving region III toward region IV.   For $j_i = 1$ A/cm^2 (which is a respectable ion current density) and $v_{i3} = 2 \times 10^5$ cm/sec (which corresponds to argon ions at an energy of the order of 1 eV),

$$n_e = n_i \cong 3 \times 10^{13} \text{ particles/cm}^3$$

Although this plasma density is relatively high, it does not appear excessive for an intense hot-cathode Penning-type discharge; this shows, therefore, the potential capability of this type of engine for achieving quite high thrust densities at a high exhaust velocity.

A few additional observations can be made about the rate of ion generation in region IV.   In the above calculations, this rate was assumed to be zero.   In some engines (e.g., Meyerand's oscillating-electron ion engine), however, this rate is finite.   The consequences of having a finite rather than a vanishing rate of ion generation in region IV are:

1. Requirement for a lower plasma density in region III than predicted by (12.11). This lowering of the plasma density, however, will not be larger, typically, than one order of magnitude.

2. Introduction of an undesirable velocity spread for the accelerated ions.

It is now possible to see the following striking analogies between the oscillating-electron engines and accelerated-ion engines:

1. Electrostatic acceleration of electrons takes the place of electrostatic acceleration of ions.

2. Minimization of electron collection by the accelerating electrode through electron focusing (electrostatic and magnetic) is the counterpart of suppression of ion interception by careful electrostatic ion focusing in the ion engine (even though electron interception here is very much higher than ion interception in ion engines).

3. Ion injection is needed here for space-charge neutralization instead of electron injection in the ion engine.

4. Prevention of ion flow to the cathode by an ion deceleration potential maximum (region II) may be compared with the prevention of electron flow to the ion emitter by a potential minimum in the ion engine.

The major basic difference between the oscillating-electron engine and the ion engine is possibly the process of momentum transfer from electrons to ions, which was discussed above.

The actual oscillating-electron engines to which the above considerations pertain differ from one another primarily in their methods of ion generation or ion injection (in region III of Fig. 12.20). Three types of ion source have been considered so far:

1. Ion generation through the oscillating electrons themselves (Penning discharge)
2. Ion generation by an arc
3. Ion generation by contact ionization of cesium

Type 1 was investigated by Meyerand et al.[27,28] in the oscillating-electron engine; type 2 was first proposed and investigated by J. S. Luce[29]; type 3 is being investigated by W. Eckhardt. This list, of course, does not exhaust the types of ion source which can be considered or which may be in a preliminary state of investigation.

THE OSCILLATING-ELECTRON ION ENGINE   The oscillating-electron ion engine as invented and investigated by Meyerand et al. is sketched in principle in Fig. 12.21; a typical axial potential distribution is also shown qualitatively in Fig. 12.21. It is seen that this potential distribution is closely similar to that in Fig. 12.20, with the difference that the well-defined potential maximum of region II in Fig. 12.20 does not exist; it is replaced by a moderate potential gradient in region III. The ions are generated by collisions between the fast electrons and neutral gas molecules or atoms admitted into region III. This device is therefore seen to be a modification of a hot-cathode Penning discharge.

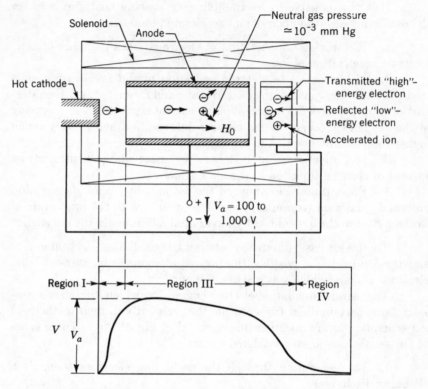

Fig. 12.21 Oscillating-electron ion engine (modified Penning discharge).

Typical performance obtained to date by an engine of this type is, according to Meyerand:

Exhaust velocity, up to 30,000 m/sec with argon
Beam power efficiency, 25 percent
Ion current density in escaping plasma beam, of the order of several amperes per square centimeter

Sputtering of the cathode (lanthanum hexaboride) was visible after several hundred hours of operation but did not interfere with normal operation of the device. Appreciably less than 50 percent of the total ion current went to the cathode.   No useful data on expellant utilization efficiency seem available yet.

The most important result from Meyerand's experiments appears to be attaining ion current densities above 1 A/cm^2 in the plasma beam, an achievement which appears rather outstanding in the present state of the art.

THE ARC-TYPE ENGINE   In the arc-type engine, ions are generated in what corresponds to region III of Fig. 12.20 by means of a high-intensity vacuum arc.   The type of arc developed by Luce[29] at Oak Ridge has been proposed for this purpose.   Because of the high ion density (up to about 10^{14} ions/cm^3) and the high degree of ionization (resulting in good fuel-utilization efficiency), this type of arc has definite merits as an ion source for an accelerated-electron engine.   The arc will be established between a hollow cathode and a cylindrical anode, the accelerated-electron beam being injected through the cathode hole.   Two possible disadvantages of this type of arc are the relatively fast rate of erosion (sputtering) of the cathode and the requirement for a relatively strong axial dc magnetic field in the arc region.   The latter, of course, is also a requirement for Meyerand's Penning-type oscillating-electron ion engine.

ACCELERATED-ELECTRON CESIUM-ION ENGINE   It is conceivable that the ions in region III of Fig. 12.20 could be provided by means of a cylindrical cesium-ion emitter, producing ions by contact ionization.   In principle, such an ion emitter could be made by a method similar to that described for the cesium-ion engine.   Possible advantages of the accelerated-electron cesium-ion engine then appear to be: (1) the ability to draw the maximum ion current density available from the ionizer, without limitations imposed by the perveance of the accelerating-ion gun electrodes, and (2) better thermal efficiency, because the thermally radiating area (approximately equal to the area of the plasma-beam cross section) can be made much smaller than the ion-emitting area.   A reduction of one order of magnitude in radiative heat loss does not seem impossible.

Further apparent advantages of the accelerated-electron cesium-ion engine over other types of oscillating-electron engines are: (1) better expellant-utilization efficiency; and (2) better control of the potentials in regions I and II to minimize the backflow of ions onto the cathode, minimize the power loss, and suppress the cathode sputtering associated with this ion backflow.

Questions still to be resolved about this type of engine are the details of potential distribution in the vicinity of the ion-injecting cylinder through which the electron beam passes in region III, the optimization of the electron-beam focusing in region III (it is possible that an appreciable dc magnetic field may be required), and a detailed understanding of the plasma potential gradient needed in region IV.   Only preliminary experiments have been performed to date on the accelerated-electron cesium-ion engine.

12.4   MAGNETIC PLASMA ACCELERATION

An alternative to electrostatic acceleration of charge particles is magnetic acceleration; Coulomb forces are used in the former, Lorentz forces in the latter.

In magnetic plasma acceleration, the accelerating force is given by $\mathbf{F} = q\mathbf{v} \times \mathbf{B}$, where q is the charge of the accelerated particle, $\mathbf{v}$ its velocity, and $\mathbf{B}$ the magnetic induction.   Because the average value of $\mathbf{F}$ can be finite even when both $\mathbf{v}$ and $\mathbf{B}$ are time-varying and quasiperiodic, the possibility exists, in principle, of accelerating a plasma with either dc or ac magnetic fields.   Both methods are being investigated and will be discussed below.

12.4.1. DC Magnetic Plasma Acceleration

A schematic of a dc magnetic plasma accelerator is shown in Fig. 12.22. An electron current I is emitted from a cathode, passed through the plasma which is to be accelerated, and collected by an anode; a dc magnetic field perpendicular to both the direction of plasma flow and of electron current flow is provided.   As a result, a volume force $\mathbf{f} = \mathbf{j} \times \mathbf{B}$ is exerted on the plasma by the flowing electron current and is available to accelerate the plasma.

Both the current flow perpendicular to a dc magnetic field and the momentum transfer from the electron current I to the ions and neutrals of the gas imply collision processes.†   Because these collisions are necessary, it also follows that a finite ohmic plasma resistivity ρ is unavoidable. Therefore, Joule heating of the plasma must necessarily accompany its acceleration by continuous Lorentz forces.   It is thus of interest: (1) to evaluate the importance of this Joule heating, (2) to compare it with the transfer of ordered kinetic energy by Lorentz forces, and (3) to determine what basic limits are imposed on this type of plasma accelerator as a consequence of Joule heating.‡

The kinetic power u_k transferred to the plasma is given by

$$u_k = T(v_2 - v_1) = IhB(v_2 - v_1) \tag{12.12}$$

---

† Devices with Hall fields in the direction of ion flow are not considered here because, in such Hall-field accelerators, acceleration is electrostatic via the Hall field, rather than magnetic via the Lorentz force.

‡ The assumption in the present considerations is that plasma heat is considered not recoverable and is therefore counted as energy loss.

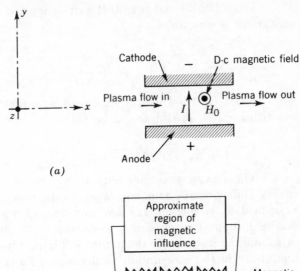

(a)

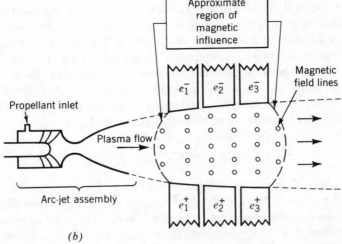

(b)

Fig. 12.22   Direct-current magnetic plasma acceleration.   (a) Principle; (b) arc-jet plasma source and dc magnetic plasma accelerator.

where T = total thrust

v_2 = exit gas velocity

v_1 = input gas velocity

I = total current flowing across magnetic field H between cathode and anode

$B = \mu_0 H$ = magnetic induction

h = distance between cathode and anode

The power lost by Joule heating of the plasma is given by

$$u_j = R_p I^2 \qquad (12.13)$$

where R_p = total plasma resistance between cathode and anode

From (12.12) and (12.13) the efficiency ε of a dc magnetic plasma accelerator is seen to be

$$\varepsilon \cong \frac{u_k}{u_k + u_j} = \frac{1}{1 + \dfrac{R_p}{hB(v_2 - v_1)} I} \tag{12.14}$$

From (12.12) I can be expressed as a function of the thrust T; substituting $I = T/hB$ into (12.14) yields

$$\varepsilon = \frac{1}{1 + R_p \, [T/h^2 B^2 (v_2 - v_1)]} \tag{12.15}$$

These expressions are only rather crude approximations. End effects, Hall currents, boundary-layer effects, power to maintain the magnetic field are all neglected; a constant channel cross section (which may not be optimum) is assumed. Nevertheless, the expressions are helpful in gaining an insight into the nature and importance of the fundamental limitations of the acceleration mechanism. To this end, it is useful to express the total plasma resistance R_p in terms of the plasma resistivity η, which is a more fundamental and more readily evaluated quantity. Under the assumption of constant plasma resistivity η throughout the region of current flow, this leads from (12.15) to

$$\varepsilon \cong \frac{1}{1 + (\eta/B^2) \, [T/V(v_2 - v_1)]} \tag{12.16}$$

where V = volume of interaction space

Equation (12.16), although approximate, is important because it shows which parameters should be optimized. For a given mission, the total thrust T will be prescribed; the exhaust velocity v_2 will also be prescribed. For an effective plasma accelerator, $v_2 \gg v_1$; hence, $v_2 - v_1 \cong v_2$. The parameters remaining available for optimization are therefore:

1. The plasma resistivity η
2. The magnetic induction B
3. The volume of the interaction space V

According to (12.16), it is desirable: (1) to minimize η and (2) to maximize $B^2 V$.

In order to evaluate the plasma resistivity η and determine how far it can be reduced, it is necessary to consider how the plasma can be injected into a Lorentz accelerator. In the devices considered and tested to date, the plasma has usually been generated by a plasma jet.[30] This implies gas temperatures of the order of 1 eV. At such temperatures, a fractional ionization of the order of several percent can be achieved by seeding the gas with an alkali metal of low ionization potential (for example, Cs, Na, or K).   This degree of ionization is high enough to make electron-neutral collisions negligible as compared with Coulomb (electron-ion) collisions.   The resistivity of such a plasma can therefore be made about as low as that of a fully ionized gas of corresponding temperature. The resistivity of a fully ionized plasma is given by[31]

$$\eta = 6.53 \times 10^3 \frac{\log \Lambda}{T_e^{3/2}} \qquad \Omega\text{-cm} \qquad (12.17)$$

where $\log \Lambda$ is a slowly varying function of density and temperature whose value is between 6 and 9 in the range of parameters of interest here and T_e (this should not be confused with those equations where T is used for thrust) is the electron temperature in degrees Kelvin.   For a temperature between 5000 and 10,000°K, this leads to a plasma resistivity of the order of $\eta \simeq 0.1$ Ω-cm.   To reduce η below this value would require a higher electron temperature.   With a plasma jet, this is not practical because of materials and life problems.   An alternative process providing additional heat to the electrons is the Joule heating itself, which results from the current flow through the interaction space of the plasma accelerator.   An upper limit to the electron temperature, however, is probably set by the heat transfer from the plasma to the electrodes with which it is in contact and by the subsequent electrode cooling and erosion problem.   The existence of sheaths at the electrodes and of temperature gradients through the plasma cross section makes difficult a more accurate evaluation of the maximum tolerable electron temperature and of the minimum corresponding plasma resistivity in the present state of the art.   An important limitation which is not taken into account in Eq. (12.16) is the effect of wall losses.   It seems that, as a consequence of these losses, dc magnetic plasma accelerators could become practical only at high power levels,[32] typically above 100 kW.   Conclusive experimental data permitting a better evaluation of the capabilities of this type of devices still appear to be lacking.

12.4.2. AC Magnetic Plasma Acceleration

In dc magnetic plasma acceleration, important limitations seem to be imposed by the presence of electrodes in direct contact with the hot plasma.   A possibility of avoiding these difficulties while still applying the concept of magnetic plasma acceleration by $\mathbf{j} \times \mathbf{B}$ Lorentz forces is to produce induced currents in the plasma by time-varying (ac) fields, thus obviating in principle the need for physical contact between plasma and electrodes.   A further advantage of currents induced by ac fields is the possibility of exerting accelerating forces on the plasma without the need for any particle collisions.   For this reason, the use of plasmas with very low resistivity is, in principle, possible.   This contrasts with dc magnetic plasma acceleration, where a finite plasma resistivity has been shown to be basically unavoidable.

TRAVELING-WAVE MAGNETIC PLASMA ACCELERATION   One class of ac magnetic fields available for plasma acceleration is traveling fields, or magnetic *traveling waves;* the other, of course, is stationary fields.

Traveling-wave devices may be subdivided† for convenience into the following categories: (1) magnetic-piston plasma accelerators, (2) traveling-wave plasma bunch accelerators, and (3) traveling-wave plasma ring accelerators.

1.   Magnetic-piston plasma accelerator.   A schematic illustration of a magnetic-piston device is shown in Fig. 12.23.   The pressure p

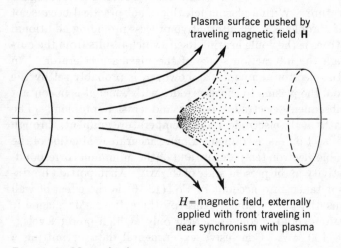

Plasma surface pushed by
traveling magnetic field **H**

$H =$ magnetic field, externally
applied with front traveling in
near synchronism with plasma

Fig. 12.23   Magnetic piston.

exerted by the magnetic field on the plasma, if the penetration of the magnetic field into the plasma is limited to a surface region, is given by

$$p = \frac{\mu_0 H^2}{2} \tag{12.18}$$

A simple derivation of this relation will be given for a one-dimensional case because it provides some insight into the physical significance

---

† This subdivision is somewhat superficial, of course, as in all three cases the force exerted on the plasma does result from the interaction between a traveling magnetic field and the currents which it induces in the plasma.

of (12.18).   Consider a slab of plasma as illustrated in Fig. 12.24.   Let a field H be established in the z direction at the left of the plane x_1.   As this field tends to penetrate into the plasma, it induces a current density

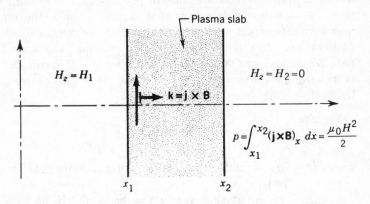

Fig. 12.24   Magnetic pressure on plasma slab.

j_y.   Then, from Maxwell's equation $\nabla \times \mathbf{H} = \mathbf{j}$, it follows that, for this one-dimensional case,

$$j_y = - \frac{\partial H_z}{\partial x} \tag{12.19}$$

The force per unit volume exerted on the plasma in the surface region is given by $\mathbf{f} = \mathbf{j} \times \mathbf{B}$,

$$f_x = j_y B_z = \mu_0 j_y H_z \tag{12.20}$$

From (12.19) and (12.20),

$$f_x = - \mu_0 H_z \frac{\partial H_z}{\partial x} = - \frac{\mu_0}{2} \frac{\partial}{\partial x} (H_z{}^2)$$

Integrating f_x between $x = x_1$ and $x = x_2$ to obtain the force per unit area yields the pressure p exerted by the magnetic field on the plasma,

$$p = - \frac{\mu_0}{2} \int_{x_1}^{x_2} \frac{\partial}{\partial x} (H_z{}^2) = \frac{\mu_0(H_1{}^2 - H_2{}^2)}{2} = \frac{\mu_0 H^2}{2}$$

if $H_1 = H$ and $H_2 = 0$.

This elementary derivation illustrates the fact that the pressure exerted by a traveling magnetic field acting as a piston on a plasma does result from the Lorentz force $\mathbf{j} \times \mathbf{B}$. A further consequence evident from this derivation is that the full magnetic pressure is exerted on the plasma only if the magnetic field is prevented from penetrating through it [otherwise, the above derivation leads to $p = (\mu_0/2)(H_1{}^2 - H_2{}^2)$, with $p = 0$ when $H_2 = H_1$!]. This means that the diffusion of the magnetic field into the plasma must be slow compared with the time available for plasma acceleration. In order to determine, in principle, the feasibility of magnetic-piston acceleration of a plasma, it is necessary to consider the process of diffusion of a magnetic field into a plasma. This process is expressed by the following diffusion equation,[33]

$$\frac{\partial \mathbf{H}}{\partial t} = D \cdot \nabla^2 \mathbf{H} \tag{12.21}$$

where $D = \eta/\mu_0$ = diffusion coefficient of magnetic field in a plasma of resistivity η

μ_0 = permittivity of vacuum = $4\pi \times 10^{-7}$ H/m

Let τ be the time available for acceleration of the plasma, l the depth to which the magnetic field is allowed to diffuse into the plasma during this time τ, and v_d the mean velocity of diffusion of the magnetic field during this time τ. The condition determining the maximum tolerable rate of diffusion of the magnetic field into the plasma is then

$$v_d \tau < l$$

Further, by definition of v_d and D,

$$v_d \simeq \frac{D}{l}$$

Hence, $$D < \frac{l^2}{\tau} \tag{12.22}$$

The time τ available for acceleration is related to the available acceleration length L and to the final plasma velocity v_p by

$$\tau = \frac{2L}{v_p} \tag{12.23}$$

which is valid under the assumption that $v_p \gg v_1$ if v_1 is the initial plasma velocity (before acceleration) and if the accelerating force is constant during the time interval τ. The following condition results, for the plasma resistivity η, from the above equation:

$$\eta \leq \mu_0 \frac{l^2 v_p}{2L} \tag{12.24}$$

For a representative numerical application, let

$$l = 10^{-2} \text{ m}$$
$$2L = 0.5 \text{ m}$$
$$v_p = 10^5 \text{ m/sec}$$
$$\mu_0 = 4\pi \times 10^{-7} \text{ H/m}$$

Then $\eta \leq 2.5 \times 10^{-5}$ Ω-m

or $\eta \leq 2.5 \times 10^{-3}$ Ω-cm

Assuming the degree of ionization of the plasma to be high enough so that Coulomb collisions will predominate over electron-neutral collisions leads, with (12.17), to the following limit for the electron temperature T_e of the plasma:

$$T_e > 6 \text{ eV (about 70,000°K)}$$

This calculation is, of course, only approximate, and it aims only at determining orders of magnitude. As such, it shows that the requirement imposed on plasma resistivity and temperatures, while not trivial, is not unrealistic. By proper preionization, or plasma injection, such plasma temperatures should, in principle, be attainable. For this reason also, plasma acceleration by a traveling magnetic piston as illustrated in Fig. 12.23 appears, in principle, possible.

In the light of the above discussion it should, however, be observed that the Lorentz body force $\mathbf{j} \times \mathbf{B}$ acts only where $\mathbf{j} \neq 0$, that is, close to the plasma surface. The material ahead of the "piston" can therefore be accelerated only by a gradient of the *particle* pressure (as in a gas-dynamic shock wave). Compression and heating can therefore not be avoided. Unless the thermal energy thus imparted to the plasma is reconverted into ordered kinetic energy by an efficient nozzle, such heating is quite undesirable.

Two modifications[34] of the device in Fig. 12.23 are shown in Figs. 12.25 and 12.26. In the device in Fig. 12.25 a "biasing" magnetic field is applied before the (traveling) field of the magnetic piston is applied. The purpose of this biasing field is to provide a magnetic insulation between the plasma and the walls. It should however be observed that a bias field parallel (in contrast to antiparallel) to the piston field leaves a "hole" in the piston and that the totality of the ionized gas therefore cannot be accelerated. In the device of Fig. 12.26 the traveling magnetic field is made sinusoidal; plasma bunches can be trapped every half wavelength in the vicinity of the axis, in the region where the magnetic field is minimum. Plasma will tend to accumulate in the dotted regions and to be carried with the magnetic field. For a high enough plasma conductivity (small rate of diffusion of magnetic field into plasma), conditions similar to those described above will prevail, with the advantage that the shape

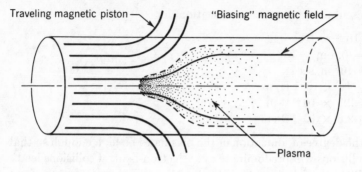

Fig. 12.25 Magnetic piston with "biasing" magnetic field.

of the lines of force of the magnetic field will tend to reduce the plasma diffusion to the walls. In the plane AA one has, with a circular cylindrical symmetry, a "cusped" field which, while it leads to a finite rate of plasma leakage, keeps this leakage below that which would exist in the absence of a magnetic field.

2. Traveling-wave plasma bunch accelerator. An alternative possibility exists for operating a device of the type shown in Fig. 12.26: if the plasma leaks through the saddle points of the magnetic field from one minimum to the preceding one, transverse azimuthal currents are induced in the plasma as a consequence of the relative motion between plasma and magnetic field. The resulting $\mathbf{j} \times \mathbf{B}$ force will still tend to accelerate the plasma, even though the force is more evenly distributed throughout the plasma volume than in the case of the magnetic piston. The advantage of this mode of operation is that it conceivably permits quasicontinuous operation, where the plasma is fed continuously into one end of the device and is dragged and accelerated by the traveling magnetic field to the other

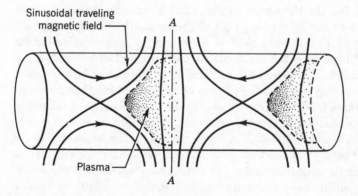

Fig. 12.26 Sinusoidal magnetic traveling-wave plasma accelerator.

end of the accelerator.   One penalty for this mode of operation, as compared with the magnetic piston, is the requirement for a higher ac magnetic field for a given average thrust density.

3.   Traveling-wave plasma ring accelerator.   The principle of a traveling-wave plasma ring accelerator is illustrated in Fig. 12.27.   It

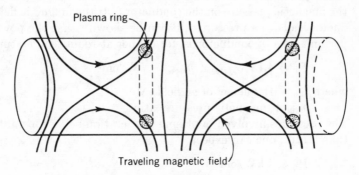

Fig. 12.27   Traveling-wave plasma ring accelerator.

does not differ basically from that of a magnetic-piston device of the type in Fig. 12.26.   Although the concept of magnetic pressure $p_m = \frac{1}{2}\mu_0 H^2$ could also be applied to the plasma ring device, it is preferable to consider directly the interaction between the magnetic field and the azimuthal current induced in the plasma ring by the change of magnetic flux enclosed by the plasma ring as a consequence of slippage between traveling magnetic field and plasma ring.   For this process of plasma acceleration to be effective, the total slippage has to be kept smaller than one-fourth the wavelength of the traveling magnetic field.   This condition is equivalent to the condition of (12.21) derived for the magnetic-piston devices and leads to similar requirements for the plasma conductivity.

Limitations and experimental results.   A feature of importance of all traveling-wave devices is that the energy stored in the accelerating magnetic field is at least of the same order of magnitude as the kinetic energy of the plasma being accelerated.   Hence, if the traveling-wave circuit were simply terminated in a matched load, this flux of magnetic energy would be lost; because it is at least equal to the flux of kinetic energy of the plasma, quite poor efficiencies would result.   One way to avoid this waste is to recuperate the magnetic energy at the output of the accelerating circuit by feeding it back through an appropriate feedback circuit to the input of the accelerating circuit; the accelerating circuit is then part of a race-track circuit.

While loss of magnetic energy at the end of the traveling-wave circuit could, in principle, be avoided, ohmic losses in the conductors of the circuit causing the magnetic field cannot be suppressed. Because of the relatively low frequencies involved (typically of the order of 1 Mc or less), these losses can be kept relatively small, at least in those cases (magnetic piston, plasma rings) where the penetration of the magnetic field into the plasma or the slippage is kept small. Indeed, in those cases, the magnetic pressure on the plasma is of the same order of magnitude as the magnetic pressure on the conductor. If the magnetic field at the conductor leads to a pressure p, it can be shown[35] that the power dissipated per unit area of conductor is, for copper at room temperature,

$$P_1 = 4.15 \sqrt{f_{Mc}} \times p_{atm} \qquad kW/cm^2$$

where f_{Mc} = frequency of ac field, Mc

p_{atm} = magnetic pressure, atm

The pressures involved in plasma acceleration being substantially smaller than 1 atm, one can expect

$$P_1 \ll 1 \ kW/cm^2$$

In comparison to this, the kinetic power transferred per unit area of accelerated plasma is

$$P_{k1} \cong p \cdot v_p \cong 10^{-2} p_{atm} v_p \qquad kW/cm^2$$

where p = magnetic pressure, N/m^2

p_{atm} = magnetic pressure, atm

v_p = plasma velocity, m/sec

It is interesting to evaluate P_1/P_{k1} from the above relations, assuming p to be about the same on plasma and conductors,

$$\frac{P_1}{P_{k1}} \simeq \frac{415 \sqrt{f_{Mc}}}{v_p} \ll 1$$

With $v_p = 10^4$ to 10^5 m/sec, it is seen that the ohmic circuit losses can be kept relatively unimportant, even if the conductor surfaces much exceeded the surface over which the magnetic pressure is exerted upon the plasma.

An additional feature common to these traveling-wave devices, where near synchronism is required between traveling magnetic field and plasma, is the necessity of increasing progressively with distance the phase velocity of the accelerating magnetic field wave; i.e., a "tapered" traveling-wave circuit is required. This applies to the magnetic-piston devices; it also applies to the traveling-wave plasma ring accelerator described above.

A further requirement common to all traveling-wave accelerators is low plasma resistivity. In all the above cases, plasma resistivity has been shown to lead to unwanted effects (diffusion of the field into the plasma, slippage of plasma rings). In other words, electron collisions with ions or neutrals are to be minimized.   It could be asked at this point how momentum is transferred to the ions in the absence of electron-ion collisions.   In all traveling-wave devices, the current flow is indeed carried by the electrons; the Lorentz force $\mathbf{j} \times \mathbf{B}$ acts, therefore, on the electrons, and the electrons alone will first be accelerated.   However, as the electrons are accelerated and leave the ions behind, a charge separation occurs and leads to a space-charge field; this space-charge field then accelerates the ions and decelerates the electrons until they all have the same average velocity.   This process is analogous to ambipolar diffusion and is effective as long as the Debye length is small compared with the dimensions of the plasma.   Returning to the requirement for minimizing electron-ion and electron-neutral collisions leads to the need for highly ionized plasmas of relatively high temperature (typically above 5×10^4 °K, as shown above).   For this reason, it does not seem practical to ionize the gas and accelerate it by means of the same electromagnetic field. Separate ionization and injection seem necessary for traveling-wave devices.   (It is indicated below that similar requirements on low plasma resistivity and separate plasma generation may also exist in stationary ac field plasma accelerators.)

This list of requirements is rather severe; even so it is not exhaustive.   In experiments performed by Penfold[36] the basic principle of this acceleration mechanism has been demonstrated.   Plasma rings with velocities of about 1.2×10^7 cm/sec ($I_{sp} \simeq 10^4$ sec) have been obtained; evidence has been found for plasma ring currents of at least several hundred amperes.   Encouraging results have also been obtained with traveling cusp plasma accelerators at AVCO.   An exhaust velocity slightly in excess of 30,000 m/sec has been obtained, with energy efficiencies of the order of 30 percent, excluding coil losses.

STATIONARY AC MAGNETIC FIELD PLASMA ACCELERATION   It is also conceivable that currents can be induced in a plasma by means of a stationary ac magnetic field and that accelerating forces can be obtained by the interactions between these currents and the inducing magnetic field.   One model of a device based upon this principle is shown in Fig. 12.28.   The ac magnetic field is produced by a single-turn coil, which is made part of a tank circuit by connection to an appropriate capacitor.   A plasma ring is produced by preionization of the gas in the vicinity of the "exit" of the coil, as shown ideally in Fig. 12.28.   The ac magnetic field induces an azimuthal current in the plasma ring; this results in a force $\mathbf{j} \times \mathbf{B}$ as shown in Fig. 12.28.   For optimum conditions the current density $\mathbf{j}$ should be in phase with the ac magnetic field $\mathbf{H}$; hence, it should be in quadrature with the induced electric field, except for the phase shift of the electrons produced by the acceleration itself.   This means that the plasma ring should

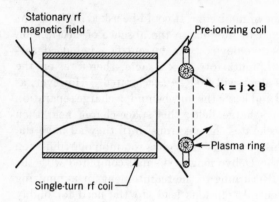

Fig. 12.28 Stationary ac magnetic field plasma
accelerator.

have only negligible ohmic resistance; in this case, as well as in that of traveling-wave acceleration, electron-ion and electron-neutral collisions are, if anything, detrimental. In the extreme, if the response of the plasma ring were purely resistive (instead of inductive) as a consequence of high ohmic resistivity, j and B would be 90° out of phase and the average force would be zero.

The requirement for minimizing collisions and plasma resistivity appears to be in contradiction to the need of collisions for ionizing the gas. For this reason, preionization of the gas or separate injection of the plasma seems necessary. This can be done on a pulse basis (where plasma rings are formed or injected periodically while the accelerating ac field is off). This procedure has been used by Miller[37] in the experiments in which he provided a preionizing coil in the vicinity of the region where the plasma ring was wanted. In Miller's experiment, the frequency (55.5 kc) and strength (up to 1 wb/m²) of the ac magnetic field were chosen of such magnitude that the plasma ring was accelerated away from the coil within about ¼ cycle. Acceleration efficiencies of the order of 5 percent have been measured, with axial exit velocities up to 10^5 m/sec; energies up to 25 J/pulse were transferred from the circuit to the axial motion of the gas.

An alternative mode of operation, still pulsed, would consist in using higher ac frequencies and lower field strengths, with the result that the plasma ring would remain several ac periods in the ac field. Because little analytical work and no experimental work seem to be available on this mode of operation, a more extensive treatment does not seem indicated at this time.

Finally, it is possible to consider continuous-wave (cw) operation. This implies continuous gas feeding and ionization in the accelerating region. Because of the apparent conflict between ionization and acceleration pointed out earlier, no satisfactory solution to the problem of cw stationary ac magnetic plasma acceleration has yet been reported. Alternating-current magnetic plasma acceleration as discussed above should not be confused with rf plasma heating and expansion through a magnetic nozzle. The latter concept, which is briefly mentioned in Sec. 12.1, Introduction, has been tested experimentally with some measure of success, overall efficiencies up to about 25 percent having been recently reported.

In conclusion, although the idea of plasma acceleration by stationary ac magnetic fields seems interesting, both the problem of proper plasma or gas injection and the problem of optimization of the circuit geometry for optimum utilization of the ac magnetic field still remain to be solved.

12.4.3. Pulsed Magnetic Plasma Acceleration

An example of a pulsed magnetic field plasma accelerator is shown in Fig. 12.29. This *hydromagnetic plasma gun*, in the form invented by

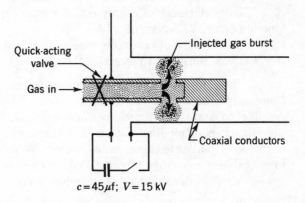

$c = 45 \mu\text{f}; \; V = 15 \text{ kV}$

Fig. 12.29 Hydromagnetic plasma gun.

Marshall,[38] consists of two coaxial conductors with provision for pulsed admission of gas burst through openings in the center conductor.   Shortly after a burst of gas is admitted between the two coaxial conductors, the power from a high-voltage capacitor bank is suddenly applied.   The gas is ionized, a large radial current flows through it, and an azimuthal magnetic field is established by the flow of the current through the central conductor. This azimuthal magnetic field interacts with the radial plasma current to produce axial $\mathbf{j} \times \mathbf{B}$ forces, which results in a magnetic pressure of the order of $\frac{1}{2}\mu_0 H^2$.   In this respect, this device is analogous to the magnetic-piston type of accelerators described above.   The essential differences are that, in the hydromagnetic plasma gun: (1) current flow through the plasma results from direct contact to the electrodes instead of being induced, and (2) the magnetic field is supported, not by a slow-wave circuit, but by a simple coaxial line.

Basic limitations to this type of plasma accelerator may result from:

1.   Difficulty of efficient energy transfer from the capacitor bank or pulsing circuit to the coaxial line

2.   Loss of magnetic energy tending to escape or be reflected from the end of the plasma gun

3.   Power loss due to physical contact between the plasma and electrodes

4.   Disadvantage of requiring a pulsed power supply

Typical characteristics quoted for the performance of such a plasma accelerator are[38]

Output gas (H_2) velocity 1.5×10^5 m/sec

Total kinetic energy of gas, per pulse, 1,200 J

Total efficiency of energy transfer from capacitor bank to directed kinetic energy, 23.5 percent

No evidence of electrode erosion observed under proper timing of the electrical pulse with respect to the gas burst

These results so far appear encouraging.   Because of the limitations suggested above, much still remains to be done before it will be possible to state how much performance can be improved.   Further investigations of modifications of the Marshall gun[39] have led so far to results very similar to those quoted above, the major difference being that lower exhaust velocities (about 60,000 m/sec) have been obtained with heavier gases, with about the same efficiency.

12.5   ELECTROMAGNETIC PLASMA ACCELERATION

In a nonuniform ac electromagnetic field, a charged particle of charge e experiences an average force given by[40]

$$\mathbf{F} = \frac{-e^2}{\epsilon_0 m_e \omega^2} \cdot \nabla E^2 \tag{12.25}$$

Although it does not appear explicitly, the effect of the ac magnetic field is also included in (12.25).   [The magnetic field has been eliminated by the use of Maxwell's equation $\nabla \times \mathbf{E} = -\mu_0(\partial H/\partial t)$, to arrive at (12.25).]

In a plasma, the force of (12.25) will be acting mainly on the electrons and will produce a force f per unit volume given by

$$\mathbf{f} = -\left(\frac{\omega_p}{\omega}\right)^2 \nabla\left(\frac{\epsilon_0 E^2}{2}\right) \tag{12.26}$$

where $\omega_p \equiv$ electron plasma frequency

$E =$ rms rf electric field

(Momentum transfer from electrons to ions is caused by space-charge forces in the same process as described above.)

In electromagnetic plasma acceleration, forces of the type given by (12.26) are produced by nonuniform rf fields and are being used to accelerate plasmas. It may be observed that when the rf field penetration inside the plasma is negligible (which it will tend to be at high plasma densities), (12.26) can be integrated to lead to a pressure $p = \epsilon_0 E^2/2$ from the rf field on the plasma surface. A more rigorous derivation would lead to

$$p = \tfrac{1}{2}(\mu_0 H^2 + \epsilon_0 E^2)$$

which may also be interpreted as radiation pressure.

A general feature of (12.25) and (12.26) is the indication that the rf fields will tend to push the plasma toward regions of minimum rf electric field. When the rf electric field is parallel to the plasma surface, the plasma can actually be pushed toward these regions where the rf vacuum field would be minimum. When the rf electric field is parallel to the gradient of the plasma density, this conclusion is no longer always true; it applies only when the frequency of the rf fields is higher than the electron plasma frequency. If the rf frequency is lower, a "peeling" effect takes place at the plasma surface and tends to draw the plasmas toward the region of maximum rf electric fields.[35]

The only attempt at electromagnetic plasma acceleration reported to date is that of Reboul et al.[41] In these experiments, fields substantially parallel to the plasma density gradient were used, as shown in Fig. 12.30. For this reason, plasma acceleration could be obtained for only the relatively low plasma densities corresponding to a plasma frequency below that (140 Mc) of the applied rf fields ($n \gtrsim 3 \times 10^8$ electrons/cm³). Under these conditions, plasma acceleration up to velocities of the order of 2.5×10^4 m/sec has been observed. Because of the low plasma density, however, resulting thrusts would be too low to be of practical interest, even if higher microwave frequencies were used.

In conclusion, the basic principle of electromagnetic plasma acceleration has been demonstrated. To make it useful for propulsion purposes, however, it will be necessary to devise means of applying the electromagnetic rf fields perpendicular to the density gradient of the plasma, avoiding rf electric fields normal to the plasma surface, and thus being

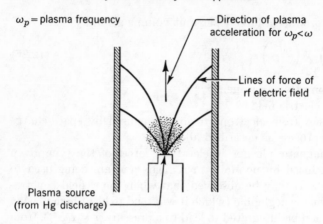

ω_p = plasma frequency

Direction of plasma acceleration for $\omega_p < \omega$

Lines of force of rf electric field

Plasma source (from Hg discharge)

Fig. 12.30   Electromagnetic plasma accelerator.

able to operate with relatively high plasma densities where the plasma frequency exceeds the frequency of the accelerating rf fields.

12.6   SPACE SIMULATION

We shall discuss here fairly briefly some of the problems associated with the testing and evaluation of performance of electrical propulsion engines. As we shall see, the simulation of space environment for an electrical propulsion engine is somewhat more complex than the usual space-simulation problem; however, adequate and detailed ground testing of these engines is an essential step in their development.   There are a number of reasons for the importance of this ground testing, including the relatively high cost of conducting space tests, the much greater detail with which the physical phenomena can be studied on the ground, the ease of making changes and adjustments in the experiment, etc.; all point toward very comprehensive ground tests prior to the real test in space.

Since electrical propulsion engines are to operate in the excellent vacuum of space, it is obvious that they must be tested under conditions where the background gas pressure will not influence the operation.   The tolerable level of background pressure depends, of course, on the mean-free path of the ions in this residual gas but will in general be of the order of 10^{-7} mm Hg.   Additional features of the physical-environment simulation which may be desirable are: (1) cold surfaces to simulate the regions of infinite space as far as radiation from the engine is concerned, (2) artificial solar radiation to simulate the power received from the sun, and (3) possibly other types of radiation such as γ rays, cosmic rays, etc.   All these physical-environment factors have been successfully simulated in various test chambers.   The most difficult aspects of space simulation, as far as electrical engines are concerned, are the electrical conditions.   It is apparent that the walls and collector of the test chamber must not act as sources of electrons; i.e., suitable grids or other barriers must be arranged so that the bombardment by ions or other particles does not cause emission of electrons which can get into the beam and influence its behavior.   A second and equally important aspect of the electrical-simulation problem arises because in the space environment the ion beam is sent into essentially an infinite region, with no boundary condition beyond the edges of the space vehicle.   In the ground-based test chamber, walls must be imposed around the engine and beam in order to provide the vacuum, and a collector must be provided to collect the ion beam; both these arrangements impose artificial boundary conditions at the edges and the end of the ion beam.   The discussion on ion-beam neutralization (Sec. 12.3.1) indicates that tests conducted with a floating collector may constitute an adequate simulation of the conditions in space, provided that secondary electron emission from the collector is avoided.   Pulsed tests can supplement the dc tests to reveal the effect of such secondary electrons, should this effect be important.

12.7 APPENDIX: MOMENTUM TRANSFER BETWEEN ELECTRONS AND IONS IN ACCELERATED - ELECTRON ENGINES

This calculation is intended to determine the rate of momentum transfer from energetic electrons to ions in a region in which a plasma potential gradient exists in the absence of electrostatic forces between plasma and electrodes.   Such a region would be the equivalent of region IV of Fig. 12.20 and will be referred to simply as region IV.   It should, however, be observed that, in the absence of electrode fields, it does not appear possible in a one-dimensional situation to maintain such a region at a fixed region in space; it appears that it should move in space (e.g., at the beam front) with the mean velocity of the accelerated ions.   Assuming, now, that the axial electrostatic forces between plasma and electrodes in region IV are negligible, the flux of momentum flowing into the plasma from the left of region IV must be equal to the flow of momentum carried by the plasma out to the right of region IV,

$$\frac{du_{in}}{dt} = \frac{du_{out}}{dt} \tag{12.27}$$

where du_{in}/dt = flux of momentum into region IV (from left)
du_{out}/dt = flux of momentum out of region IV (with plasma beam to right)

Assuming most electrons to be reflected in region IV and assuming the ion energy in region III to be much smaller than the electron energy leads to the following expression, for the flow of momentum du_{in}/dt into region IV,

$$\frac{du_{\text{in}}}{dt} = n_{e3}m_e\overline{v_{e3}^2} \qquad (12.28)$$

where n_{e3} = electron density in region III

m_e = mass of electron

v_{e3} = axial electron velocity in region III

$\overline{v_{e3}^2}$ = mean-square axial electron velocity in region III

(Note that, because of the assumption of near total reflection, $n_{e3}/2$ electrons/cm³ will have a velocity directed toward region III and will experience a change of momentum equal to $2m_e v_{e3}$ upon reflection; also, the velocity distribution of electrons flowing in the $+x$ and $-x$ directions will be nearly identical.)

Assuming, at the exit of region IV, the electron energy to be much smaller than the energy of the accelerated ions leads, for the flux of momentum leaving region IV, to

$$\frac{du_{\text{out}}}{dt} = n_{i3}m_i\overline{v_{i3}v_{i4}} \qquad (12.29)$$

where n_{i3} = ion density in region III

m_i = ion mass

v_{i3} = axial velocity of ions flowing toward region IV

v_{i4} = axial velocity of ions leaving region IV

Equations (12.28) and (12.29) say that the momentum flux into region IV is essentially carried by the electrons, while the momentum flux out of region IV is essentially carried by the accelerated ions [in effect, the latter assumption means that $(m_i + m_e)v_{i4} \simeq m_i v_{i4}$].

By using (12.27) to (12.29), it is possible to relate the oscillating-electron current density in region III to the ion current density emerging in the plasma beam to the right of region IV. To this effect, the following quantities are defined:

j'_{e3} = density of electron current flowing in $+x$ direction (away from cathode) in region III

j''_{e3} = density of electron current flowing in $-x$ direction in region III

$j_{e3} = j'_{e3} + j''_{e3} = |j'_{e3}| - |j''_{e3}|$

j_i = density of ion current flowing in $+x$ direction (no ions reflected toward $-x$ direction exist with the potential distribution shown in Fig. 12.21)

By using (12.27) to (12.29) the following relation is found between

j'_{e3} and j_i,

$$\frac{-j'_{e3}}{j_i} = \frac{1}{2}\sqrt{\frac{m_i}{m_e}} \cdot \sqrt{\frac{\bar{v}_{i4}}{\bar{v}_{i3}}}$$

with the simplifying assumptions, in (12.28) and (12.27), that $\overline{v_{i3}v_{i4}} \cong \bar{v}_{i3} \cdot \bar{v}_{i4}$ and $\overline{v_{e3}^2/v_{e3}} \cong \bar{v}_{e3}$.

Under these conditions,

$$\bar{v}_{i4}/\bar{v}_{i3} \cong \sqrt{W_{i\,out}/W_{i\,in}}$$

where $W_{i\,out}$ is the average energy of ions leaving region IV (to the right) and $W_{i\,in}$ is the average energy of ions in region III (at the entrance to region IV). Hence,

$$-j'_{e3} \simeq \left(\frac{1}{2}\sqrt{\frac{m_i}{m_e}} \cdot \sqrt[4]{\frac{W_{1\,out}}{W_{i\,in}}}\right) \cdot j_i \tag{12.30}$$

REFERENCES

1. Goddard, R. H.: Notebook (dated Sept. 6, 1906), *Astronautics*, **4**, 24–27 (April, 1959).
2. Oberth, H.: "Wege zur Raumschiffahrt," R. Oldenbourg, Munich, Berlin, 1929.
3. Langmuir, D. B.: "Space Technology," pp. 9–18, Wiley, New York, 1959.
4. Fay, C. E., A. L. Samuel, and W. Shockley: On the Theory of Space Charge between Parallel Plane Electrodes, *Bell System Tech. J.*, **17**, 49–79 (January, 1938).
5. Ramo-Wooldridge Research Laboratory Staff: Electrostatic Propulsion, *Proc. IRE*, **48**, 477–491 (April, 1960).
6. Wells, W. H.: *Jet Propulsion Lab., Pasadena, Calif., Tech. Release* 34-118, 1960; H. Mirels: On Ion Rocket Neutralization, "Progress in Astronautics and Rocketry," vol. 5, Electrostatic Propulsion, Academic, New York, 1961; W. Eckhardt: Private communication.
7. Buneman, O.: Dissipation of Currents in Ionized Media, *Phys. Rev.*, **115**, 503 (1959).
8. Bernstein, W.: Oscillations in Synthetic Plasma Beams, *Phys. Fluids*, **6**, 1032 (1963).
9. Knechtli, R. C., and J. Y. Wada: Collisionless Thermalization of Electrons in a Highly Ionized Plasma, APS Meeting, San Diego, Calif., Nov. 6–9, 1963; *Bull. Am. Phys. Soc.*, **9**, 341 (1964).
10. Etter, J. E., et al.: Neutralization of Ion Beams, in "Electrostatic Propulsion," pp. 357–372, Academic, New York, 1961.
11. Brewer, G. R., J. E. Etter, and J. R. Anderson: Design and Performance of Small Model Ion Engines, presented at the American Rocket Society Meeting, Los Angeles, Calif., May 9, 1960; *Paper ARS* #1125-60.

12. Sellen, J. M., et al.: Beam Diagnostic Techniques, in "Electrostatic Propulsion," p. 451, Academic, New York, 1961; J. M. Sellen and R. F. Kemp: Cesium Ion Beam Neutralization in Vehicular Simulation, presented at the American Rocket Society Meeting, Los Angeles, Calif., June 13–16, 1961; *Paper* 61-84-1778.

13. Ward, J. W., and R. A. Hubach: Neutralization of Ion Beams from Engines of Annular Geometry, *Am. Rocket Soc. J.*, **32,** 1730 (1962).

14. Taylor, J. B., and I. Langmuir: The Evaporation of Atoms, Ions and Electrons from Cesium Film on Tungsten, *Phys. Rev.*, **44,** 423–458 (September, 1933).

15. Husmann, O.: These data were obtained from a comprehensive study of surface ionization being carried out by Dr. O. Husmann at the Hughes Research Laboratories, Malibu, Calif.

16. Pierce, J. R.: Rectilinear Electron Flow in Beams, *J. Appl. Phys.*, **11,** 548–554 (March, 1940).

17. Brewer, G. R.: Note on the Determination of Electrode Shapes for a Pierce-type Electron Gun, *J. Appl. Phys.*, **28,** 634 (May, 1957).

18. Kaufman, H. R.: An Ion Rocket with an Electron Bombardment Source, Lewis Research Center, Cleveland, Ohio, *NASA Tech. Note* D-585, 1960.

19. King, H. J., W. O. Eckhardt, W. Ward, and R. C. Knechtli: Electron-bombardment Thrustors Using Liquid Mercury Cathodes, AIAA Meeting, San Diego, Calif., March 7–9, 1966; *Paper* 66-232.

20. Sohl, G., R. C. Speiser, and J. A. Wolters: Life Testing of Electron-bombardment Cesium Ion Engines, AIAA Meeting, San Diego, Calif., March 7–9, 1966; *Paper* 66-233.

21. Luce, J. S.: Ion Erosion of Accelerating Electrodes in Space Vehicles, presented at the American Rocket Society Meeting, Los Angeles, Calif., May 9, 1960; *Paper* 1159-60.

22. Eckhardt, W. O., J. Hyman, G. Hagen, C. R. Buckey, and R. C. Knechtli: Research on Ion Beam Formation from Plasma Sources, AIAA Meeting, New York, Jan. 20–22, 1964; *Preprint* 64-8.

23. Von Ardenne, M.: Electronenphysik, Ionenphysik und Ubermikroskopie, *VEB deut. Verlag Wiss.*, 1956, p. 554.

24. Hayes, R. J., C. A. Heubner, and J. M. Glassmeyer: The Duoplasmatron Ion Rocket, M.S. thesis, Massachusetts Institute of Technology, Cambridge, Mass., June, 1960.

25. Burton, B. S.: The Duoplasmatron: Theoretical and Experimental Studies, presented at the Electrostatic Propulsion Conference, Monterey, Calif., November, 1960; "Electrostatic Propulsion," p. 21, Academic, New York, 1961.

26. Bond, A. F., et al.: Summary Report on Bombardment Ion Engine Development, *NASA Rept.* ER-5180, Contract NAS 8.42, Thompson-Ramo Wooldridge, October, 1963.

27. Rose, D.: Acceleration of a Neutralized Ion Beam, *M.I.T. Research Labs. on Electronics*; *RLE Quart. Rept.* 53, p. 3, Apr. 15, 1959; F. Salz, R. G. Meyerand, et al.: Electrostatic Potential Gradients in a Penning Discharge, *Phys. Rev. Letters*, **6,** 523–525 (Nov. 10, 1961).

28. Davis, J. W., et al.: Theoretical and Experimental Descriptions of the Oscillating Electron Ion Engine, presented at the National IAS-ARS Joint Meeting, Los Angeles, Calif., June 13–16, 1961.

29. Luce, J. S.: Intense Gaseous Discharge, *Proc. Second UN Internatl. Conf. on Peaceful Uses of Atomic Energy*, Geneva, 1958, vol. 31, pp. 305–314.

30. Giannini, G. M.: The Arc Jet, presented at the Second Symposium on Advanced Propulsion Concepts, Boston, Mass., Oct. 7–8, 1959.

31. Spitzer, L.: "Physics of Fully Ionized Gases," p. 139, Interscience, New York, 1962.

32. Janes, G. S.: Magnetohydrodynamic Propulsion, *Avco-Everett Research Lab. Research Rept.* 90, August, 1960.

33. Cowling, T. G.: "Magnetohydrodynamics," p. 4, Interscience, New York, 1957.

34. Clauser, M. U.: The Magnetic Induction Plasma Engine, *Space Technol. Lab., Los Angeles, Calif., Rept.* STL/TR-60-0000-00263, Aug. 19, 1960. The writer has here proposed devices of the type shown in Figs. 12.23 and 12.25.   R. X. Meyer: Magnetic Plasma Propulsion by Means of a Traveling Sinusoidal Field, in S. W. Kash (ed.), "Plasma Acceleration," p. 37, Stanford, Stanford, Calif., 1960.   The writer has here proposed devices of the type shown in Fig. 12.26.

35. Dow, D., and R. C. Knechtli: Plasma Containment by R.F. and D.C. Field Combinations, *J. Electronics Control*, **7**, 316–343 (November, 1959).

36. Penfold, A. S.: Experimental Results Concerning the Electromagnetic Acceleration of Plasma Toroids, presented at the Fourth Air Force Office of Scientific Research Contractor's Meeting on Ion and Plasma Acceleration, Beverly Hills, Calif., Apr. 20–21, 1961.

37. Miller, D. B.: Measurements on an Experimental Induction Plasma Accelerator, *Am Rocket Soc. J.*, April, 1962, pp. 549–552.

38. Marshall, J.: Hydromagnetic Plasma Gun, in S. W. Kash (ed.), "Plasma Acceleration," pp. 60–72, Stanford, Stanford, Calif., 1960.

39a. Garovity, B., and P. Gloersen: Experimental Performance of a Pulsed Gas Entry Coaxial Plasma Accelerator and Applications to Space Missions, presented at the American Rocket Society Meeting, New York, December, 1960; *Paper* 1535-60.

39b. Gooding, T. J., B. R. Hayworth, and R. H. Lovberg: Development of a Coaxial Plasma Gun for Space Propulsion, *Gen. Dynamics Astronaut. Rept.* GD/A63-0454, 1963.

40. Boot, H. A. H., et al.: Containment of a Fully Ionized Plasma by R.F. Fields, *J. Electronics Control*, **4**, 434–453 (April, 1958).

41. Reboul, T. T., C. D. Gordon, and G. A. Swartz: Plasma Acceleration by a Quasistatic RF Electric Field Gradient, presented at the ARS Meeting, Washington, D.C., Dec. 5–8, 1960; *Paper* 1532-60.

13

CONTROLLED THERMONUCLEAR FUSION

PYLE

ROBERT V. PYLE, *Senior Physicist, Lawrence Radiation Laboratory, and Lecturer in Nuclear Engineering, University of California, Berkeley, California*

13

13.1 INTRODUCTION

The controlled thermonuclear fusion program is a worldwide effort to produce useful amounts of energy from exothermic reactions between the nuclei of light elements. This program utilizes and depends on the most advanced techniques from many branches of modern science and engineering, but the main emphasis at present is of necessity directed toward the understanding of the physics of hot plasmas, in particular stability.

Although the present effort is devoted mainly to understanding and manipulating plasma phenomena, the feasibility of a *useful* fusion reactor also depends critically on nuclear and atomic cross sections and on engineering considerations such as achievable magnetic field strengths and field shapes, vacuums, neutron blankets, and the like. The few preliminary engineering studies that have been made suggest cautious optimism if instabilities can be suppressed.

In the present chapter we mention only a few of the considerations that are involved and refer the reader to more exhaustive treatments in books and journals.[1] The presentation is oriented toward the prevailing philosophy that a useful reactor must operate in the steady state, or nearly so, with the plasma fuel stably confined by magnetic fields. However, it should be noted that some scientists argue that long-time confinement may not be necessary or even possible. In the latter case it may be possible

454

to produce power in bursts during the relatively short time that a very dense and hot plasma remains assembled.   This might be accomplished by a powerful magnetic compression, or by firing high-velocity particles of solid fuel at a target, or by heating fuel with an intense burst from a laser. It is hard to evaluate and perhaps premature to consider the eventual usefulness of most of these techniques.

13.2 NUCLEAR REACTIONS

The necessary plasma parameters of a power reactor are strongly influenced by the properties of the nuclear-fusion reactions.   The rapid increase of Coulomb repulsion with increasing nuclear charge of the fuel means that only the lightest elements can be considered.   Of these the d-d, d-t, and d-He^3 are the most promising.   The first has the advantage that the fuel is abundant in nature, the second that the reactions occur with the smallest energy input, and the third that the reaction products both carry a charge and can be deflected by magnetic fields.   The energy release per particle, in million electron volts, is indicated in parentheses.

$$d + d \rightarrow n(2.45) + He^3 (0.82)$$
or
$$\rightarrow p(3.02) + t(1.01)$$
$$d + t \rightarrow n(14.1) + He^4(3.52)$$
$$d + He^3 \rightarrow p(14.7) + He^4(3.7)$$

It is obvious from the above reactions that useful fusion devices will be potent neutron sources and will require about as much personnel shielding as fission reactors do.   (This is true even for d-He^3 reactors because d-d reactions in the fuel will be unavoidable.)   For example, a power plant equivalent to an automobile engine would produce at least 10^{17} neutrons/sec and would require several feet of concrete for personnel shielding.   For this and other reasons it seems likely that a fusion power source will be a large installation.

Fusion cross-section data have been combined by Arnold et al.[2] and others to give semiempirical expressions of the general barrier penetration form $\sigma = (a/W) \exp(-b/\sqrt{W})$.   Here W is the relative particle energy, and a and b are constants.   The advantage of large energy is apparent from the exponential behavior.

Somewhat more complicated expressions, valid over a wider energy range, have been published recently,[3]

$$\sigma_{dd} = \frac{1.55 \times 10^7}{W} (e^{-31.4/\sqrt{W}})[(W - 300)^2 + 6.09 \times 10^4]^{-1}$$

$$\sigma_{dt} = \frac{4.51 \times 10^7}{W} (e^{-34.4/\sqrt{W}})[(W - 48.8)^2 + 1.71 \times 10^3]^{-1}$$

$$\sigma_{dHe3} = \frac{3.33 \times 10^8}{W} (e^{-68.8/\sqrt{W}})[(W - 183)^2 + 2.04 \times 10^4]^{-1}$$

Here W is the kinetic energy in the center-of-mass system in kiloelectron volts and the cross sections are in barns. These fit the data to 5 percent for center-of-mass energies up to about 100 keV. Reaction rates can be calculated from expressions of this kind if the velocity distribution functions of the ions are known. The results are usually given as a rate parameter,

$$\langle \sigma v \rangle = \int f_1(\mathbf{v}_1) f_2(\mathbf{v}_2) v \sigma(v) \, d\mathbf{v}_1 \, d\mathbf{v}_2$$

where $\mathbf{v}_1$ and $\mathbf{v}_2$ are the velocities of the two interacting species and v is the relative velocity. The reaction rate per unit volume is then $R = n_1 n_2 \langle \sigma v \rangle$ (or $R = \frac{1}{2} n^2 \langle \sigma v \rangle$ if there is only one species), where n is the ion density.

The distribution functions $f(v)$ are generally assumed to be Maxwellian; this assumption gives $\langle \sigma v \rangle$ values that are approximately correct for any reasonable distribution provided that the mean particle energies are sufficiently high. Cases of interest to power production fall in this category. However, at lower energies the reaction rates are strongly dependent on the shapes of the distribution functions. For example, if the mean energy of ions in a deuterium plasma is 1 keV, then $\langle \sigma v \rangle$ for an isotropic distribution in which all the ions have the same energy is about 100 times smaller than for a Maxwell-Boltzmann distribution.

For Maxwell distributions the rate-parameter expressions obtained from the cross-section data are given in Ref. 3 and Fig. 13.1.[4]

13.3 GENERAL PLASMA PARAMETERS

A rough idea of plasma density and temperature suitable for a quasi-steady-state fusion reactor can be obtained by noting that the shielding and complexity of the apparatus will be as great as in a fission reactor, and we can expect that similar power densities will be required for competitive operation, give or take an order of magnitude or so. Economic considerations suggest that it will be hard to operate at temperatures much above 100 keV and impossible at temperatures much below, say, 20 keV. At 100 keV the power densities, in order of magnitude, are $10^{-35} n^2$, $10^{-33} n^2$, and $10^{-34} n^2$ W/m³ for the d-d, d-t, and d-He^3 reactions, respectively. The exact numbers depend on whether reaction products are also burned, velocity distributions, etc. The *maximum* particle densities that can be contained by magnetic fields at a given temperature are given by $nkT = B^2/2\mu_0$, and if we assume that $B = 10$ Wb/m² is as high a magnetic field as is technologically and economically feasible in a large volume, then power densities of 10^6 to 10^8 W/m³ are consistent with temperatures of the order of 100 keV and densities of 10^{20} to 10^{22}/m³, depending on the fuel mixture. (The mass densities are accordingly very small, $\sim 10^{-6}$

kg/m³, corresponding to room-temperature pressures of about 1 to 100 millitorr.) The optimization of density and temperature of course depends on many considerations; e.g., closed toroidal devices may be able to operate at considerably lower temperatures than open-ended mirror machines.

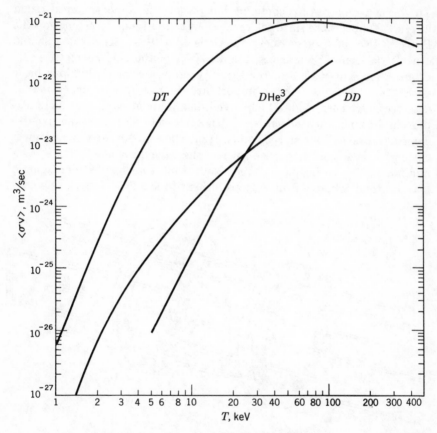

Fig. 13.1 $\langle \sigma v \rangle$ for fusion reactions assuming Maxwellian distributions. [*Adapted from Ref.* 4. *J. L. Tuck, Nuclear Fusion,* **1**, 201 (1961).]

It is worthwhile to estimate how long an average ion must be contained if the energy expended in heating the fuel is to be recovered from the nuclear-reaction products. One estimate considered Coulomb bremsstrahlung to be the only loss mechanism.[5] After optimizing the ion temperature to give the most advantageous situation, Lawson found that containment times of $t > 10^{22}/n$ and $10^{20}/n$ are required for the d-d and d-t reactions respectively. Here n is the ion density/m³. These estimates increase as other loss mechanisms are taken into account; roughly speaking, times of the order of seconds appear necessary for steady-state devices.

13.4 PLASMA CONFINEMENT

The basic geometries for confining hot plasmas have not changed much in general appearance since the revelations of the First Conference on the Peaceful Uses of Atomic Energy at Geneva in 1958.   The reader should consult the general references[1] for drawings of the field geometries and Chap. 5 for a discussion of the stability properties of the various basic geometries.   However, later important theoretical and experimental advances have been made which have brought about significant modifications in the magnetic field shapes.   A notable example is the experimental demonstration by Ioffe and coworkers[6] that a plasma can be contained much better if it is created in a region where the vacuum magnetic field has a minimum.   The principle has been known for a long time: diamagnetic plasmas tend to move from regions of high to low field.   Desirable con-

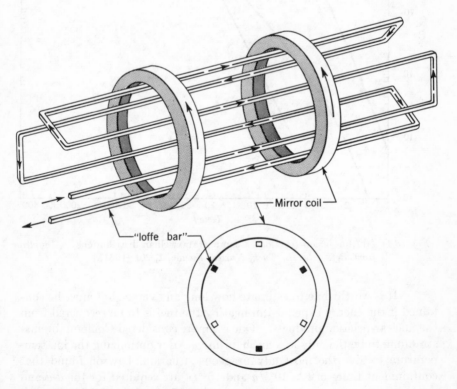

Fig. 13.2 The Ioffe minimum-*B* geometry.

figurations exist in earlier proposed geometries, e.g., cusp, picket fence, and Astron, but the interest in such field shapes has increased enormously since Ioffe's demonstration.

Ioffe's combined mirror and cusp geometry is sketched in Fig. 13.2. This example has the advantage over a simple mirror geometry that the magnetic field strength increases away from the plasma to give macroscopic stability and over the simple cusp configuration in that the field does not go to zero within the plasma, and consequently adiabatically invariant motion can be retained.   The application of the "minimum B" principle has substantially increased the lifetime of plasma in a high-energy injection mirror experiment.[7]   There are a number of geometries with the desirable properties mentioned above.   A variation is the "tennis-ball seam" coil of Larkin,[8] which offers advantages for neutral beam injection (Fig. 13.3).

The creation of a minimum-B field throughout the volume of a toroidally confined plasma is not possible, but considerable advantage is

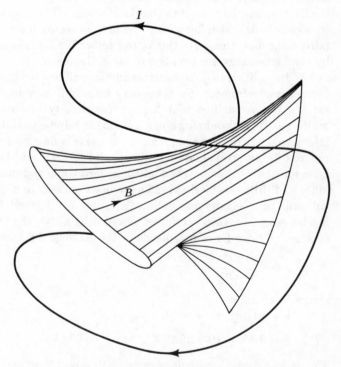

Fig. 13.3   Field lines of a "tennis-ball seam" winding. (F. M. Larkin, A Single Loop Conductor for Producing a Magnetic Well, CLM-R 37, 1964.   Unpublished.)

obtained if a minimum average-B situation exists.[9]   Tuck has shown how such a situation can be created by rigid conductors within the plasma,[10] and Furth and Rosenbluth have demonstrated that this "maximum-$\int dl/B$" condition can be satisfied with purely external conductors.[11]   It should be noted that none of these configurations guarantees microscopic stability.   In most of the present higher-density experiments, operation regimes having apparent macroscopic stability still exhibit particle loss rates much larger than are calculated from classical theory.   In some cases the loss rates appear to correspond to those predicted by Bohm diffusion (Chap. 4).   It is therefore important to calculate upper limits to plasma loss due to nonmicroscopic instabilities; such studies have been started by Fowler.[12]

We see a trend toward more complex and more precisely controlled magnetic field configurations, the latter brought about by observations in several types of apparatus that changes of 0.1 percent or less in the magnetic field distributions can have profound effects on the confinement times. In an actual thermonuclear reactor it will probably be necessary to have a fairly thick layer between the plasma and the field coils for cooling of the liner and moderation of the hard radiation.   This may make it difficult to do exotic field shaping within the plasma region unless the latter is of fairly large diameter.   (In the Astron device the field shaping is produced by the high-energy electron layer within the plasma.)

In addition to quasistatic magnetic confining fields, electric or high-frequency electromagnetic fields may be useful in reducing plasma loss rates.   The possibility that high-energy-density electromagnetic fields may be useful for confining a plasma and/or helping to stabilize it is under intensive study in the Soviet Union.   Some encouraging results have been reported,[13] but no evaluation of the possible technological importance can be made at this time.   When all the macroscopic and microscopic instability and drift loss mechanisms have been reduced to a minimum, there will still be a maximum confinement time set by end losses in linear machines and by classical diffusion and drift across the fields in toroidal devices.   The order of magnitude of the classical diffusion time is

$$\tau \approx \frac{(R/2.4)^2 \, B^2}{nkT \, \eta} \tag{13.1}$$

where R = radius

η = plasma resistivity,

$$\eta \approx 1.65 \times 10^{-9} \ln \Lambda T_e^{-3/2} \Omega\text{-m} \qquad T_e \text{ in keV} \tag{13.2}$$

This time may be hundreds of seconds in power reactors.

Mirror machines lose particles out the ends by Coulomb scattering into a loss cone.   The characteristic time is approximately the time for scattering through a large angle.   From Eq. (4.10) we can obtain [e.g., see eq. (8.24) of Ref. 1A4]

$$\tau_{ii} \approx 6.4 \times 10^{17} T_i^{3/2} (n_i \ln \Lambda)^{-1} \qquad \text{sec}$$
$$\text{for deuterons, } T_i \text{ in keV} \quad (13.3)$$

This expression must be multiplied by $[1 - (1 - 1/R)^{1/2}]^{-1}$, where R is the mirror ratio, to get the actual loss time.   The result may be a few seconds for reactor parameters.   One might ask why this additional loss mechanism should be tolerated in a fusion experiment; i.e., why not stick to toroidal devices?   The answer is that mirror machines offer several geometrical advantages to the experimenter, allowing him to make diagnostic measurements and perform some plasma-physics experiments in the easiest way.   It is even possible that they will turn out to be more stable than toroidal devices, although the reverse generally has been expected.

Charge exchange is another phenomenon that causes loss of hot-fuel ions.   It may not be important when thermonuclear conditions are reached, but it can seriously inhibit the buildup of the plasma and is in fact the dominant loss mechanism in some of the present high-energy injection experiments.   The charge-exchange cross sections of deuterons in typical gases are shown in Fig. 13.4.   The lifetime of an ion in a system that is dominated by charge exchange is $1/n_0 \sigma_{10} v$, where n_0 is the neutral atom density, σ_{10} the electron-capture cross section, and v the ion speed. Using the data of Fig. 13.4, we find, for example, that 20- and 100-keV deuterons in molecular nitrogen at a pressure of 10^{-8} torr have lifetimes of only 0.01 and 0.03 sec, respectively.   So it is necessary to start with very good vacuums or remove the impurities early in the plasma cycle.

13.5 PLASMA PRODUCTION AND HEATING

Methods for producing plasma have been described in Chap. 10.   The ways that are most applicable to thermonuclear experiments are (1) the ionization of cold neutral gas within the confinement region by pulsed or high-frequency electric fields or by high-energy electrons, (2) the injection of a "blob" of high-density plasma from some sort of gun, and (3) the gradual buildup of plasma within an evacuated region by ionizing or dissociating a high-energy beam.   Each method has advantages and disadvantages.

The ionization *in situ* of cold neutral gas is the oldest and easiest way to build up a high-density plasma.   However, the plasma must then be heated by several orders of magnitude to reach thermonuclear temperatures, and the plasma may be driven unstable during this process or may accumulate impurities.   The injection of blobs of high-density high-energy plasma at least partially solves the heating problem, but the pres-

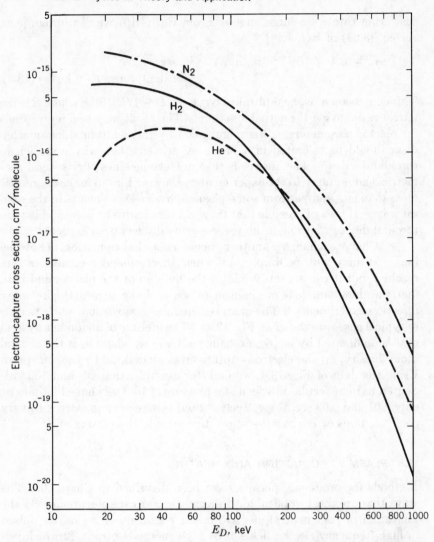

Fig. 13.4   Electron-capture cross sections of deuterons in H_2, He, and N_2. [*Adapted from C. F. Barnett and H. K. Reynolds, Phys. Rev.*, **109**, 385 (1958); *P. M. Stier and C. F. Barnett, Phys. Rev.*, **103**, 896 (1956).]

ent guns produce fairly large quantities of impurities, and trapping of the plasma has so far often been inefficient. The high-energy-beam injection method produces a plasma with ions of any desired mean energy. Multiple sources of different energies or modulated sources could be used to produce an approximately Maxwellian distribution if desired. The required plasma buildup time is very long with present ion sources, so that ideal confinement situations are required, but trapping becomes more efficient as the density increases, and there is reason to expect that the first plasmas of thermonuclear character may be produced by the high-energy injection technique.

In order to trap a high-energy beam efficiently in a static field the charge-mass ratio of the particles must be changed within the containment region.   This can be accomplished by interactions with a background neutral gas, or a high-density arc, or a previously injected plasma.   A number of examples of the first two methods are given in the general references.   They suffer from the disadvantage that the charge/mass changing centers also lead to charge exchange of trapped ions or to scattering out of mirrors.   These methods have not yet been applied to toroidal devices.   There is a slower but nevertheless useful method for trapping high-energy beams that does not require collisions and therefore minimizes losses.   It has been discussed by Hiskes[15] and is called *Lorentz ionization* or *dissociation* because particle motion with velocity **v** through a magnetic field **B** is equivalent to motion in an electric field $\mathbf{E} = \mathbf{v} \times \mathbf{B}$. Atoms or molecules in an electric field have ionization or dissociation lifetimes that decrease rapidly with increasing excitation level of the particle and with increasing electric field.   For example, a 100-keV deuterium atom excited to a principal quantum number of $n = 10$ has an ionization lifetime of ~ 1 sec in a 1.0 Wb/m^2 magnetic field and a lifetime of $\sim 10^{-12}$ sec in a 5 Wb/m^2 field.   The respective lifetimes of an $n = 12$ atom are both less than 10^{-12} sec.   The disadvantage of this trapping technique is that only a small fraction of the high-energy beam, perhaps 0.1 percent, is in a useful excitation state, although collisionally induced cascades to higher quantum levels will increase the effect as the plasma density builds up.   Two present high-energy injection experiments rely entirely on Lorentz ionization.[7,16]

The heating of a plasma to the desired fusion temperature also can be approached by various routes, some of which are closely tied to the method of creating the plasma.   No further heating is required in the high-energy injection methods just mentioned.   If the plasma is produced by other methods, ohmic heating, magnetic compression, absorption of electromagnetic radiation, or "magnetic pumping" can be tried.   Plasma can also be born hot[17] or heated[6] by pulsing on a large electric field that is perpendicular to the confining magnetic field.   In the Astron, the multimillion-electron-volt electrons that create the plasma and the confining magnetic field are also intended to heat the ions.[1]

Ohmic or Joule heating is in principle suitable for heating electrons to temperatures of a few hundred electron volts, although there is evidence that the heating currents cause electrostatic instabilities that in turn lead to rapid diffusion across the confining magnetic fields.   The resistivity decreases with increasing temperature as $T_e^{-3/2}$ so that the rate of power input at constant current falls rapidly as the electron temperature rises. The most vigorous attempt to heat a plasma by strictly ohmic means is the Russian Tokamak[18] series of experiments.   They use the largest possible longitudinal magnetic fields in their toruses so that very large heating currents can be used without causing macroscopic instability.   A rough expression for T_e as a function of time is given by

$$T_e \approx 10^8 (B^2 t n^{-1} R^{-2})^{2/5} \qquad \text{keV} \tag{13.4}$$

where R = major radius of the torus
This expression applies when the energy losses are classical diffusion and radiation.   According to this expression temperatures of several kiloelectron volts should be obtainable with their apparatus.   On the other hand, if Bohm diffusion is the main energy-transport mechanism, the maximum electron temperature turns out to be

$$T_e \approx 10^5 (B^2 r^2 n^{-1} R^{-2})^{2/7} \qquad \text{keV} \tag{13.5}$$

where r = minor radius of the plasma
For their parameters this expression gives an electron temperature of several hundred electron volts; they observe a maximum of 150 eV and suspect enhanced diffusion.   In heating of this kind the ions must gain energy from the electrons and consequently are at somewhat lower temperatures.

The electrons can also be heated by electromagnetic radiation at the cyclotron frequency.   Large amounts of microwave power are required, but apparently there is not much of a problem in coupling the energy to the plasma as long as the resonance condition is met somewhere within the confinement region.   Plasmas with electron temperatures of tens of kilovolts have been produced in this way.[19]   The ions of the plasma can be heated by radiation near their cyclotron frequency, i.e., in the megahertz region.   The coupling of energy of this relatively low frequency into a plasma is a difficult matter.   Baker, Boley, and coworkers at Berkeley have put large amounts of power into a plasma in a mirror geometry through electrodes near one of the mirrors;[20] apparently in this case end effects limited the ion temperature to a few tens of volts.   A more fruitful but more difficult approach is vigorously pursued by the Stellarator group, where the ion-cyclotron frequency power is propagated into the plasma across the magnetic field.[14]   Ion temperatures in the kilovolt range have been produced in small volumes, and we can expect that this technique, perhaps in combination with heating at the electron-cyclotron frequency, will produce even more interesting plasma temperatures in the near future.

At an even lower frequency, energy can be put into the ions by "magnetic pumping."[14] In this method a large amplitude modulation is applied to the confining magnetic field in a localized region. Ions passing through this region alternately gain and lose energy, but there is a slow net increase. The coupling is so weak that very large amounts of power are required for significant results. The technology is being developed, and calculations show that the technique is capable of producing ion temperatures of several kilovolts in the present C Stellarator.

In addition to the quasi-steady-state methods mentioned above, there are vigorous research programs to investigate heating by magnetic compression, interdiffusion of magnetic fields, and $\mathbf{E} \times \mathbf{B}$ effects.[14,21] An interesting and apparently efficient nonlinear approach is "turbulent heating" produced by deliberately induced instabilities.[22]

Finally we must mention the method of plasma heating that suggested the possibility of a self-sustaining plasma and direct energy recovery: the heating of the electrons by charged nuclear-reaction products. It is unknown at present what fraction of the energy will be transferred before the fast ions escape from the system or whether or not it is even desirable to contain them.[14,4]

Energy will be transported out of the confinement region by the previously mentioned processes of classical diffusion, enhanced diffusion, macroscopic drifts and instabilities, charge exchange, and scattering into loss cones. In a well-designed and well-behaved reactor the ion cooling should be chiefly by energy transfer to cooler electrons. The electrons in turn will lose energy by Coulomb and magnetic bremsstrahlung, by excitation of incompletely ionized impurity atoms, and in some cases perhaps by thermal conduction. All these processes have been discussed in previous chapters.

13.6 NUMERICAL EXAMPLE

The purpose of this section is to remind the reader of some of the formulas and ideas developed in previous sections of the book. For a simple numerical exercise we choose a steady-state mirror machine fueled by the injection of high-energy neutral hydrogen isotopes. This is but one of the possible approaches to fusion power, and reference should be made to the general references for the others. No attempt will be made to optimize parameters.

Assume that it is desired to fill a volume of 1 m^3 with 10^{20} ions having a mean energy of about 100 keV. If the ion distribution can be approximated as a three-dimensional Maxwellian, the ion temperature will be $\sim$67 keV. From the reaction yields and the data of Fig. 13.1 we find that about 5 MW of power is generated if the fuel is d-t and about 0.5

MW if it is pure deuterium.   This power will be recovered as heat and after conversion to electricity must supply the magnet, the injector, and the cooling system; hopefully there will be something left over to sell. From $nkT \leq B_0^2/2\mu_0$ we find that the *minimum* confining magnetic field is about 1.5 Wb/m²; for stable mirror confinement it will probably have to be at least 5 Wb/m².   The mirror fields must be considerably larger, say three times this value.   It will be essential to minimize the magnet power by using very cold conductors or superconducting windings.   This will mean that a considerable barrier must be placed between the hot plasma and the coils.

For a mirror ratio of 3 the minimum average time for scattering into a loss cone is

$$T \approx \frac{1}{2}\left[1 - \left(1 - \frac{1}{R} \right)^{1/2} \right]^{-1} 6.4 \times 10^{17} T_i^{3/2}[n(\ln \Lambda)]^{-1} \approx 1 \text{ sec}$$

Thus about 10^{20} high-energy neutral particles must be injected each second.   This is equivalent to a current of 6 A of H° atoms, which could be obtained from 30 A of H_2^+ ions.   Aside from the technical difficulties of producing beams of this size, the primary beam power of 6 MW jeopardizes this particular reactor design even in the *d-t* case.   During the time that the ions are confined, they transfer energy to the colder electrons at the rate [see Ref. 144, eq. (8.39)]

$$\frac{dT_i}{dt} \approx -10^{-19}n \left(1 - \frac{T_e}{T_i} \right) T_e^{-3/2}T_i \ln \Lambda \qquad \text{keV/sec} \qquad (13.6)$$

where all temperatures are in kiloelectron volts. T_e is unknown, although it can be solved for when the various electron energy-loss rates are known. Let us assume that it is about one-third the ion temperature; then the characteristic time for energy transfer from ions to electrons is a little over 1 sec.   The injection energy would have to be raised to maintain the 67-keV ion temperature that was assumed in the reaction-rate calculation.

The electron temperature will tend to be reduced by several processes: electrons can escape through the mirrors and be replaced by cold electrons, they can radiate, and they can excite impurity ions.   It was shown in Chap. 4 that electrons should scatter out of the mirrors at a faster rate than the ions; assume that the plasma potential will adjust or can be adjusted to a value such that the rates are equal.   The Coulomb bremsstrahlung loss rate per electron obtained from Eq. (8.39) in Chap. 8 [see also Ref. 144, eq. (11.17)] gives

$$\frac{dT_e}{dt} = -3 \times 10^{-21}nT_e^{/2} \qquad \text{keV/sec}$$

This works out to about 2 keV/sec in this example.   The rate of cooling by cyclotron radiation would be [see Ref. 1A4, eq. (11.83)]

$$\frac{dT_e}{dt} \approx -0.4B^2 T_e \qquad \text{keV/sec with } T_e \text{ in keV}$$

if all the radiation escaped.   The fraction that actually escapes through the plasma surface depends on the geometry; in our case the potential loss of 50 keV/sec given by the above expression might be reduced to around 2 keV/sec.   This loss can be even further reduced by reflection of most of the radiation back into the plasma.   If the wall reflectivity is r [$r > 0.9$ for cyclotron (infrared) frequencies], then the cyclotron radiation loss can be reduced by the factor $1 - r$.

This unoptimized and non-self-consistent example shows that the major energy loss is by the escape of particles through the mirrors.   The loss rate is somewhat greater than the nuclear energy-production rate for a d-d fuel, but less than the production rate for d-t.   Post has made a careful but necessarily approximate study of energy balance in mirror machines, including magnet and accelerator power consumption, and has found that net power production is possible with d-t fuel if the system is large enough.[23]   One such example has the following parameters: central field = 6 Wb/m^2, mirror ratio = 3.3, vacuum-chamber radius = $\frac{1}{2}$ m, length = 50 m, $n = 10^{20}$m^{-3}, $T_i = 67$ keV, coils of Na at 10°K, and a net electrical power output of 100 MW.   Power for minimum-B coils was not included but would probably be quite large.   Toroidal devices should have a more favorable power balance, but they may, on the other hand, turn out to have loss problems peculiar to themselves.

In a real calculation there are a number of considerations that are very hard to include in a quantitative way: the length of time that nuclear-reaction products remain within the plasma, the effect of impurities, collective interactions, plasma potential, the maximum β consistent with stability, and so forth.   If d-t fuel is required, even the regeneration of the tritium is not a trivial problem, although it appears to be feasible.

It is for these reasons that very few "economics" studies have been made in the past few years—it is first necessary to learn how to predict and control the behavior of hot plasmas.   Very substantial progress is being made in this direction.

REFERENCES

1. Some general source material on plasma research related to controlled fusion.
   A. Books
      1. Bishop, A. S.: "Project Sherwood—The U.S. Program in Controlled Fusion," Addison-Wesley, Reading, Mass., 1958.

2. Artsimovich, L. A.: "Controlled Thermonuclear Reactions," Russian, Fizmatgiz, Moscow, 1961, English, Gordon and Breach Science Publishers, New York, 1964.
3. Glasstone, S., and R. H. Lovberg: "Controlled Thermonuclear Reactions," Van Nostrand, Princeton, N.J., 1960.
4. Rose, D. J., and M. Clark, Jr.: "Plasmas and Controlled Fusion," M.I.T. and Wiley, Cambridge, Mass., and New York, 1961.

B. Review Articles
1. Post, R. F.: *Revs. Modern Phys.*, **28,** 338 (1956).
2. Post, R. F.: *Ann. Revs. Nuclear Sci.*, **9,** 367 (1959).

C. Journals of Current Research
1. *Journal of Nuclear Energy*, part C, Thermonuclear Research, Pergamon, New York.
2. *Nuclear Fusion, Journal of Plasma Physics and Thermonuclear Fusion*, International Atomic Energy Agency, Vienna.
3. *Physics of Fluids*, American Institute of Physics, New York.

D. Miscellany
1. Second United Nations International Conference on the Peaceful Uses of Atomic Energy, United Nations, Geneva, vol. 31, Theoretical and Experimental Aspects of Nuclear Fusion, vol. 32, Controlled Fusion Devices.
2. "Plasma Physics and Controlled Nuclear Fusion Research," Intern. Atomic Energy Agency, Vienna, 1966 (*Proc. Second Intern. Conf. on Controlled Fusion, Culham, England*, 1965).

2. Arnold, W. R., J. A. Phillips, G. A. Sawyer, E. J. Stovall, Jr., and J. L. Tuck: *Phys. Rev.*, **93,** 483 (1954).
3. Kozlov, B. N.: *Atomnaya Energiya*, **12,** 238 (1962); *Soviet Atomic Energy*, **12,** 247 (1962).
4. Tuck, J. L.: *Nuclear Fusion*, **1,** 201 (1961).
5. Lawson, J. D.: *Proc. Phys. Soc. (London)*, **B70,** 6 (1957).
6. Baiborodov, Yu. T., M. S. Ioffe, V. M. Petrov, and R. I. Sobolev: *J. Nuclear Energy*, **C5,** 409 (1963).   Gives earlier references.
7. Damm, C. C., J. H. Foote, A. H. Futch, A. L. Gardner, and R. F. Post: *Phys. Rev. Letters*, **13,** 464 (1964).
8. Larkin, F. M.: A Single Loop Conductor for Producing a Magnetic Well, CLM-R37, 1964.   Unpublished.
9. Rosenbluth, M. N., and C. L. Longmire: *Ann. Phys.*, **1,** 120 (1957).
10. Tuck, J. L.: *Nature*, **187,** 863 (1960).
11. Furth, H. P., and M. N. Rosenbluth: *Phys. Fluids*, **7,** 765 (1964).
12. Fowler, T. K.: *Phys. Fluids*, **8,** 459 (1965).
13. Osovets, S. M.: *Kurchatov Atomic Energy Inst. Rept.* IAE-665, Moscow, 1964, English translation *U.C.R.L. Trans.* 1133(L); S. M. Osovets: *J. Nuclear Energy*, **C6,** 421 (1964).
14. Barnett, C. F., and H. K. Reynolds: *Phys. Rev.*, **109,** 385 (1958); P. M. Stier and C. F. Barnett: *Phys. Rev.*, **103,** 896 (1956).
15. Hiskes, J. R.: *Nuclear Fusion*, **2,** 38 (1962).

16. Sweetman, D. R.: *Nuclear Fusion Suppl.*, pt. 1, p. 279, 1962.   This reference contains many short articles on the state of fusion research as of 1962.

17. Halbach, K., W. R. Baker, and R. W. Layman: *Phys. Fluids*, **5,** 1482 (1962).

18. Artsimovich, L. A., S. V. Mirnov, and V. S. Strelkov: *Atomnaya Energiya*, **17,** 170 (1964) and *Kurchatov Atomic Energy Inst. Rept.* IAE-684, Moscow, 1964, English translation *U.C.R.L. Trans.* 1132(L).

19. Dandl, R. A., A. C. England, W. B. Ard, H. O. Eason, M. C. Becker, and G. M. Haas: *Nuclear Fusion*, **4,** 344 (1964).

20. Boley, F. I., J. M. Wilcox, A. W. deSilva, P. R. Forman, G. W. Hamilton, and C. N. Watson-Munro: *Phys. Fluids*, **6,** 925 (1963).

21. *Journal of Nuclear Energy*, part C.   Contains an extensive bibliography of current articles pertaining to the heating and containment of plasmas in thermonuclear experiments.

22. Babykin, M. V., P. P. Gavrin, E. K. Zavoyskiy, L. I. Rudakov, and V. A. Skoryupin: *Zhur. Eksptl. i Teoret. Fiz.*, **47,** 1597 (1964); E. K. Zavoiskii: *J. Nuclear Energy*, **C5,** 381 (1963).

23. Post, R. F.: *Nuclear Fusion Suppl.*, pt. 1, p. 99, 1962.

NAME INDEX

The page numbers in *italics* are those of references at the ends of chapters.